The Patrick Moore Practical Astronomy Series

Series Editor

Gerald R. Hubbell, Mark Slade Remote Observatory
Locust Grove, USA

The Patrick Moore Practical Astronomy Series is a treasure trove of how-to guides for the amateur astronomer. The books in this series are written for hobbyists at all levels, from the enthusiastic newcomer to the veteran observer. They thus go far beyond more general, popular-level books in both scope and depth, exploring in detail the latest trends, techniques, and equipment being used by amateur astronomers around the world.

You will find herein a diverse list of books on constellations, astronomy catalogues, astrophotography, eclipse chasing, telescope equipment, software, and so much more. All books in the series boast full-color images as well as practical sections for putting your newfound knowledge to use, including star charts and target objects, glossaries, hands-on DIY projects, troubleshooting walk-throughs, and a plethora of other helpful features.

Overall, this series bridges the gap between the many introductory books available and more specialized technical publications, providing digestible, hands-on guides for those wishing to expand their knowledge of the night skies.

Jane Clark

Photographing Galaxies from Light Polluted Skies

A How-to Guide for Amateur Astronomers

Springer

Jane Clark
Risca, UK

ISSN 1431-9756 ISSN 2197-6562 (electronic)
The Patrick Moore Practical Astronomy Series
ISBN 978-1-0716-4505-5 ISBN 978-1-0716-4506-2 (eBook)
https://doi.org/10.1007/978-1-0716-4506-2

This Springer imprint is published by the registered company Springer Science+Business Media, LLC, part of Springer Nature.
The registered company address is: 1 New York Plaza, New York, NY 10004, U.S.A.

To two people who died tragically young
while this book was in production:
My sister Susan Gregory, aged 56
My astronomer friend Gregory
Titley, aged 36

Preface

I have set this book out using the case study method. That is, I think I can best show you how to photograph galaxies in a suburban environment by discussing actual galaxies that I imaged from my back yard.

I think you are likely to trip over many of the same problems as me: stray reflections caused by neighbors who shine lights where they have no business to, humid air which reflects back the ever-increasing amount of light people send up into the sky, the switch from orange street lights to while-light LEDs, and so on.

Even if you lived in a desert hundreds of miles from the nearest town, you would still have to figure out how to focus and collimate, how to capture and store images and all the other things astrophotographers do. These issues are common to all of us, and I will discuss them.

Before getting into galaxies, I was a solar system astronomer. I always knew I wanted to move on from the Solar System. I did not do this in a rational, thought-out way. I just started, and became hooked. My first results weren't great. They weren't even good. I used the wrong hardware, the wrong software and used them both in the wrong way. The day job did not help one tiny bit with spending long night hours in the observatory. Eventually, two years ago, I reached retirement age and was no longer constrained to be wide awake during the day.

I still wasn't so committed to galaxy photography. I thought I was going to make a mathematical model of colliding galaxies, and began a program of reading to figure out how to do this. But, like the lead character Buck in Jack London's *Call of the Wild*, I was drawn away from the comfort of my desk to the discomfort of my unheated observatory.

With the persistence that would allegedly get a determined monkey to type out great literature, I tried this and that. In among the Facebook "likes" I received some occasionally blunt feedback from people who knew what they were doing. It occasionally stung, but the critics were rarely wrong, and they were far more useful to me than people who will "like" anything you load onto Facebook.

Eventually, fifteen months ago, it all came together and I could reproducibly make halfway decent images of galaxies. This was the time to write up, while I still remember the learning curve. I now had to get a set of images ready for the submission date I agreed with Springer. This hitherto informal project became much more formal and focused. My haphazard learning style would no longer do. The result is that I am better now than I was fifteen months ago. I have done things I would never have believed possible.

The cases I have chosen are not the images that were straightforward to produce. I have a good collection of those. I had fun producing them, but I didn't really learn much about photography from them. No, the ones that initially flummoxed me are the ones I have written about, so that you can glean knowledge of ways out of difficulty.

Some external factors helped. The rise of CMOS astrophotography cameras is as big a leap forward as the rise of CCD cameras was from film astrophotography. I gradually traded my way up to a really good camera.

Another step forward was the rise of artificial intelligence (AI) for image processing. AI may yet break our civilization. I'm not blind to that risk. But it has moved image processing software on a lot. In our highly automated smartphone cameras, we are not aware of the AI. But astrophotography is much less automated. Astrophotographers still have to do a lot of the image processing manually. The day may come when it's as automatic as photographing your kids or cats in cute poses. It may arrive quicker than we realize. But it's not here yet.

The telescopes and mounts have got better. We are seeing better optical and mechanical quality than ten or fifteen years ago.

I think in a way the Covid pandemic helped too. Astronomy was a great lockdown activity. Some great software was written by hobbyists. Online communities grew a lot stronger. The various social media may also destroy our civilization one day, but they have their uses.

We therefore now live in a golden era for amateur astronomy. I have expensive hardware and software now. There is cheaper kit and free software around that will work. Stuff for a quarter the price of my equipment is better than a quarter as good. You pay more and more for marginal improvements.

Another development which is rapidly advancing is the use of remote observatories. I dedicate my final chapter to my adventures with a remote observatory company to enable me to explore the galaxies in the Southern Hemisphere I can't see, using telescopes of which I can only dream. Did you ever think you would process images taken with a 24-inch telescope? Observatory ownership may not be the way forward for everyone. There are ways to have the fun without incurring crippling costs.

I have focused on the best software for astro image processing, PixInsight. It isn't free. You pay a one-off fee, which entitles you to updates. You can go quite a long way with Siril for free if you really can't afford PixInsight, but I have no regrets about investing the money or the time to master PixInsight. It is still developing rapidly. It may have changed by the time you read this.

In Chap. 1 I show how the ideas we now have about galaxies emerged as we began to realize that we live in one. Then I jump in with both feet to discuss how we use telescopes, and how we begin to process images. Once we have that background information, I move on to my case studies. The bulk of the book consists of those case studies.

Chapters 4 and 5 are much more pedagogical than the chapters that follow them. By the time you get to Chap. 6, I figured that there would be much less value to repeating every step every time. There is also some repetition. I thought that this was a sin, but a lesser sin than interrupting the narrative while you find cross-referenced pages would be. I hope I'm right.

My aim is to give you some tools and techniques so that you can explore the night skies for yourself, and to save you from spending years floundering like I did. I welcome any feedback at jane.clark@finerandd.com.

In the interest of full disclosure, I should state that I have only commented on equipment I own. In the case of Telescope Live, I pay a full subscription. No-one has sponsored me, and none of the suppliers knew I was writing the book until two weeks ago when I double-checked with Telescope Live that they were OK with me publishing the images. Therefore, I have no financial interest in plugging any equipment or service.

Risca, Wales, UK Jane Clark
May 2024

Acknowledgments

My first thanks go to my astronomer friends from whom I learned a lot about astrophotography. I often picked up ideas from people I barely knew, but I would particularly mention Lee Pullen, Chris Lee, Ryan Parle, Dave Bennett, Andy Smith, Bob Stuart and Rob Davies, and the late Mike West, all of whom have asked questions or made comments that made me think. Many other astronomers have encouraged me such as Alison Camacho, Mel Rigby, Fiona Lambert, Stephen Price, Richard Mansfield, Pete Quin, Andrew Grasemann, Edward and Theresa Cooper, Greg Titley, Aaron Cuthbertson, Tiffany Kew, Phill Wallace, Mike Bradley, Ray Dent, Tony Cook, Sue Walker, Bob Bowen, and Carolyn Alexander. If I have offended someone by omission, I apologize.

I would also like to thank my daughter and son-in-law for their encouragement, and my dear friend Pauline Thomas for encouragement and support of my strange hobby.

I gratefully acknowledge Alar and Juri Toomre, Min Yun, Kathryn Kreckel, Wendy Freedman, and the late Greg Titley for permission to use their images.

Finally, I would like to thank my editors: Michael Maimone for encouraging this project when it was little more than a mad idea, Hisako Niko, and Nivetha Moorthi on the production side.

Competing Interests The author has no competing interests to declare that are relevant to the content of this manuscript.

Contents

About the Author

Jane Clark is a British amateur astronomer who earned her living as an engineering physicist before retiring in 2022. She has a PhD in Physics and an MBA from Warwick University. She completed 2 years of postdoctoral training at Case Western Reserve University in Ohio before returning to England to begin an industrial career. She became interested in both astronomy and photography as a teenager in the 1970s, photography much more seriously, although as her career progressed and family commitments increased, both interests lapsed. She acquired a telescope in 2006, shortly after completing her MBA, and quickly became hooked on observing. This experience made her realize that astronomy is a lot more fun than business administration. In 2017 she achieved her ambition of having an observatory in her back yard. She is a member of Bristol, Cardiff and Newtown Astronomical Societies, and was a founder member of West Norfolk Astronomy Society. Jane gives talks to astronomy clubs, and other societies as diverse as the cub scouts, the University of the Third Age, and church wives' groups.

Chapter 1
How We Became Aware of Galaxies

Telescopes became good enough to detect what we know to be galaxies around the time of Messier, though of course some people claim to be able to see the nucleus of the Andromeda galaxy without a telescope. Charles Messier first published his catalogue of nebulae in 1774, with 45 objects; and eventually built it up to 102 objects. More have since been added: there are now 110.

He lived a long life, during which he discovered 13 comets, a good haul for anyone. He worked as an astronomer for the French Navy; and was honored in his lifetime by several foreign scientific societies.

Messier's motive was that his catalogue was his list of "I'm not going to be fooled by that one again" objects. There does not appear to have been much knowledge of what they were. Of the 110, we now know 40 to be galaxies.

So how did we get there?

Hubble's discovery that the Andromeda Nebula M31 is outside our galaxy is the stuff of legend. It took me a while to realize that in order to know that, he had to have a concept that our own galaxy exists, and of what our galaxy is. This took a while.

The first piece of progress was to realize that the Sun is a star. It is alleged that certain Greek scholars had this idea, though their cosmology was radically different from the ideas of modern science according to J.L.E. Dreyer, an astronomer and historian whom we will meet again [1]. I therefore treat these claims with some skepticism. Aristarchus of Samos was said by later writers to have "placed the Sun among the Stars" [1], a rather vague statement which could mean anything.

A more plausible claim goes to a philosopher & monk named Giordano Bruno (1548–1600) whose ideas about many things were unorthodox. He believed that the stars were other suns, and thought it likely that intelligent life could exist on some of them. This was unlikely to be popular with a Catholic Church which was losing ground to the Reformation and therefore feeling threatened and retreating into extreme conservatism. Bruno was executed in a particularly barbaric way, more probably for his exotic theological views than for his radical cosmology.

J. Clark, *Photographing Galaxies from Light Polluted Skies*, The Patrick Moore Practical Astronomy Series, https://doi.org/10.1007/978-1-0716-4506-2_1

Christiaan Huygens, a contemporary of Isaac Newton, made the assumption that the star Sirius would have the same brightness as the Sun. On the basis of its brightness, he estimated its distance. Looked at the Sun through ever-smaller pinholes until he thought that the Sun was as bright as he remembered Sirus to be [2]. Scottish astronomer James Gregory tried the same idea, but comparing Sirius to a planet during the night [3]. Hughens estimated 30,000 AU to Sirius; Gregory estimated 83,000 AU. Neither realized that Sirius is much brighter than the Sun, and that it is in fact over half a million AU away. But it was a start. Buth the idea existed that stars were well outside the Solar System; and that the Sun was a star.

In 1750, English astronomer Thomas Wright published *An original theory or new hypothesis of the Universe* [4]. In this book he asserts that the Milky Way is planar. He also suggests that some nebulae may be similar star systems, which we may perhaps never have telescopes good enough to see.

This idea inspired the philosopher Immanuel Kant, who had studied Newton's Principia and was numerate enough to follow it. He was a good enough scientist to deduce the idea that the Moon's orbit could slow the Earth's rotation. In his book "The Universal Natural History and Theory of Heaven" in 1755. He writes (and I have used Google Translate here) [5]

> Because it is more natural and understandable that they are not individual stars that are so large, but systems of many, the distance of which is represented in such a narrow space that the light, which is imperceptible from each of them individually, forms into one given their immeasurable number pale glow. The analogy with the star system in which we find ourselves, its shape, which is exactly as it should be according to our teaching concept, the weakness of the light, which requires an assumed infinite distance: everything agrees perfectly, these elliptical figures such world orders and, so to speak, Milky Ways, the constitution of which we have just developed; and if conjectures in which analogy and observation completely agree to support one another have the same worth as formal proofs, then the certainty of these systems will have to be taken for granted.

For a long time afterwards, investigators were unable to move beyond the idea of stars of equal brightness. There's no hiding from the fact that the idea of so-called "island universes" was around in the 1750s.

The observational evidence backing this up was at this point rather thin. People had an idea that the Milky way encircles the Earth, and that there are more stars along this circle than elsewhere in the sky. And the existence of nebulas was known. I wonder if people focus on these ideas of the 1750s using the lens of hindsight. Crazy theories were no doubt as common then as now. We have just forgotten about those of old.

So now we will follow the slow accumulation of data capable of supporting, or not supporting, the "island universes" idea.

The next great advances were made by William Herschel, who did so much more than discover the planet Uranus. I was told during a recent visit to the Herschel Museum in Bath, England that he was an obsessive telescope mirror grinder, so much so that his sister spoon-fed him while he worked. He made the best telescopes of his day, and put them to good use. This sister, Caroline, was in fact his co-worker for much of his observation, and is credited with the discovery of eight comets, a very respectable total.

One of the things Herschel did was to study binary stars in great detail, on which subject he published papers over several years [6, 7, 8]. He was therefore well aware that pairs of stars varied in brightness.

Yet he did not break away from the assumption that all stars are equally bright. He assumed that first magnitude stars were at a certain distance, second magnitude stars were at double that distance, and so on, doubling the distance for every drop in magnitude [9]. Considering that he was writing 53 years before the first successful parallax measurement of the distance to a star [10], it is hard to criticize him for making this assumption.

People often make assumptions they know not to be the whole truth in order to make scientific progress. For example, Isaac Newton knew full well that his law of gravity implied that the stars get pulled together, but he suspended enough disbelief to work out the details of the Solar System. He had no way to investigate the mutual gravitational attraction of stars, even though he knew that there was an issue there.

As is well-known, Herschel started life as musician and composer. His music is by no means unpleasant. There's a YouTube playlist at https://youtube.com/playlist?list=OLAK5uy_kM2hNgNOJ3JPOm3rrbQ_gLn3_Ufuj79IU&si=zn0D2gIHoFmVgsm2. I was also impressed to note when reading one of his papers that he must have read Newton's Principia. He was a very bright guy.

After considering various scenarios, dominated by the mutual gravitational attraction of stars, Hershel goes on to state that from his observations of the distribution across the sky, and brightnesses (his proxy for distance),

> I shall now proceed to shew that the stupendous sideral system we inhabit, this extensive stratum, and its secondary branch, consisting of many millions of stars, is, in all probability a *detached nebula* [9].

The italics are Herschel's. This is the first claim, backed up with evidence, that we live in a region of space that is unusually highly populated with stars, outside which there aren't so many stars. Nowadays we would call that isolated region of stars a galaxy.

He goes on to describe the number of stars he saw [9]:

> In order to go upon grounds that seem to me to be capable of great certainty, they being no less than an actual survey of the bounds of our sidereal system, which I have plainly perceived, as far as I have yet gone round it, every where terminated, and in most places very narrowly too, it will be proper to shew the strength of my sounding line, if I may so call it, that it may appear whether it was sufficiently long for the purpose.
>
> In the most crowded part of the milky way I have had fields of view that contained no less than 588 stars, and these were continued for many minutes, so that in a quarter of an hour's time there passed no less than 116000 stars through the field of view of my telescope.

He is clearly not tracking the apparent motion of the sky, but, rather, letting stars pass through the view of a static telescope. Because he had the best telescopes of his day, funded by his patron King George III,[1] his audience would not have been famil-

[1] King George III of Great Britain, who was also the monarch of Hershel's native land Hannover, gets a bad press, particularly in the USA. The fact that he funded the top astronomer in his domains

iar with the sheer numbers of stars he could see. It was a very different world, in which nobody got to see the results of the great spaceborne observatories almost as soon as they were released. There wasn't even photography. Herschel's son John would go on to devise the process of 'fixing' a photographic image with sodium thiosulphate, a process which lasted until the rise of digital photography in the late twentieth century. We will hear more of him in due course.

As a result of all this, and of dividing the sky into regions to search, his map of the star-containing region is shown in Fig. 1.1. There is one bold black dot in the center. This is his position of the Sun. So his universe is heliocentric.

A final point about Herschel is that he found and catalogued far more nebulae than Messier. It is curious that history has forgotten this, and remembered Messier as the great cataloguer.

In 1833 his son John Herschel published a massive 156-page paper entitled *Observations of Nebulae and Clusters of Stars, Made at Slough, with a Twenty-Feet Reflector, between the Years 1825 and 1833* [11]. The twenty feet refers to the focal length, not the aperture. The drawings in this document are a wonder to behold.

Just as comparison, Fig. 1.2 shows his drawings of globular clusters M13 & M5. The high quality of his draftsmanship and observational ability are unmistakable.

Of M64, Fig. 1.3, top right, he remarks that his father, William Herschel, noticed the feature we now call the 'black eye".

M51 is the only one we would not nowadays immediately recognize. He was also puzzled by this one, and wrote:

> M51 – This very singular object is thus described by Messier: *Cloudy without stars…One cannot see it but with difficulty with an ordinary 3½-foot telescope…It is double, each part having a bright center, the one separated from the other by 4' 35". The two atmospheres touch each other.*[2] By this description it is evident that the peculiar phenomena of the nebulous ring which encircles the central nucleus had escaped his observation, as might have been expected from the inferior light of his telescopes. My Father describes it in his observations of MESSIER'S nebulae (which are not included in his catalogues,) as a bright round nebula, surrounded by a halo or glory at a distance from it, and accompanied with a

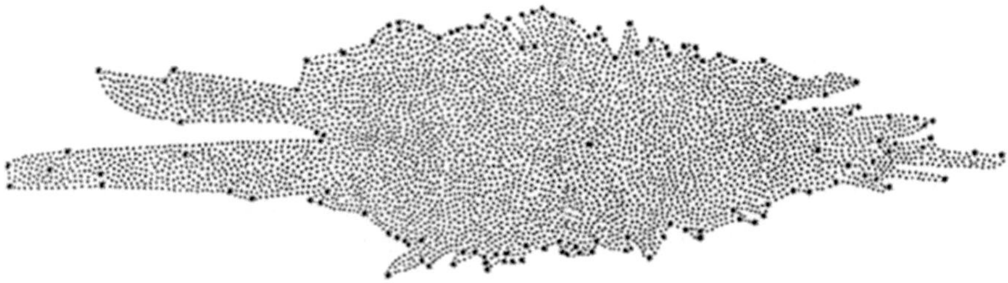

Fig. 1.1 Herschel's map of the Milky Way. The bold black dot in the middle is the position of the Sun. (Image from Reference [9], copyright expired)

shows that he can't have been all bad.

[2] I have translated the words in the italics from the French in which Herschel quotes Messier.

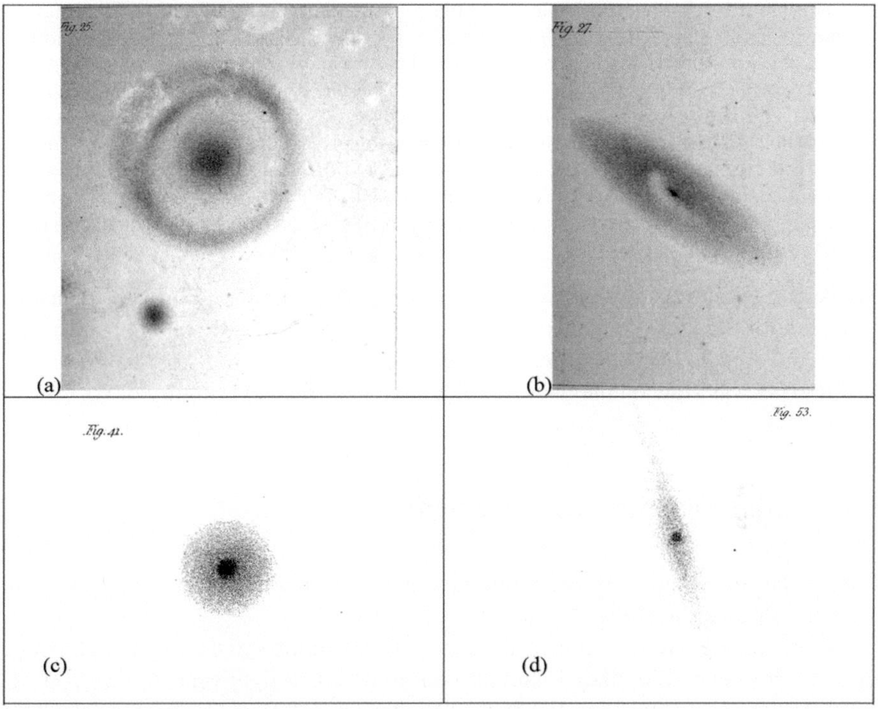

Fig. 1.2 John Herschel's drawings of galaxies. (a) M51, (b) M64, (c) M94 & (d) M65. (Images from Reference [11], copyright expired)

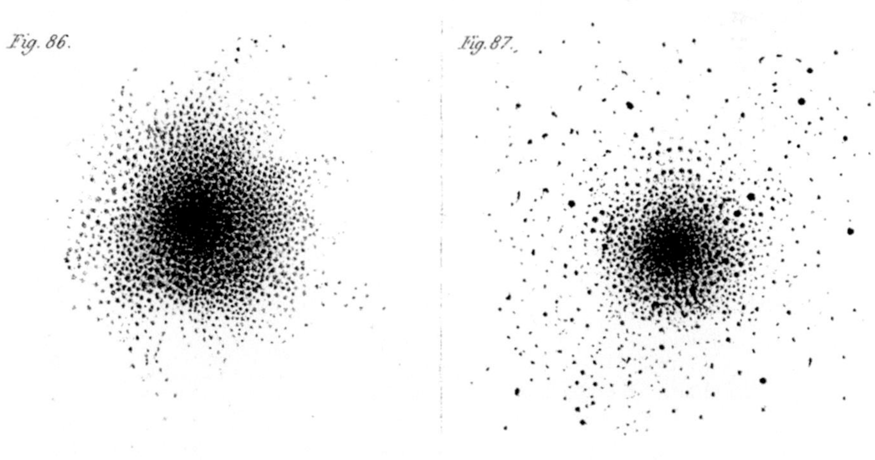

Fig. 1.3 John Herschel's drawings of globular clusters. Left: M13. Right: M5. (Images from Reference [11], copyright expired)

> companion; but I do not find that the partial subdivision of the ring into two branches throughout its south following limb was noticed by him. This is, however, one of its most remarkable and interesting features. Supposing it to consist of stars, the appearance it would present to a spectator placed on a planet attendant on one of them ex-centrically situated towards the north preceding quarter of the central mass, would be exactly similar to that of our Milky Way, traversing in a manner precisely analogous the firmament of large stars, into which the central cluster would be seen projected, and (owing to its greater distance) appearing, like it, to consist of stars much smaller than those in other parts of the heavens. Can it be, then, that we have here a brother-system bearing a real physical resemblance to our own?

So we have the remarkable insight for the year 1833 that we could very plausibly call a hint of the idea of a galaxy separate from the Milky Way. Given his difficulty seeing the object, this is an especially clever idea.

Progress was slow on the question of galaxies for the next century or so.

Interstellar Distance Measurement

During the nineteenth century, progress was slowly made on developing ways to measure distances to stars.

Geheiman-Rath & Bessel measured the parallax to the star 61 Cygni, publishing their result in late 1838: they found this star to be 10.28 light years or 657,700 AU away. The modern value reported by the planetarium software *Cartes du Ciel* is 11.4 light years, the value found by the Hipparcos space probe. This was the first reasonably accurate value of a stellar distance to be published. The principle of parallax is shown in Fig. 1.4.

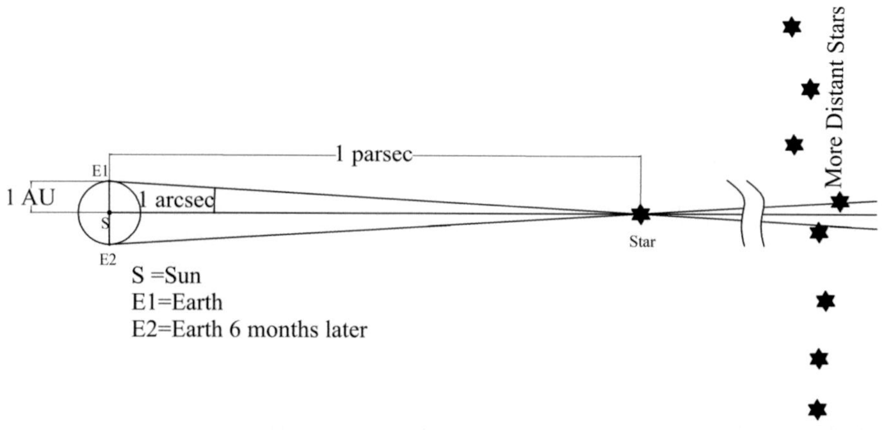

Fig. 1.4 The principle of parallax. The position of a star against the much more distant background is noted at times six months apart. If the angle is 1 arcsecond, the distance is defined as 1 parsec or 1 pc. If the angle were 0.1 arcseconds, the distance would be 10 pc. 1 pc = 3.26 light years. (Image by the Author)

At about the same time, John Herschel spent time from 1834–8 in what is now South Africa mapping the skies of the Southern Hemisphere.

Nineteenth century technology did not allow the parallax measurement of many stellar distances. The progress made with these difficult measures is shown in Table 1.1.

In order to map out a galaxy, it is necessary to measure the distances to many, many stars: we now know that there are of the order of 100 billion stars in the Milky Way.

Therefore a method other than parallax was required to determine stellar distances: there just weren't enough parallax measurements to do the job.

Henrietta Swan Leavitt of Harvard University, was given the job of analyzing large numbers of variable stars in photographs of the Magellanic Clouds. Each cloud could be taken to be at a fixed distance from the Earth. In the Small Magellanic Cloud, she identified 25 Cepheid variables [13], a type of star that pulsates regularly, and noticed a simple relationship between the time it took them to go through once cycle of pulsation and their brightness (Fig. 1.5). This at last gave a "standard candle": a way to measure distances by comparing the periods and brightnesses of any Cepheid variables that can be observed. Thus, there was a distance measurement that did not depend on parallaxes. For this, Miss Leavitt was nominated for a Nobel Prize. Unfortunately, she had died prematurely of cancer, and so was ineligible to win the prize.

A decade later, Harlow Shapley reported the distance of the Large Magellanic Cloud as 34.5 kpc, and that of the Small Magellanic Cloud as 31.6 kpc [14].

At about the same time as Leavitt's law was published, a way was found to relate the brightness of stars to their spectra, in particular to their temperatures deduced from their spectra. A history of the origins of this diagram, with references, is given by Nielsen in 1964 [15]: it is now known after its two independent discoverers as a Hertzsprung-Russell diagram, or HR diagram. These two astronomers were working independently on different continents, but were in correspondence with one another, and seem to have enjoyed friendly relations.

Hertzsprung plotted the diagram for the well-known open clusters the Pleiades (M45) and the Hyades (C41) and Russell plotted bright nearby stars for which the parallaxes were known in 1911.

They both discovered what we now call the main sequence, which contains the vast majority of stars, and the giant branch, which is nowadays subdivided. The branch for white dwarfs was discovered later.

Table 1.1 Progress in measuring stellar parallaxes 1840–1912

Year	Number of reliable stellar parallaxes measured
1840	3
1880	20
1912	163

Source: Reference [12]

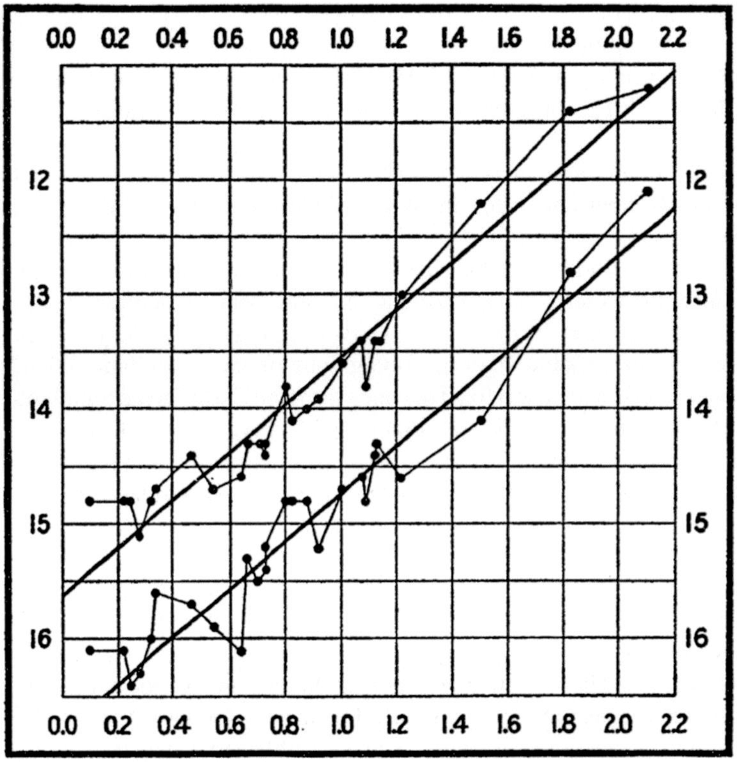

Fig. 1.5 Leavitt's Law for Cepheid variable stars in the Small Magellanic Cloud. The vertical axis is brightness expressed as magnitude, and the horizontal axis is the logarithm of the pulsation period. The upper line is the maximum brightness; the lower line is the minimum brightness. (Image: E. C. Pickering & H. C. Leavitt. ©American Astronomical Society. Source: Reference [13])

There are two key points for our story. First, stars cannot occupy just any position on the HR diagram. Figure 1.6 shows this. This has important implications for the theory of stellar evolution, which is outside the scope of this book. Second, a new way to estimate distances to stars now appeared: from the spectrum, or a proxy measure such as the difference between blue and visible magnitudes,[3] the absolute magnitude can be estimated, and the distance obtained. This distance determination method is called spectral parallax. It has a much longer range than geometric parallax, and can even be used to estimate the distances to stars in other galaxies, via a method called the tip-of-the-red-giant-branch method.

[3]These are measured using the so-called Bessell filters, B for Blue and V for Visible, which peaks in the green. Such filters are sold by Baader Planetarium, see e.g. https://www.baader-planetarium.com/en/baader-ubvri-bessel-filter-set-%E2%80%93-photometric.html. This website shows the transmission curves for these filters.

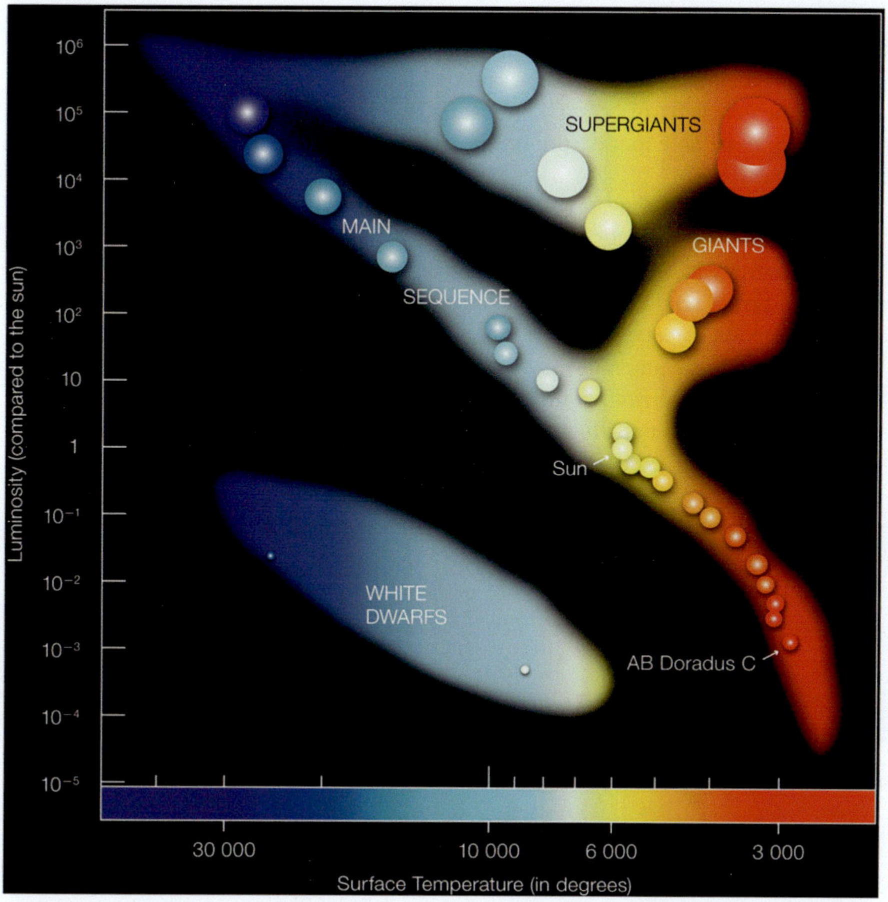

Fig. 1.6 An example of a Hertzsprung-Russell diagram plotted for 23,000 stars. The vertical axis, the luminosity, is given as a multiple of the luminosity of the Sun. The Horizontal axis is given as temperature in Kelvin. (Image Credit: ESO, source https://www.eso.org/public/images/eso0728c)

So now we have moved from the rather vague ad-hoc estimates of distance by Seeliger and Kapteyn to a much more precise method. The HR diagram acquired an underlying scientific basis once the theory of stellar evolution was developed.

Another method of distance method was developed that works only for clusters of stars if they move though space together as a unit & don't rotate. This method is useful because, first, it gave astronomers a way to cross-check their geometric parallax distance measurements. Since the geometric parallaxes are tiny and notoriously easy to get wrong because of atmospheric effects, this was useful. The second useful thing was that it gave a way to obtain absolute magnitudes for the HR diagrams for clusters of stars. This was especially important for the Hyades, which was the first open cluster to be close enough for geometric parallax methods, for the unsurprising reason that it is the nearest.

The method was first devised to work out the direction and speed of travel of the Sun through the field of nearby stars, which is towards the constellation of Hercules. The idea first came from a geometer named Bravais[16], who is best known for his work on the geometry of crystal lattices. He was ahead of his time in both fields. The first person to be able to measure the sun's travel was Kapteyn [17], whom we have already mentioned.

The method of Bravais & Kapteyn seems to have been first used by Lewis Boss in 1908 [18]. The principle of the method is shown in Figs. 1.7 and 1.8.

In the same way that parallel, straight railroad lines on flat plane appear to converge to a point, so do the proper motions of the stars in a co-moving cluster in a three-dimensional space also appear to converge to a point. The cunning trick is that we also know these stars are moving either towards or away from us because the Doppler effect can be seen in their spectral lines.

The Doppler effect is illustrated in Fig. 1.9. Red light has a longer wavelength than blue light. Movement away from us makes light appear redder, while movement towards us makes light appear bluer. (The same thing happens if space is expanding, making the distance to the star increase. In that case, the effect is called the cosmological redshift.)

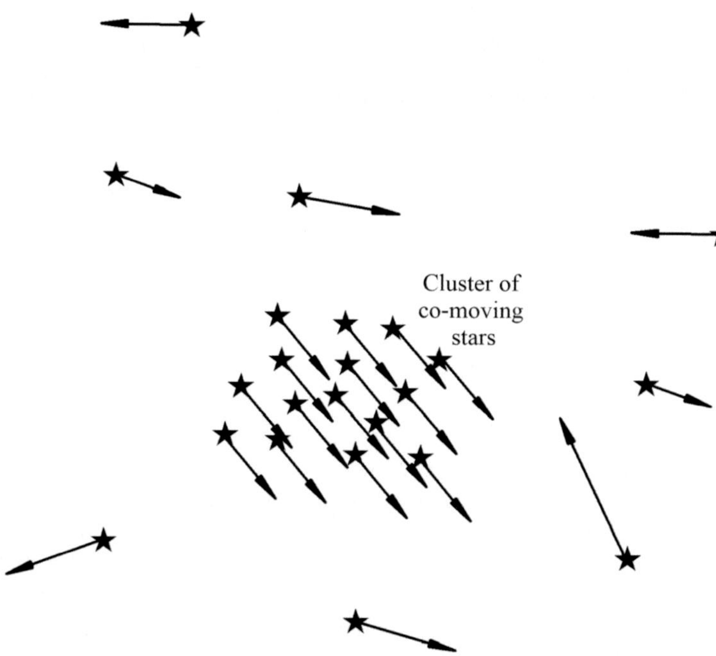

Fig. 1.7 In two dimensions, a cluster of co-moving stars would have parallel proper motions. (Figure by the Author)

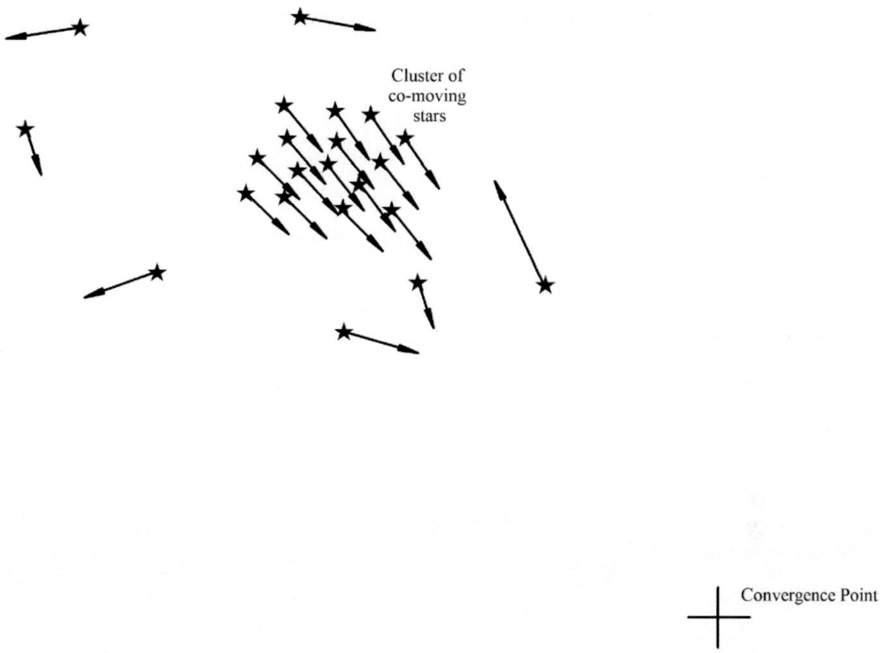

Fig. 1.8 In three dimensions, the apparent motions of a cluster of co-moving stars are not parallel. Instead they appear to converge to a point. (Figure by the Author)

By mathematical rearrangement, the distance to the cluster can be calculated from the above information. The distance thus obtained is known as the moving cluster parallax. There was a time when this method was thought to be of historic interest only, but unfortunately data from a space probe called the Hipparcos satellite showed a very different distance to the Pleiades from that obtained by other methods. This embarrassment gave a new lease of life to the moving cluster method, because it was one of the ways people showed that the reported distance was wrong [19].

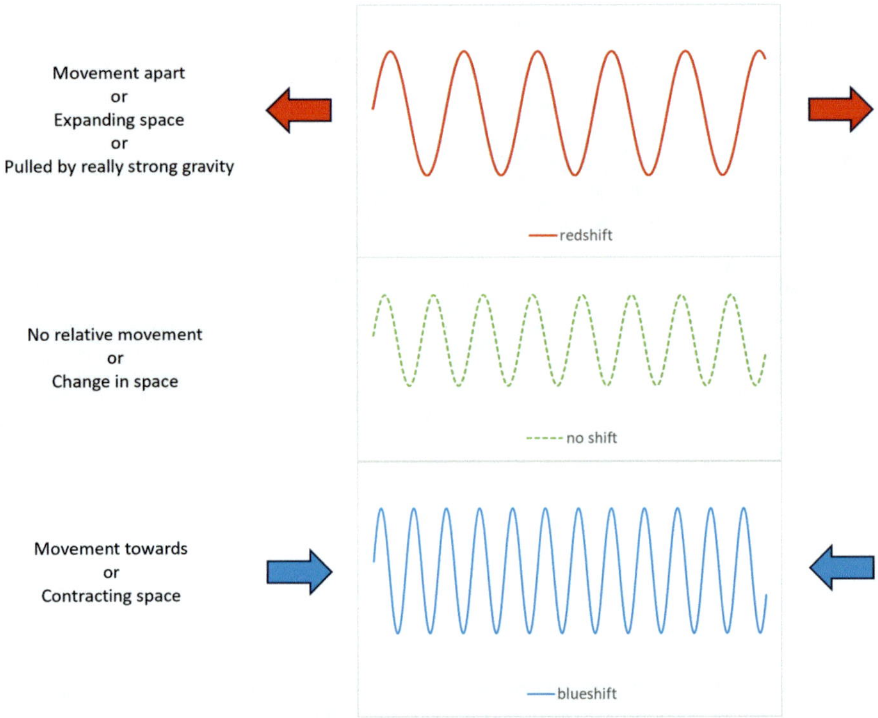

Fig. 1.9 In the doppler effect, the light from a star moving away appears to be redder if the star is moving away (top) and bluer if it is moving towards the observer (bottom). Red light has a longer wavelength than blue light. (Image by the Author)

Models of the Universe by 1922

J. C. Kapteyn, 1922

Kapteyn's modus operandi was to form strong relationships with observers who worked in climates with clearer skies than the Netherlands, and to analyze their data. He had a small team of people doing this in his university in Groningen. This was not only forced on him by the need for clear skies, but also by the fact that his university could not afford a top-class observatory.

He produced an approximate model in 1922 [20], the year he died, probably of cancer [21]. His model was not intended to be exact; but with the computational power then available, it was meant to be tractable. The contours of star density are shown in Fig. 1.10. For the first time, the Sun is shown offset from the center of the universe, by a distance of almost 2 kiloparsecs (kpc). Seeliger had earlier produced a not dissimilar model, with the Sun at the center [22].

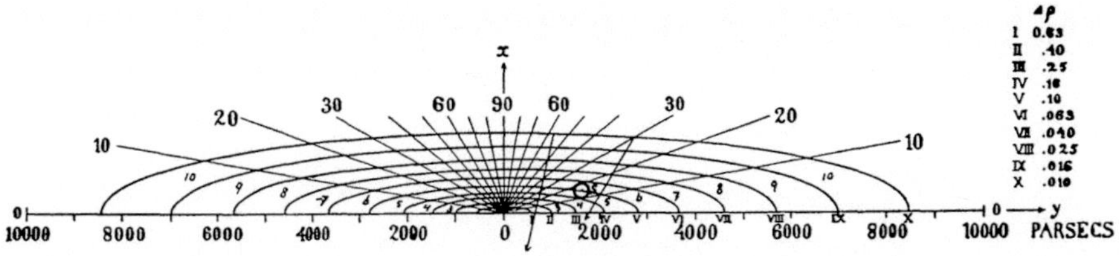

Fig. 1.10 Kapteyn's Universe. Source: Reference [20]. He shows the universe to be made of shells, each with an average star density listed in Table 1.2. For the first time, the Sun is not at the centre of the system. It is close to the centre, at the point shown to the right of centre where Shells 5 & 6 meet. (Image: J. C. Kapteyn, Reference [20] © American Astronomical Society)

Table 1.2 Average number of stars per cubic parsec in the shells shown in Fig. 1.10

Shell	Average density of stars per cubic parsec
1	0.0358
2	0.0226
3	0.0143
4	0.00900
5	0.00568
6	0.00358
7	0.00226
8	0.00143
9	0.000900
10	0.000568

Source: Reference [20], © American Astronomical Society

For Kapteyn, his demise came at a particularly unfortunate time, as he did not live to see the great leaps forward made in the later 1920s.

Eddington [23] and Seeliger [24] produced similar models of the universe, though Seeliger's was heliocentric.

H. Shapley

Shapley discovered a much bigger universe than did Kapteyn. His method was to use variable stars, mainly Cepheid variables, to measure the distances to globular clusters during the late 1910s. After an immense amount of work, he found that the globular clusters were much further away than the size of Kapteyn's universe; and that they appear in the sky at points around the constellation Sagittarius. They are not at all evenly distributed.

Globular clusters are distinct entities with a characteristic appearance (Fig. 1.11). They contain hundreds of thousands of stars. Cepheid variables and RR Lyrae variables are common.

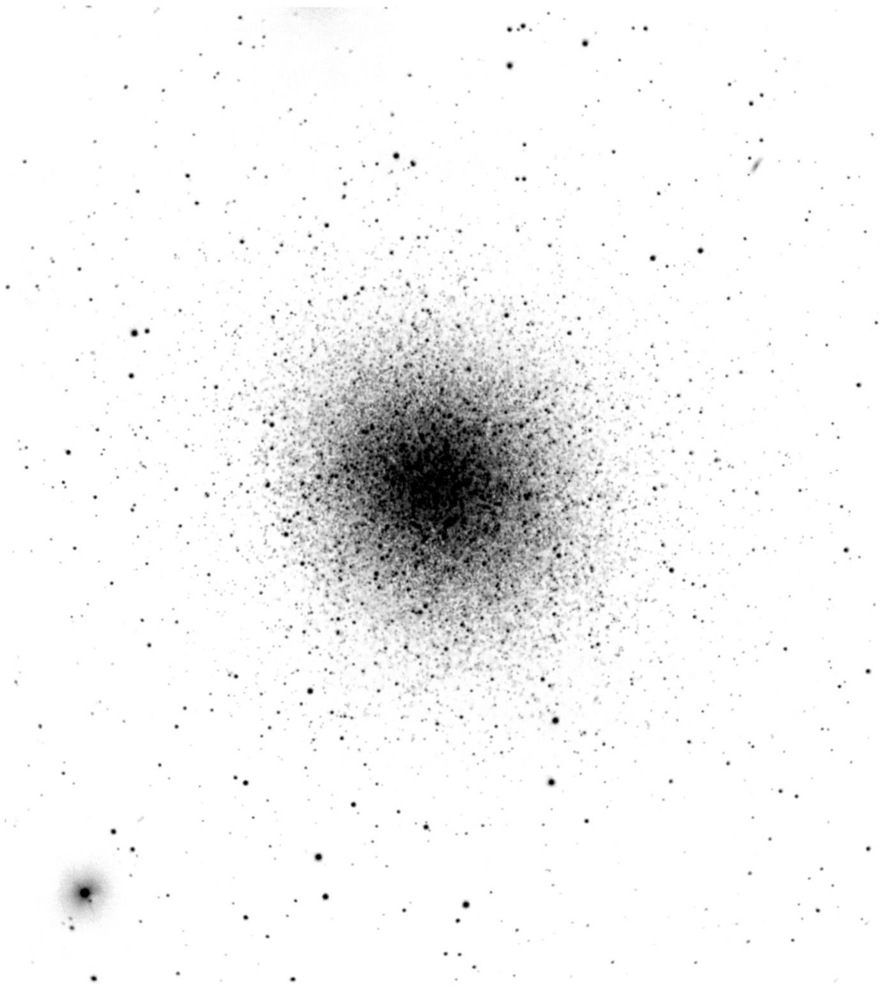

Fig. 1.11 A globular cluster, M13. Celestron C11 at f/6.3, 83×60s. (Image by the author)

By this time, astronomers were well aware that the Milky Way has a bulge in Sagittarius. It is not visible from the United Kingdom, but I got a clear enough view of it during a trip to Australia. You can't miss it.

Shapley proposed that the globular clusters orbit around this bulge, and put it at their center [25]. This required quite a leap of the imagination, especially since the distance to the central bulge was not then known. For this to be true, and for the Sun to be part of the Milky Way, Shapley proposed a galaxy rather like the one shown in Fig. 1.12. Shapley put the size of the Milky way at about 300,000 light years, or roughly 100 kpc.

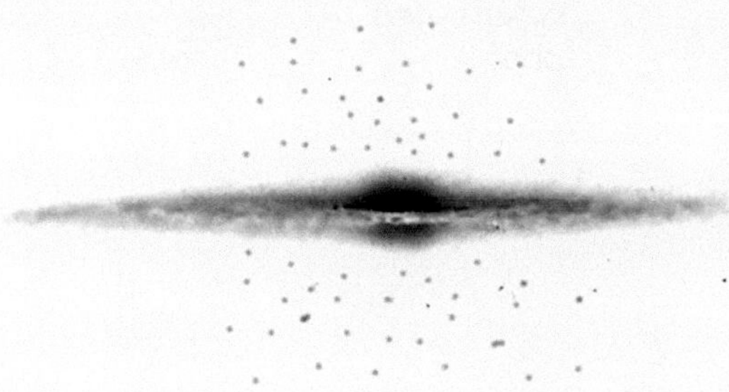

Fig. 1.12 Shapley's view of the universe, with globular clusters orbiting the center. (Image by the Author. (I cheated slightly, and used a photo of the needle Galaxy, NG4565, and drew in the globular clusters using the image processing software Gimp))

I have shown the galaxy edge-on on purpose. The details of the plane were not then known. Shapley believed that the stars were not evenly distributed, but he did not know enough to fill in the details; and wisely kept his counsel.

Hubble: The Andromeda Nebula M31 Is Outside the Milky Way

Not long afterwards, in about 1924, Edwin Hubble, by then a staffer at the Mount Wilson Observatory, with its newly commissioned 100-inch telescope, discovered a Cepheid variable, the celebrated V1, while looking for novae in M31. This showed that M31 was ten times as far away as the Small Magellanic Cloud, where Leavitt had discovered her law. He went on to confirm this result with over forty such variables, and published his results in a 1929 review paper [26]. In Fig. 1.13, I have plotted Hubble's 1929 data on Cepheid Variables, and compared them with Leavitt's data from Fig. 1.5.

This bombshell result was a game-changer. People could now think in terms, not so much of using our galaxy to understand the spiral nebulae, as to use the spiral nebulae to understand our galaxy.

Although people had proposed a spiral structure for the Milky Way, such as Cornelius Easton in 1900 [27], this structure was not confirmed until after World War II, when radio astronomers were able to detect the spiral structure [28].

Sometimes the Milky Way is referred to as 'the Galaxy' with a capital G, whereas lower-case galaxy is any galaxy.

The bar in the Milky Way was postulated as early as 1964, but unambiguously observed in 1991 [29]. The size of the Galactic bar was found in 2005 to be larger than thought [30].

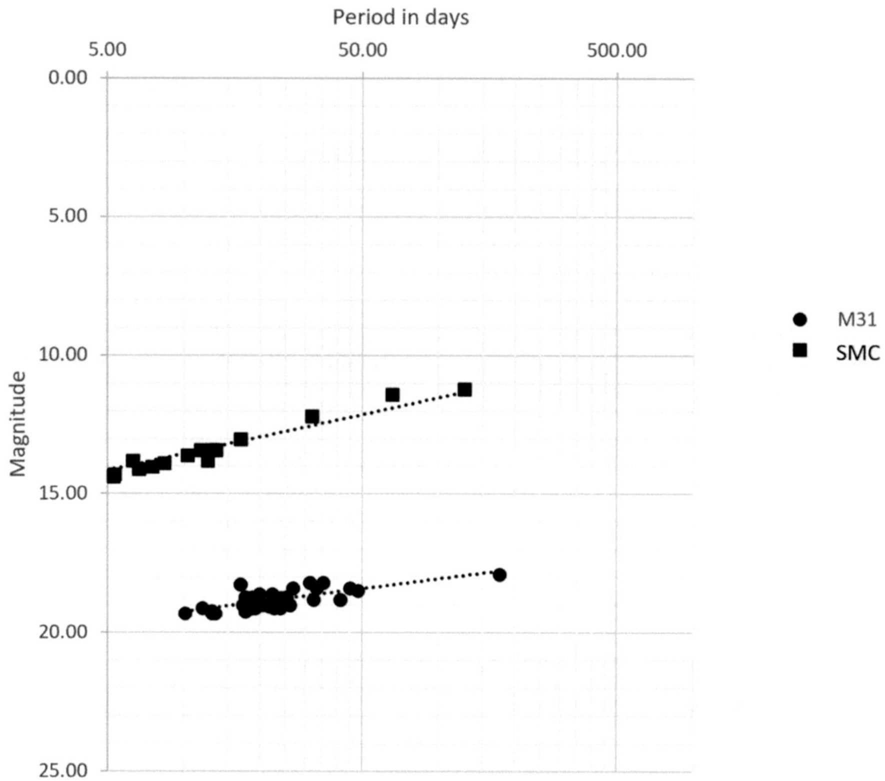

Fig. 1.13 Hubble's and Leavitt's data for Cepheid variables in the Small Magellanic Cloud (SMC) and M31, showing that the variables in M31 are roughly five magnitudes dimmer than those in the SMC, corresponding to a factor of 10 difference in distance. (Image by the Author, data from References [13, 26])

Hubble's Classification System & the Expanding Universe

In 1927 Hubble published his classification of galaxies [31], which was subsequently turned into the 'tuning fork' diagram (Fig. 1.14). You will still see people talk about galaxies to the left of this diagram as "early type" galaxies, and those to the right as "late type" galaxies, even though no-one now seriously believes that such an evolutionary path exists. I suppose it's a bit like the way metallurgy is still full of alchemists' jargon: jargon seems to outlive its usefulness. Hubble examined something like 400 galaxies to derive this classification.

Hubble was one of a few people who noticed that the spectra of galaxies are redshifted. He worked out that the redshift is roughly proportional to distance, and thereby discovered the expansion of the universe, as shown in Fig. 1.15. His work was published in 1929 [33]. He had an incredibly busy and productive decade in the 1920s.

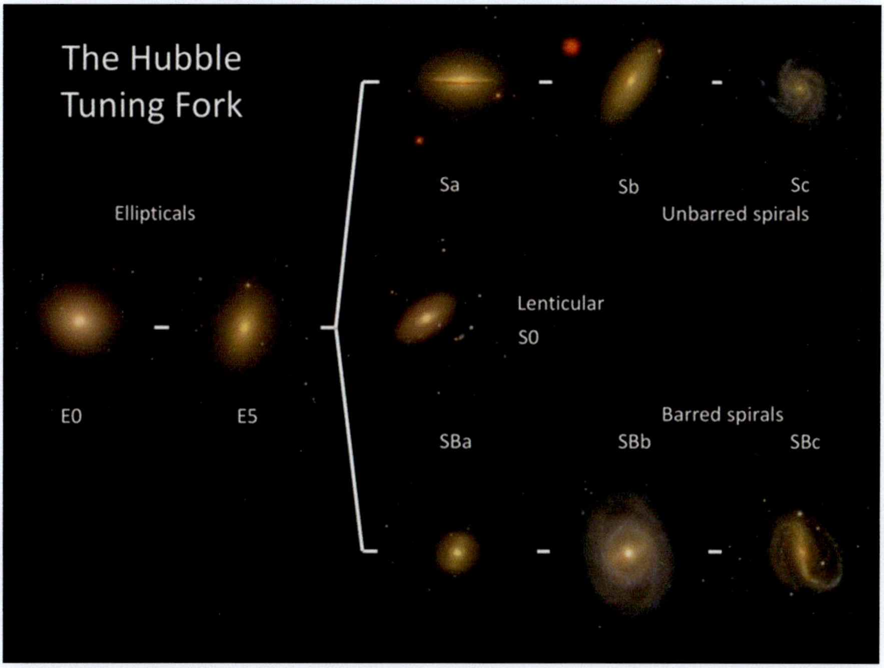

Fig. 1.14 A modern version of the 'tuning fork' diagram, which Hubble never actually used in his original paper, to illustrate his classification of the Galaxies [31]. (Source: Reference [32], made available under a CC by 4.0 license from https://ras.ac.uk/news-and-press/research-highlights/citizen-scientists-re-tune-hubbles-galaxy-classification)

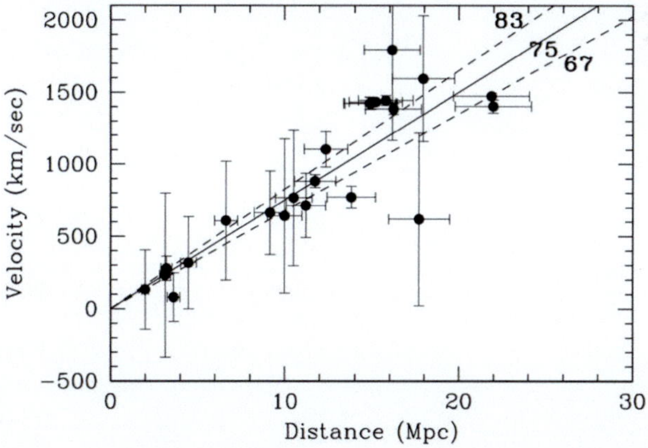

Fig. 1.15 A modern version of the redshift of the local universe as found by Hubble. (Source: Freedman et al. [34], used with permission. Numbers in the top right show different values of the Hubble constant for different curves fitted through the data, in units of km s^{-1} Mpc^{-1})

The galaxies that these astronomers and astrophysicists discovered are there for all to see. It is not easy for amateurs to photograph them, but it can be done, even from urban skies. In the rest of this book, I will show you how I learned to do so.

Chapter 2
At the Telescope

You could photograph galaxies with a DSLR or mirrorless camera and along with a long telephoto lens on a tracking mount, but I have never heard of anyone doing this, except for an unsuccessful attempt I once made, discussed in Chap. 5. Most people I know, however, use telescopes with cameras.

How big a telescope do you need? At the moment I have two: an 11-inch Schmidt Cassegrain telescope and a 3-inch (actually 73 mm) apochromatic refractor. The three-inch scope works just fine, though it is designed for wide-angle astrophotography. Therefore you don't need a huge telescope.

The mount is the more critical component. You need a mount that can handle your payload of scope, accessories & camera with room to spare. The familiar rule of thumb for astrophotography is that your mount should handle twice the payload you actually put on it. I think that's probably a little bit conservative, but you should bear this rule in mind when you choose your mount.

Mounts

The purposes of a mount are to hold your telescope steady and track the apparent motion of the heavens as the Earth rotates. The apparent motion of the heavens would look simple if you & I were at the poles or the equator. Elsewhere, it looks more complex. Away from the equator, some constellations never set. They are said to be *circumpolar*. In the United Kingdom, the constellation *Ursa Major*, colloquially known variously as *The Plough* and *The Big Dipper*, is circumpolar. The densely populated parts of North America are quite a bit further south than the UK, which lies roughly between 50°N and 60°N. In North America, Ursa Major may not be circumpolar for you.

There are two types of constellation behaviors your mount will need to cope with: circumpolar and non-circumpolar. As Figs. 2.1, 2.2, and 2.3 show, the

J. Clark, *Photographing Galaxies from Light Polluted Skies*, The Patrick Moore Practical Astronomy Series, https://doi.org/10.1007/978-1-0716-4506-2_2

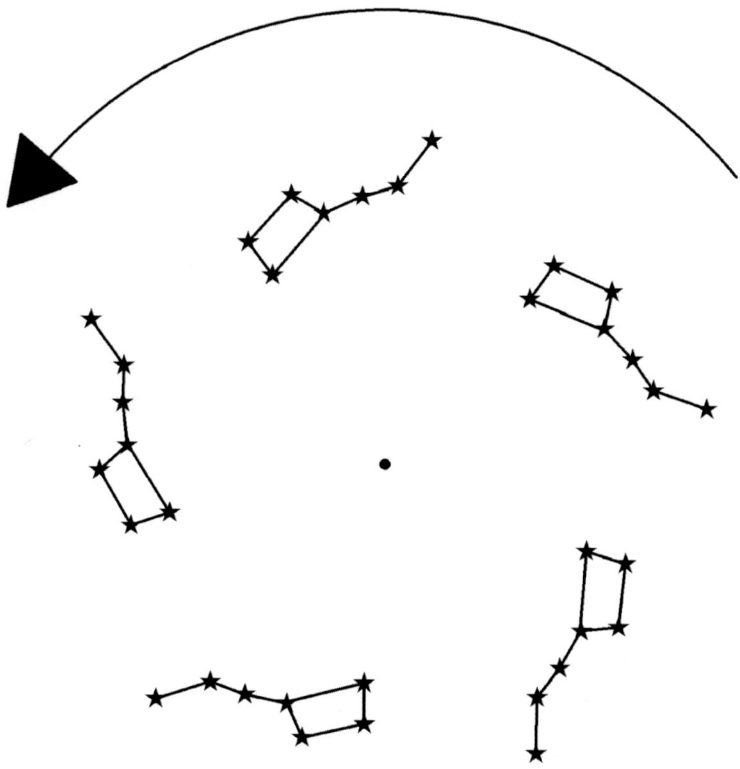

Fig. 2.1 The apparent movement of a circumpolar constellation as the Earth Rotates. The positions are 4 hours 48 minutes apart. The constellation appears to move anticlockwise. The main part of Ursa Major is circumpolar above about 41°N. A circumpolar constellation in the Southern Hemisphere would appear to rotate clockwise. (Figure by the Author)

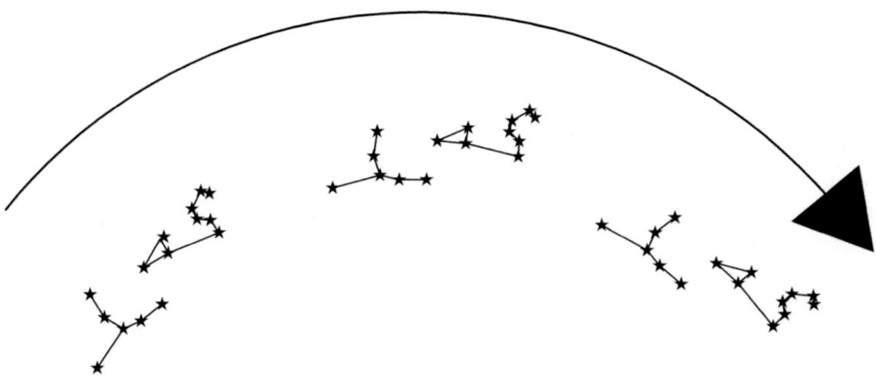

Fig. 2.2 The apparent movement of the non-circumpolar constellations Leo & Virgo as the Earth Rotates, as seen from the Northern Hemisphere. The constellations appear to move from left to right if you look south. The Virgo Cluster of galaxies lies between these two constellations. (Figure by the Author)

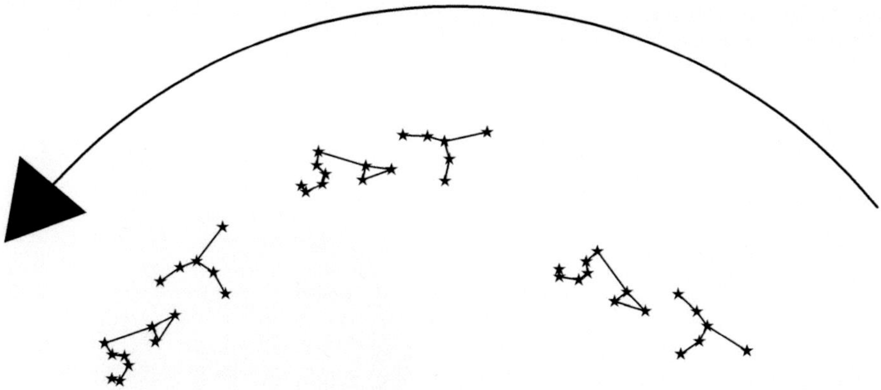

Fig. 2.3 The apparent movement of the non-circumpolar constellations Leo & Virgo as the Earth Rotates, as seen from the Southern Hemisphere. The constellations appear to move from right to left if you look north. The Virgo Cluster of galaxies lies between these two constellations. (Figure by the Author)

constellations don't just appear to move: they appear to rotate. This apparent rotation is known as *field rotation*. The field of view in your camera will appear to rotate if you do not correct for field rotation.

Since the root cause of this is that the Earth is rotating, and since galaxies and stars are so immensely far away that parallax effects are negligible, the ideal solution is to rotate your scope about an axis parallel to the Earth's axis, but in the opposite direction, i.e. from east to west, as the Earth rotates from West to East. This idea is shown in Fig. 2.4. A mount that achieves this is called an equatorial mount.

The situation shown is an idealization. In practice the alignment and rotation speed are never perfect. Actions required to deal with this will be detailed shortly.

First, I want to tell you a little bit about mounts.

The mount shown in Fig. 2.5 is a so-called "German" equatorial mount. Its Right Ascension axis can be adjusted until it points directly at the celestial pole. This axis needs to be motorized to track the apparent movement of the heavens, per Fig. 2.4.

A newer variant on the equatorial mount theme is a harmonic drive mount. These rely on a type of flexible gear called a harmonic or strain-wave gear. They do not have counter-weights. At the time of writing, they are not offered for large telescopes. That may change.

Not all mounts are Equatorial. Fork Mounts of the kind shown in Fig. 2.6 are common, but they are not equatorial. Unlike equatorial mounts, they do not correct for field rotation.

If you own one, all is not lost. You can either buy a Wedge to tilt the azimuth angle, i.e. the base of the fork mount, until it is parallel to the Earth's axis, or you can attach a camera rotator in front of your camera to compensate for field rotation. The latter solution is something of a bodge. I would only really recommend it if, for

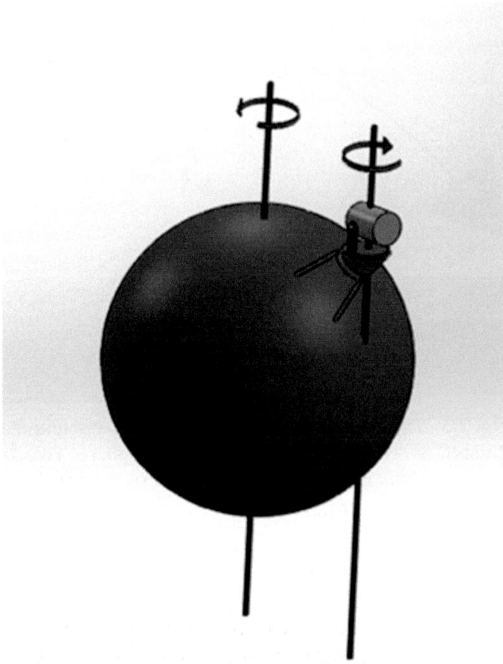

Fig. 2.4 If your telescope has an axis of rotation parallel to the Earth's axis of rotation, the heavens will appear to be still in your field of view. An axis of rotation perpendicular to this would enable you to slew the scope to different declinations. (Image by the Author © Jane Clark)

example, you have a rolling-roof observatory building with insufficient headroom to insert a wedge.

Most professional telescopes nowadays are fork-mounted on wedges.

With a camera rotator, don't try to counter-rotate the camera in real time. That won't end well. Instead, shoot a few frames and then rotate the camera. The stacking process will compensate for small amounts of field rotation. Repeat until all data is captured.

The wedge would look like the one shown in Figs. 2.4 and 2.7. Meade and Celestron both sell them.

German equatorial mounts have their load-carrying limits. They become inefficient above a certain weight, though they will work fine for telescopes up to about 14 inches in diameter, provided that they are adequately specified. For larger telescopes, including modern professional ones in fixed observatories, a design similar in principle to that in Fig. 2.7 is used.

For amateur-sized telescopes, the downside of a German equatorial mount is that when the target passes the meridian, or passes through due north (in the northern hemisphere) for a northerly galaxy, you have to perform a meridian flip, which can be time-consuming, infuriatingly so if you live in a climate where the weather is unlikely to stay clear. This means that the telescope and the counter-weights have to swap sides. The camera also needs to rotate through 180° if you want to continue stacking.

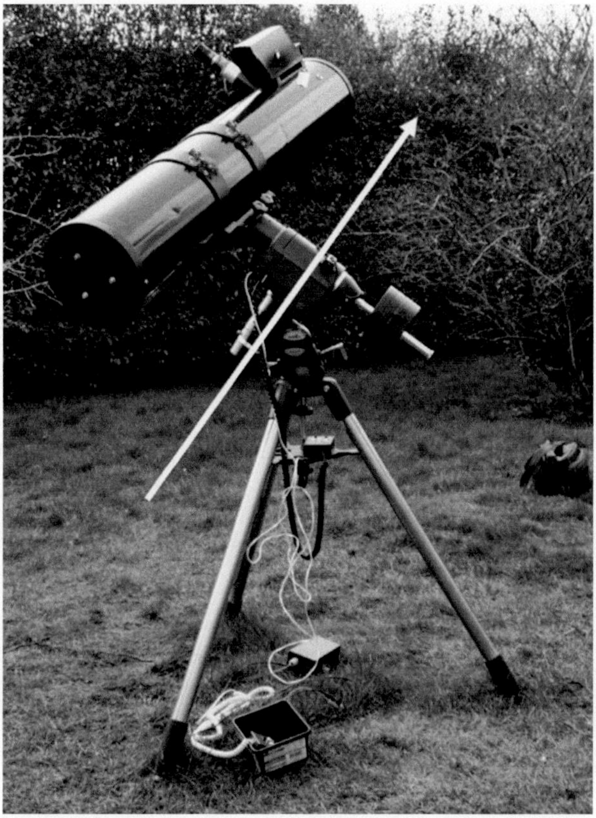

Fig. 2.5 A German-type equatorial mount. The axis shown by the white arrow is adjusted so that it lies parallel to the Earth's axis. That means pointing the telescope north in the Northern Hemisphere or south in the Southern Hemisphere. (Image by the Author © Jane Clark)

Wedge-and-fork mounts don't need to perform meridian flips, but they can be significantly harder to keep balanced as you add and remove equipment. You need a good supply of small counter-weights to add & remove.

All loads on mounts should be so well balanced that, with the clutches off, you can rotate the telescope with one finger.

Commercial mounts for amateurs have to be designed to sell into a mass market, where the customers will live at locations with a wide range of latitudes. Therefore, mechanisms for adjusting the latitude have to be built in. That constraint does not apply to giant, bespoke professional telescopes.

Fig. 2.6 This is a telescope with a fork mount. This mount will not correct for field rotation, which greatly reduces its usefulness for deep sky imaging. If you own one, the best remedy is to buy a wedge to tilt the azimuth angle until it is parallel to the Earth's axis of rotation. (Image by the Author © Jane Clark)

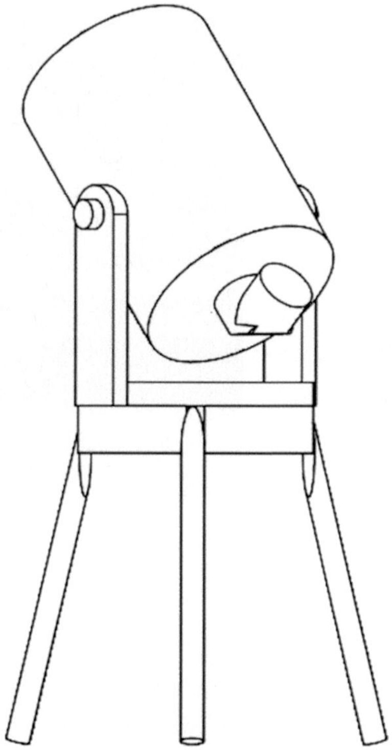

Fig. 2.7 This is a telescope with a fork mount, like the one in Fig. 1.6, but with a wedge to tilt the azimuth angle of rotation until it is parallel to the Earth's Axis. (Image by the Author © Jane Clark)

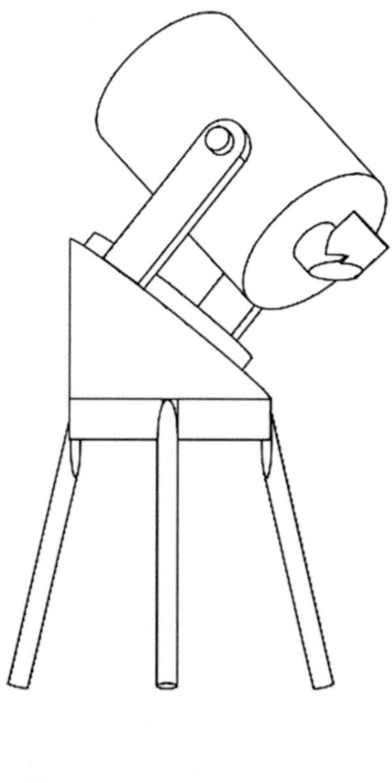

Compensating for Imperfect Alignment of Equatorial Mounts

It goes without saying that the better you can align your mounts, and the better you balance the weights on the two axes, the easier your life will be.

The technique of choice for compensating for alignment imperfections is to "guide" your mount. This is a technique whereby you send commands to your mount to keep a star in the center of crosshairs on a computer screen, fine tuning the tracking at frequent intervals. The overwhelmingly dominant software for doing this on a PC has been a piece of freeware called PHD2 for at least the last decade. It can be found at https://openphdguiding.org.

A screenshot of PHD2 in action is shown in Fig. 2.8. Once you have set the camera up and loaded its particulars, including guide scope focal length, into PHD2, you use the buttons along the bottom left. The USB icon connects the guide camera. The circular arrows icon starts capturing images from this camera. The star-and-magnifier icon selects the guide star. The cross-hairs Icon, which shows green when it is not greyed out, is the one to click to start guiding.

If the set-up is not already calibrated, it will spend 10–15 minutes calibrating the guiding. You really only want to do this once, so on subsequent runs, you navigate

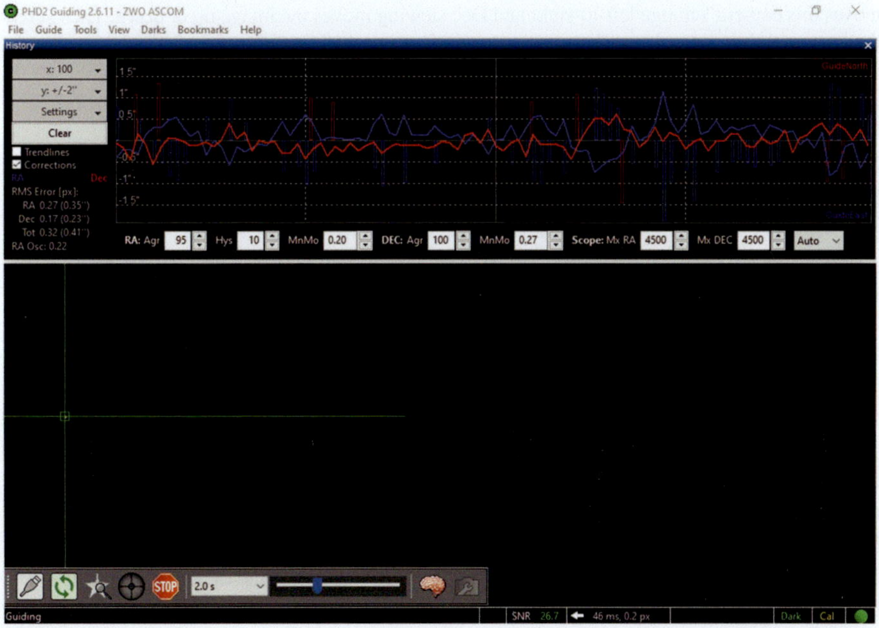

Fig. 2.8 A screenshot pf PHD2 in action. The image is captured with a second camera. Crosshairs are placed at a star chosen by clicking on the star-and-magnifying-glass icon at the bottom left. The software sends commands to the mount to slew slightly to keep the star at the center of the green crosshairs. I have chosen to display the history of the corrections made. In the graph, blue is right ascension & red is declination. (Image by the Author)

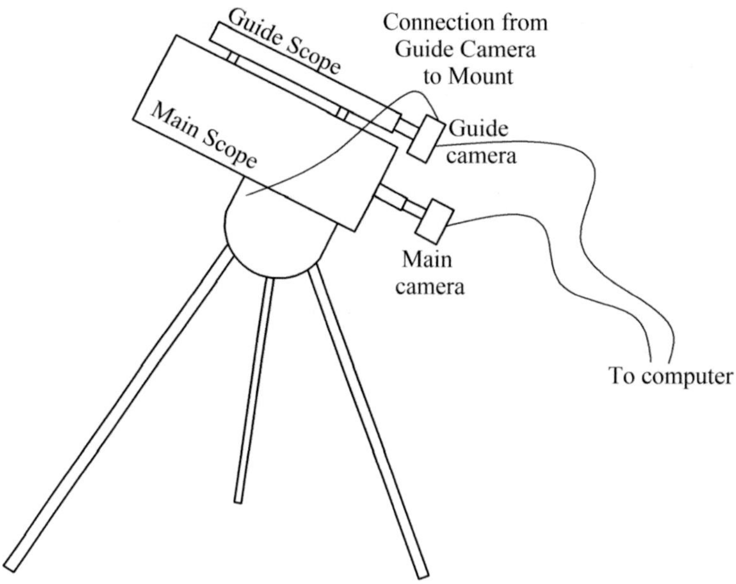

Fig. 2.9 The principle of guiding with a guide scope. A separate telescope is attached to your main telescope. A camera is fitted to it. You should focus this carefully, and rotate it until RA slewing moves this camera's image left-to-right and Dec slewing moves the image up or down. Then screw the camera tightly into place. Thumbscrews are often supplied for this purpose. The guide camera is connected to the mount and to your computer. You can then tell PHD2 what it needs to know to use this auxiliary scope and camera to guide

Tools → Restore Calibration Data. If you have a German Equatorial Mount, and perform a meridional flip, you can navigate Tools → Flip Calibration Now, and your calibration data will be good to go.

Another piece of setting up that you have to do the first time is to take a set of dark frames. The camera will do this for all possible exposures up to about 30 seconds, which takes a while, so take care not to lose your dark frames. The easiest way to do this is by selecting the wrong camera as your guide camera. PHD2 will see all the cameras that are connected to your PC. Take care to select the right one.

You can then use one of three methods to guide.

The method I use is to have a secondary telescope attached to my main telescope for guiding. The scope and camera do not need to be of the same quality as your main imaging scope. This arrangement is shown in Fig. 2.9. You can either attach the guiding camera to your mount or to your computer. You have to tell PHD2 which of these two methods you have chosen. I attach to the mount rather than the PC because that frees up a port on the PC.

A variant on this theme is to use a guide scope with enough built-in intelligence to bypass your PC. Celestron offer the Celestron StarSense Autoguider™.

The main alternative is to use an off-axis guider. The principle is shown in Figs. 2.10 and 2.11. A ring in the train of rings holding the main camera to the

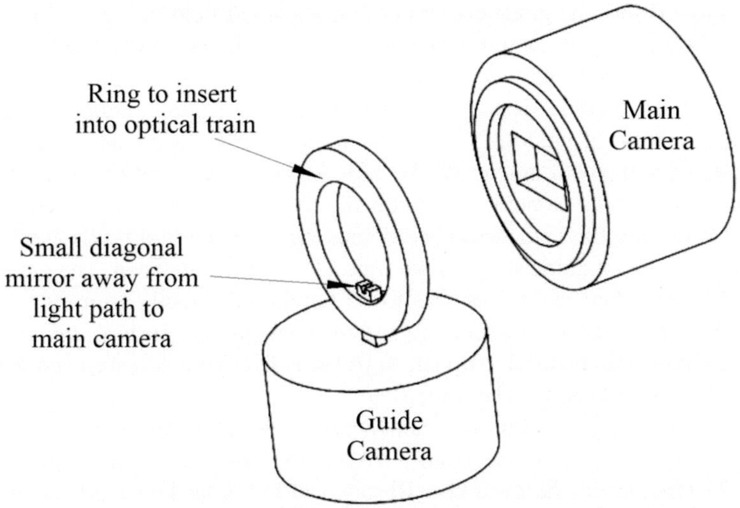

Fig. 2.10 The principle of off-axis guiding. A ring in the train of rings holding the main camera to the telescope contains a small 45°-angled mirror to send light that would not reach the main camera chip to a secondary, guiding camera. You can then tell PHD2 what it needs to know to use this auxiliary camera to guide. (Image by the Author)

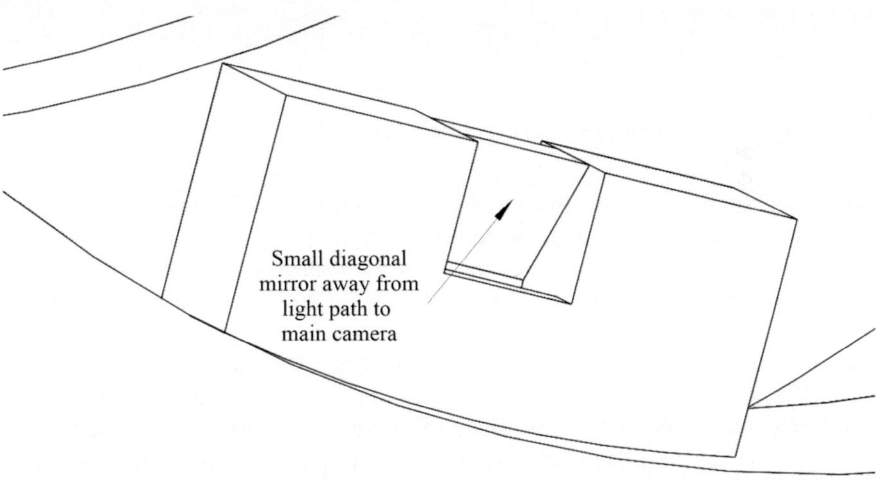

Fig. 2.11 The principle of off-axis guiding. Close-up showing the small 45°-angled mirror to send light to the secondary, guiding camera. (Image by the Author)

telescope contains a small 45°-angled mirror to send light that would not reach the main camera chip to a secondary, guiding camera. You can then tell PHD2 what it needs to know to use this auxiliary camera to guide.

The advantage of this method is that the image guide camera has the same focal length and magnification as the main image, leading to more accurate guiding. The disadvantage is that the guide camera has only a small field of view, so it may not see a star bright enough to guide on. Many off-axis guiders allow you to rotate the mirror around the circumference of the primary camera to find a guide star.

With a separate guide scope, you almost never have trouble finding a guide star.

A third method of guiding, which I have not tried, is to use an on-axis guider. This contains a diagonal mirror which sends visible light to your main camera and IR light to your guide camera. You guide and record from the same image, only at different wavelengths. This method combines the accuracy of off-axis guiding with the ease of finding a guide star of a separate scope. It is not cheap. The only equipment I know of that can do this has to be purchased directly from https://www.innovationsforesight.com/product-category/on-axis-guider. It has the additional advantage that it can be used to update your focus live: your scope can cool enough that it contracts and needs to be refocused.

If portability is your aim, you can bypass a PC altogether and capture your images & control your scope from a ZWO ASIAIR Mini™ or ZWO ASIAIR Plus™. The underlying hardware is a Raspberry PI™. This device is very small and can be used with a ZWO Harmonic Equatorial Mount. You need a smartphone to control the ASIAIR. Combining this with a ZWO mount and a modest sized telescope will take care of your guiding, polar alignment, mount control, focusing and electronic filter wheel control, all provided that you use ZWO equipment, except for your telescope. I have not tried this because I have a fixed observatory, but I have friends who swear by it.

Telescopes for Photographing Galaxies

Deep sky photographers of nebulae[1] often favor refracting telescopes, because they have smaller magnification and a wide field of view.

With one or two notable exceptions, galaxies appear to be much smaller than nebulae, and will look rather small through a typical apochromatic refractor. If you have a camera with small pixels – typically less than 4 μm – you can crop your image and still obtain something respectable from a refractor.

Really, however, I would think about a telescope with a longer focal length, and hence higher magnification. In this section I discuss telescope types, and the trade-offs made. In the 17 years I have been doing astrophotography, better telescopes have appeared to the point where the choice is bewildering. The vast majority of us have to live with limited budgets. I'm sure that this list is not exhaustive, but it covers the kinds of telescopes I have seen used. I've omitted one or two types of very fine telescope, which are not really very useful for photographing all but the 'Local

[1] The plural of *nebula* is *nebulae*. In Classical Latin, this would have been pronounced "neb-you-lye". You will hear this pronunciation from time to time. The more common pronunciation in English is "neb-you-lee". It is acceptable to use *nebulas* as the plural. But it is wrong to pronounce *nebulae* as "neb-you-la".

Group'[2] of galaxies, such as Rowe-Ackerman Schmidt Astrographs, known as RASA scopes.

Refractors

This is the original kind of telescope, and is shown in Fig. 2.12. You can buy very cheap (and nasty) versions. A more sophisticated one will correct for spherical aberration by having aspherical lenses, often concave on the inside and convex on the outside. *Achromatic* telescopes bring two colors of light into sharp focus at once, typically red & blue, usually by having two types of glass, often crown glass and flint glass. You can add a third type of fluoro-crown glass, to correct for three wavelengths, red, green & blue, instead of two. Such a lens is called *apochromatic*. This gives a significantly sharper image with less color fringing. Unfortunately, fluoro-crown glass is expensive, making apochromatic telescopes quite a bit more expensive than achromats.

The advantages of refracting telescopes are that they are light-weight, and give more image contrast than mirror-based telescopes. They also never need to be collimated. The principal limitation is that it is not practical to make large refractors. They would need very long focal lengths, making them unwieldy, and they are not price-competitive with reflecting telescopes.

The purpose of the field flattener is to bring off axis parts of the image (i.e. most of the image) into common focus on a flat camera chip. I certainly would not buy a refracting telescope without a field flattener. The distance between a field flattener and the camera chip is critical: the flattening effect only works at one distance, which the telescope's manual should tell you.

I own an apochromatic refractor: a William Optics Zenithstar III Z73, with a 73 mm (almost 3 inches) f/5.9 objective lens. It is a low-magnification,

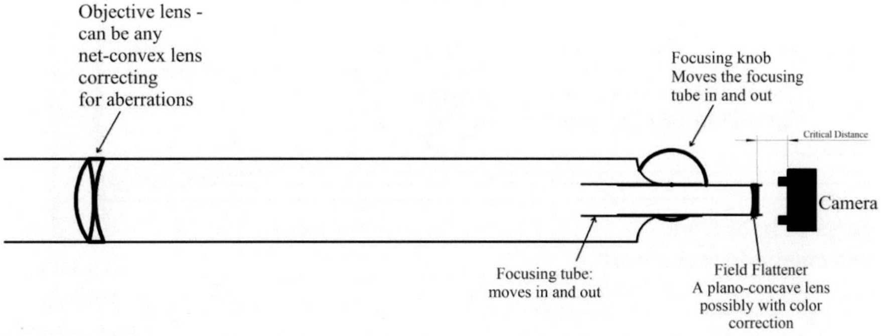

Fig. 2.12 The basic layout for photography of a refracting telescope. (Image by the Author)

[2] M31, M33, the Milky Way, and their smaller satellites.

wide-field-of-view instrument, for photographing objects with large angular diameters. I used the field-of-view calculator at https://astronomy.tools/calculators/field_of_view to make sure that it would fit the Andromeda Galaxy M31 in one shot. It will – just. It has some other nice features, like a Bahtinov mask built into the lens cap, and a dual-speed focuser. I have never found it to be necessary to motorize the focuser: it is very easy to use, and has minimal backlash. Admittedly I use it on a very rigid mount, piggybacked on my big Schmidt-Cassegrain telescope. You will be able to see the results I obtain with the refractor in various places in Chaps. 5 and 6.

Newtonian Telescopes

This was invented by Sir Isaac Newton because in his day there was no cure for chromatic aberration. It is a very practical design that sells well to this day. Over the years I have owned two, and have always vaguely regretted selling them when I upgraded, so good were they. Alas I only have so much space – and money! The layout of a Newtonian telescope is shown in Fig. 2.13.

Advantages are that dewfall is much less of a hazard if the tube is closed, as opposed to an open truss structure, and that larger apertures are cheaper to achieve than with refractors.

Downsides are that the mirrors tarnish because they are open to the elements, and that the focuser on larger models can become quite high up, necessitating the use of step ladders. Using step ladders in the dark isn't the best idea ever, especially if the ground isn't very flat.

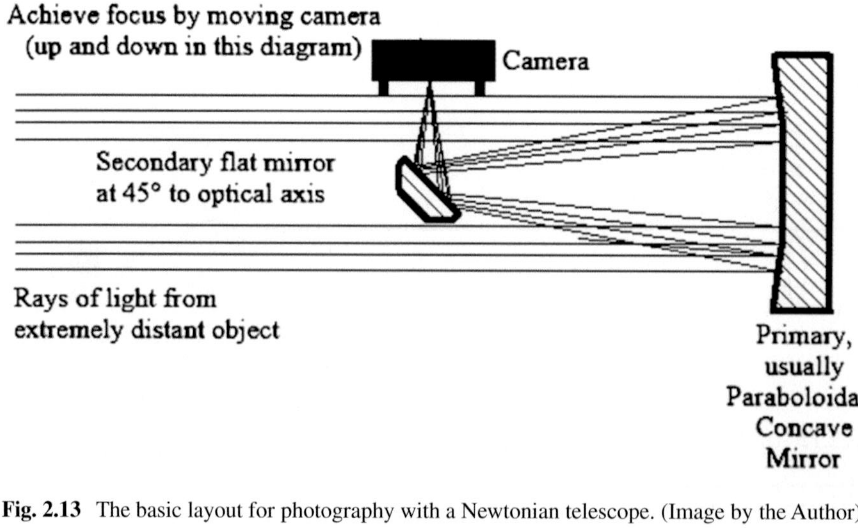

Fig. 2.13 The basic layout for photography with a Newtonian telescope. (Image by the Author)

They require regular collimation, most commonly of the secondary mirror. There are plenty of instruments to do this, such as laser collimators and cheshire collimators. The first time I collimated my Newtonian took ages, but I soon reduced it to a 2-minute task.

The 'spider' arms holding the secondary mirror in place are the items that deform, leading to loss of collimation. They also cause diffraction spikes in your image, which is why they aren't made big & strong.

A further disadvantage of most Newtonian telescopes is that there isn't enough focus travel to accommodate a DSLR camera.

Schmidt Cassegrain Telescopes (SCT)

The layout of an SCT is shown in Fig. 2.14.

The advantages of this telescope type are that it is compact, and that you don't need to climb a ladder in the dark to adjust the focus or the camera. The compactness also means that you can manage with a smaller observatory building.

The downsides are firstly that the glass corrector plate needs to be protected from dewfall, with heater strips, dew shields or an observatory dome. I use both heater strips and the observatory dome. I don't then also need a dew shield, although I do occasionally use one to keep out stray light from inconsiderate neighbors.

The second downside is that the primary mirror is moved to achieve focus. In larger basic models, this mirror can be ever so slightly floppy, which degrades your sharpness as the telescope tracks through the night. In more expensive models, there are mirror locks to minimize this effect. Even in basic models, this focusing mechanism is expensive.

Some manufacturers offer more sophisticated variants. Meade offer an Advanced Coma-Free (ACF) version with hyperboloidal secondary mirror and a different

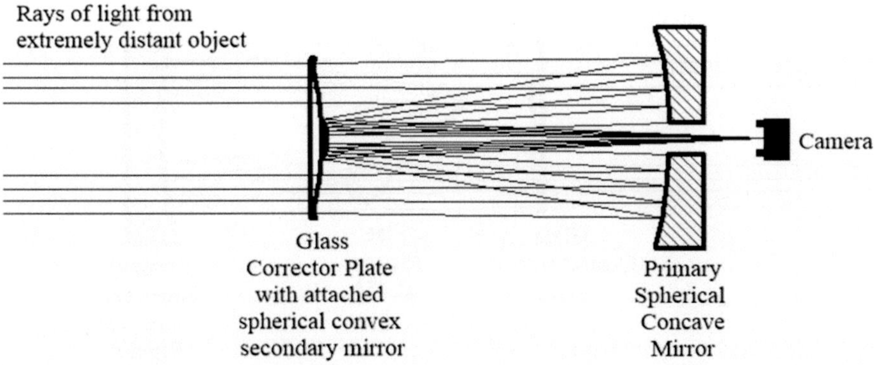

Fig. 2.14 The basic layout for photography of a Schmidt-Cassegrain telescope. (Image by the Author)

corrector plate. Celestron offer an "Edge HD" version with a plano-concave field-flattener lens close to the opening in the primary mirror. Both these models offer locking mechanisms for the primary mirror once it is focused. For most of my work in this book, I owned a Celestron 11-inch SCT, but late in the project I upgraded to a used Celestron Edge HD 1100, the 11-inch model, and sold the basic SCT. They are both good telescopes, but the Edge HD does give sharper images.

Of course, the weakest link in your optical train is the pesky atmosphere. If it is uncooperative, no amount of wizzo optics can fix that. That's not quite true. I look forward to the day when someone invents an adaptive optics system for amateurs, like the ones on professional telescopes. Meanwhile, back at the crummy seeing…

Schmidt-Newtonian Telescopes

This is an attempt to cut the cost of SCTs by having a flat secondary mirror, inclined at 45° like in a Newtonian. The spherical primary mirror, made possible by the corrector plate, is also a way to save cost. As with many hybrids, it has the worst disadvantages of both telescope types, but if you are on a low budget, the cost saving may be a benefit that outweighs the disadvantages. When I was at the parent stage of life, my budget was much tighter than it is now. I sympathize with cost concerns. Figure 2.15 shows a schematic layout.

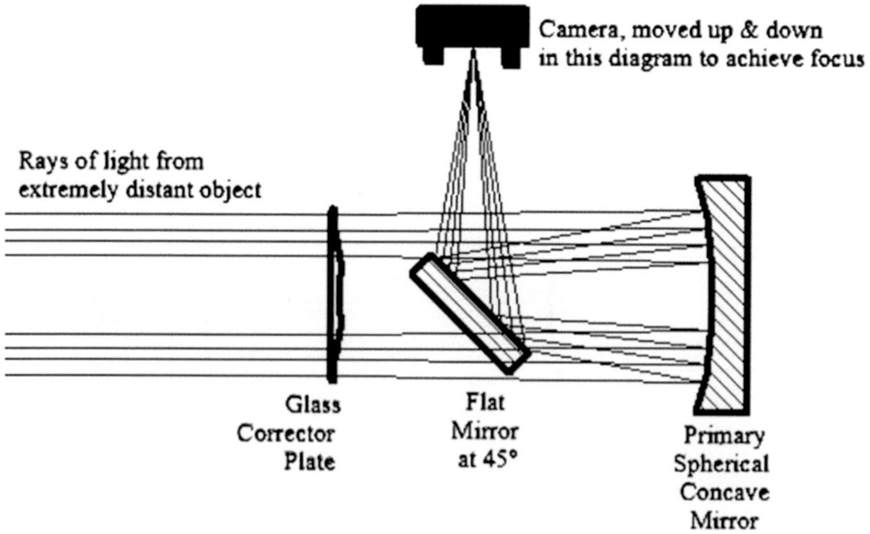

Fig. 2.15 The basic layout for photography of a Schmidt-Newtonian telescope. (Image by the Author)

Maksutov Telescopes

Maksutovs are a variant on SCTs with a heavy concave corrector plate. They only really work at long focal lengths and at modest apertures, up to about 7 inches. The biggest ones are sold as 180 mm telescopes, but 150 mm (6 inch) 127 mm (c.5 inch) models are also common. The optical layout is shown in Fig. 2.16.

They have high f/ numbers, typically f/13-f/15. This high magnification feature is attractive for photographing galaxies. They are often recommended for planetary photography for the same reason.

Focus is attained by moving the primary mirror.

Typically, the secondary mirror cannot be collimated, but since it is attached to a heavy corrector lens, this is not much of a problem. The downside of this heavy corrector lens is that it can be slow to cool down at night. It would generally be best to store such a telescope in a cool place.

Dall-Kirkham Telescopes

A Dall-Kirkham telescope is like an SCT, only with no correcting lens and an ellipsoidal primary mirror, not a spherical one. The secondary mirror is spherical. This only works at long focal lengths, above f/15. In practice a set of lenses is placed fairly close to the camera chip to correct the aberrations and enabling f/8 to be achieved. The manufacturers understandably play their cards very close to their chests about the nature of the correcting lenses. The basic layout is shown in Fig. 2.17.

These are very high-end telescopes for amateurs. They are very expensive. Apart from the 12.5-inch model, I am tempted to say that they are aimed at the professional market.

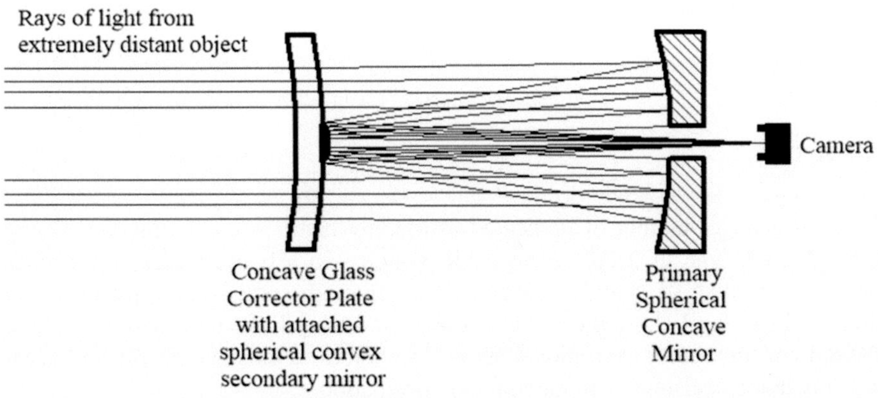

Fig. 2.16 The basic layout for photography of a Maksutov telescope. (Image by the Author)

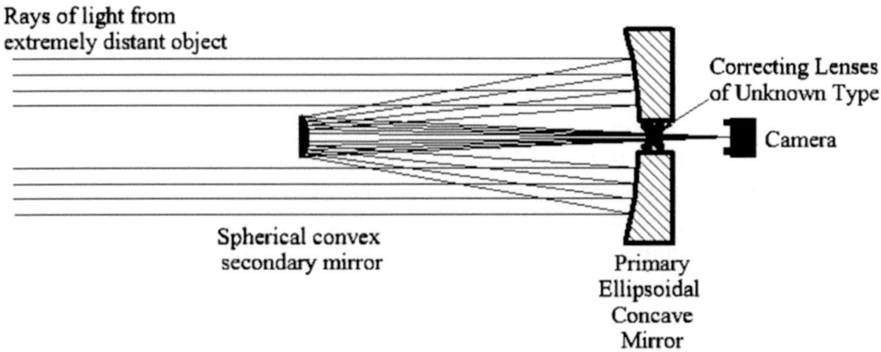

Fig. 2.17 The basic layout for photography of a Corrected Dall-Kirkham telescope. (Image by the Author)

Perhaps I should also add that my cartoon depiction of the primary mirror is unfair to PlaneWave, a leading manufacturer of Corrected Dall-Kirkham telescopes, who have gone out of their way to reduce the weight of their primary mirrors.

Automatic Tabletop Telescopes

Small telescopes are beginning to appear with built-in cameras, that are controlled from a smart phone. They are very automated, and will produce photos of deep-sky objects which are better than just recognizable. Investing in these systems is not going to buy you a long-term product. New ones are coming onto the market at a rapid rate. But there is promise that they could be a big step in democratizing astronomy. In my opinion, that has to be a good thing, because more and more people will have fun and learn to appreciate the evils of light pollution. Instead of having a handful of hardcore observers in your astronomy club, most of them could leave their armchairs and take pictures. I intend to keep a weather eye on developments here.

Cameras

There is a huge number of dedicated astronomy cameras on the market. Going through each type in detail seems a bit pointless in a book because electronics becomes obsolete so quickly. Instead, I'll go through some basic principles, and then explain why I chose the cameras I own, and what I like most and least about them. I am sometimes as irrational about my choices as the next person, and occasionally made decisions with incomplete information.

How Many Bits?

You typically find cameras described as 8-bit, 10-bit, 12-bit, 14-bit or 16-bit. Table 2.1 shows how many shades you get with different bit depths.

Does it matter? Yes. For most domestic photos, 8-bit jpeg files do the job. If I want to stretch a cloudscape – astronomers have to deal with cloudy weather somehow! – to capture the detail, I often find that 8 bits aren't enough. You end up with the transitions between shades looking like a contour map. So, for this, I switch my digital single lens reflex camera to 14-bit mode, and can get better detail.

In astrophotography, you often end up doing pretty extreme stretching of the image. So, the more bits your camera can handle, the better your chance of getting a good image after stretching.

The price you pay for this is of course that bit depth costs dollars, and you end up with bigger files of your images, in formats other than jpeg.

Differences Between Immature and Mature Technologies

Any form of technology has a life cycle. This has been true ever since the first stone was used to attack an animal somebody wanted to eat.

In Fig. 2.18, for any technology, there is an era of little or no development, followed by an era when the technology is being developed, and improving. Eventually the improvement stops, usually because it gets more and more expensive to make only marginal improvements. For example, if you buy or lease a car that will go between service intervals without breaking down, there's no point in increasing its mechanical quality. More realistically, perhaps 0.1% of cars of that type break down in between services. To reduce that fraction of dissatisfied customers from one in a thousand to one in ten thousand is very expensive in development costs. It may be cheaper either to put up with losing the odd customer, or even to give them a new car. So quality no longer improves: there is no business benefit to the manufacturer.

It may even be impossible to improve the technology much. For example, you could never make a laptop computer that uses thermionic valve technology instead of transistors. Yet in the 1950s, that was the only way to make an electronic computer.

Table 2.1 The number of shades available for different bit depths

Number of bits	Number of shades per channel
8	256
10	1024
12	4096
14	16,384
16	65,356

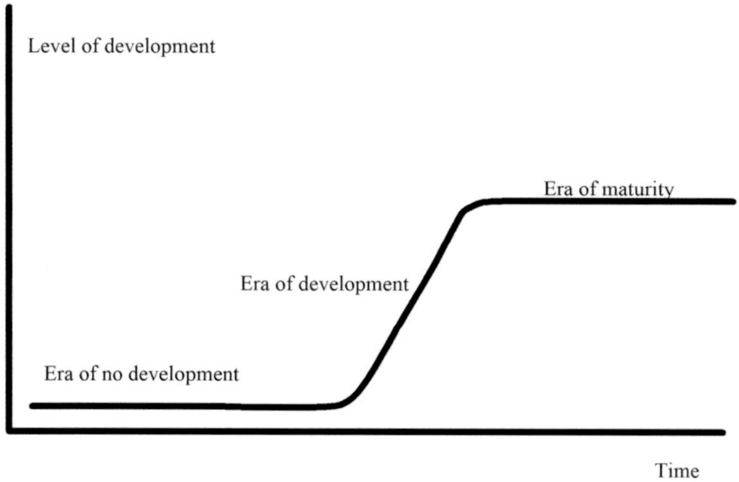

Fig. 2.18 Every technology has an era of no development, an era of rapid development and an era of maturity, when the only way to develop it further is to introduce new technologies. (Image by the Author)

The risk to the supplier of not improving their technology is that another technology may come along and disrupt their business.

An example which is well known to astrophotographers is the switch from film to digital photography. One of the main players in the film photography industry was Kodak. Even though their technical staff actually made the first hand held digital camera in 1975, they failed to make the transition to digital photography quickly enough, and were outsmarted by Asian competitors, eventually filing for bankruptcy in 2012. They are now a niche player in photography, not the dominant company that they once were [35].

Figure 2.19 shows what may happen when a new technology comes along. For a while, the new technology is inferior to the old. When I bought a good single-lens reflex camera back in 1995, my real motive was to buy a camera with automatic focus because I had a small child whose development I wanted to record. I enquired about digital cameras, but discovered that they were not yet as good as film cameras. So I didn't switch to digital, being more concerned about image quality than convenience. By the time I did buy a digital single-lens reflex (DSLR) camera in 2008, the situation was reversed: the quality was much better, and the infrastructure of PCs to store and manipulate the images was a lot better.

My mother made the opposite decision. She wasn't nearly as into photography as I was, and valued the convenience more than the quality. She switched to digital in about the year 2000. In other words, the benefits of the new technology were there for her even before the image quality was very good. That happens a lot: the new technology offers benefits the old one did not, even before it becomes better than the old technology.

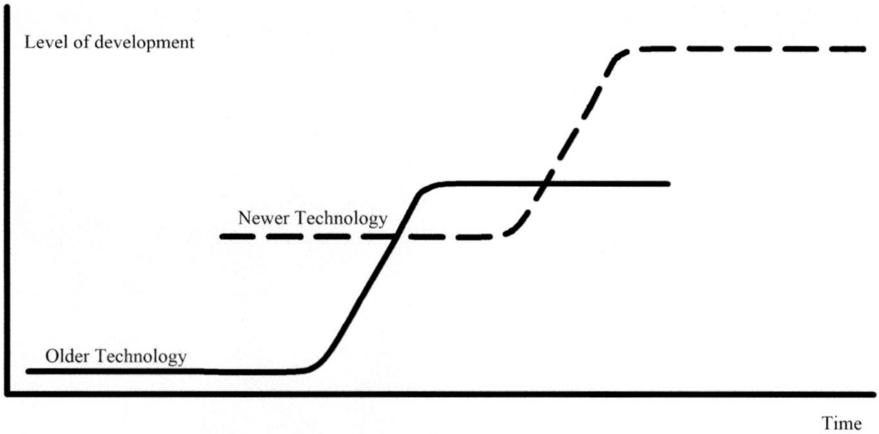

Fig. 2.19 A newer technology may come along and eventually replace the older technology before itself maturing. (Image by the Author)

Applying These Ideas to Cameras

DSLR Cameras

DSLR cameras have been capable of high-quality astrophotography for many years. They have CMOS chips, which is probably essential if they are to download images onto the memory chip in a reasonable time, and are always one-shot color cameras. They are a very good choice if your budget does not stretch to owning a separate astrophotography camera. My DSLR is shown in Fig. 2.20.

They have three disadvantages for astronomy. First, the chip is not cooled or temperature controlled. This means that you need to take your dark, flat & bias frames at the same time as your light frames (the frames you actually want). These will be explained in Chap. 3. Second, they block IR light, which may restrict their use. There is a small cottage industry that can modify a DSLR to remove the IR-blocking filter, but the camera cannot then be used for domestic photography. Thirdly, space has to be created inside the camera for the mirror to flip. The mirror is shown in Fig. 2.21. This means that the chip is a long way behind the lens flange, the big metal ring shown in Fig. 2.21. Not all telescopes can focus onto a DSLR chip. Newtonian telescopes are notorious for this, so please be sure to try before you buy.

Nowadays, DSLRs are being replaced be mirrorless cameras, with a digital image visible through the viewfinder, rather than a live image. In principle, this should obviate the need to allow space for the mirror. Unfortunately, in practice they need to be backwardly compatible with users' investment in DSLR lenses, so the same problem often exists with mirrorless cameras.

Back focus is not a problem with dedicated astronomical cameras. They will all work with most telescopes.

By default, DSLRs and mirrorless cameras produce 8-bit jpeg files. You can set them to produce RAW files, which are 14-bit in the case of Canon, 12-bit in the case

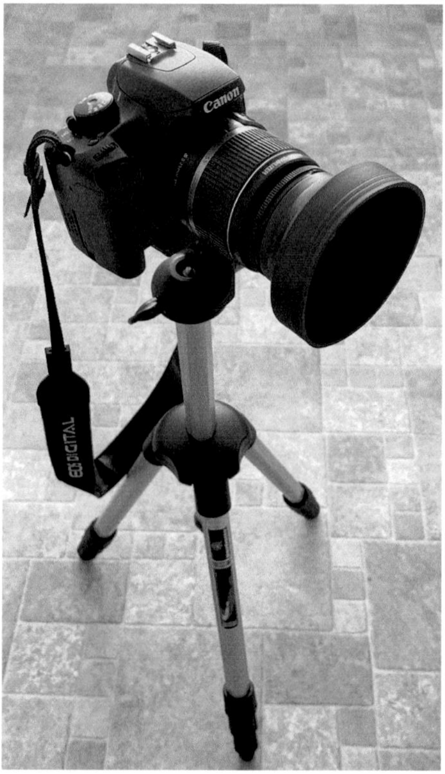

Fig. 2.20 My DSLR camera. (Image by the Author)

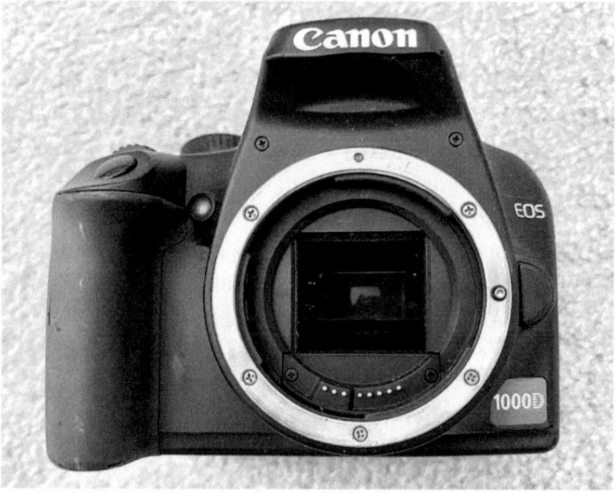

Fig. 2.21 Inside a DSLR is a mirror, which reflects to the viewfinder, except when the shutter is open. Then it moves out of the way. (Image by the Author)

of Nikon and 14-bit in the case of Fujifilm. RAW files are also proprietary: Canon produces files with extension .CR2; Nikon produces files with the .NEF extension; and Fujifilm uses the .RAF extension. The manufacturers don't publish the specifications for these files, but good image-processing software can read them. It is worth doing this for astrophotography.

CCD Dedicated Astrophotography Cameras

These were the cameras of choice if you were rich from about 2005 to 2018. They revolutionized astrophotography.

Cheap ones such as the Philips ToUcam took over from film for planetary photography. Very good they were too. CCD cameras for deep sky photography were quite a bit more expensive, but really excellent.

You could buy cooled cameras (see e.g. Fig. 2.22). The ability to shoot at a constant temperature greatly reduced the frequency with which you had to take dark, flat & bias frames.

What were the disadvantages of CCD cameras? Cost was undoubtedly one. They download the image one row of pixels at a time. This can lead to contamination of the other pixels in the row if one of them contains a brighter image than the others. So you can find linear defects in the image, which can be tedious to repair at the processing stage.

The method of downloading is slow. This matters not a whit for long exposures, such as taking 60-second frames, but it is killing for fast planetary photography. More seriously for deep sky photography, this feature tends to limit the chip size.

I used to own a monochrome Atik 414EX CCD camera. The only thing that irritated me about it was the small chip size. In those days I was not very good at finding galaxies, and wasted a lot of time trying to find them. I never had any complaints about the image quality.

I bought it because I also bought a spectrograph, and it was the camera they recommended. By 2019 when I bought it, it was already obsolete. I now have a CMOS camera that the spectrograph manufacturer recommends.

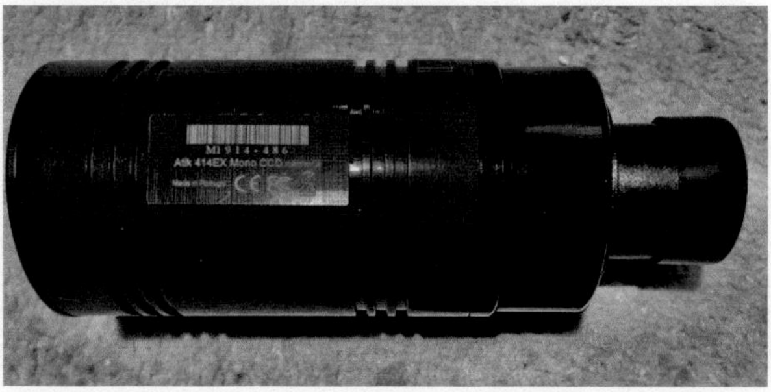

Fig. 2.22 My old ATIK 414EX camera. It required 12 V DC power and communicated with my PC via a USB 2.0 B lead. (Image by the Author)

The advantage of CCD over CMOS cameras was that the CCD ones are less prone to noise. I didn't really notice that when I was using mine, but with hindsight I can see that it was true. The model I had was a 16-bit camera.

CMOS Dedicated Astrophotography Cameras

These were available from about 2010 onwards, but the early models were not all that good.

That changed in about 2017 with the release of the cooled ZWO ASI 294MC and similar cameras with the Sony IMX 294 back-illuminated CMOS sensor. Now CMOS cameras are very useable for deep sky astrophotography.

I bought a AWO ASI 294MC camera, largely because I wanted a one-shot color camera. Its main other benefit for me over the ATIK camera was a larger image chip. This made it easier for me to find galaxies. You can't see most of them in a finder scope in an urban or suburban environment, so you have to find them with the camera attached.

But the one-shot color capability was also a benefit because I live in a place with a cloudy climate. I could make every frame count, whereas I was never sure of having the opportunity to shoot luminosity, then R, G & B frames. If I lived in a desert climate, I probably would have stuck with monochrome. But I don't; and I don't miss shooting separate L, R, G & B frames.

This camera could download about 7 full frames a second, and could download smaller frames a lot faster, making it good enough for planetary imaging. It is a 14-bit camera.

It is shown in Fig. 2.23.

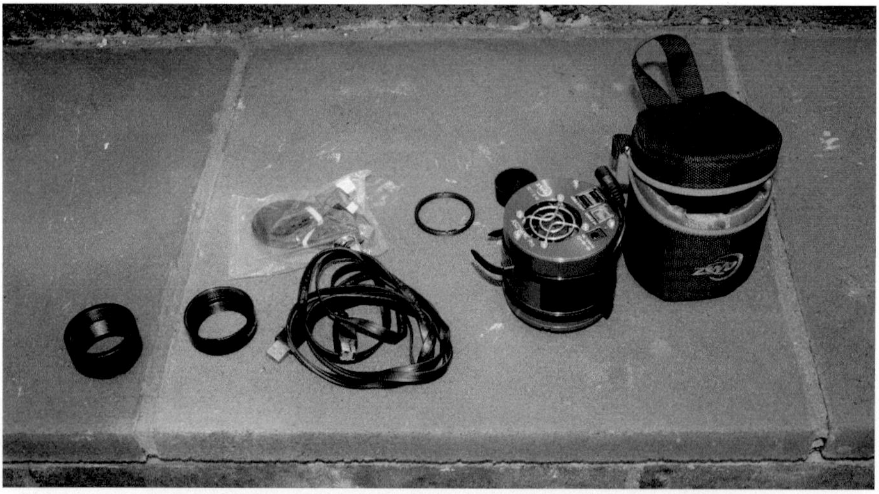

Fig. 2.23 My ZWO ASI294MC camera. It required 12 V DC power and communicated with my PC via a USB 3.0 lead. I added an external heater to combat dewfall on the cover of the cooled chip. In practice this was only a problem in the British summer. (Image by the Author)

I did find it noisy, which meant I had to expose for a long time to get rid of noise. Noise removal software has improved massively in the last few years, so that would be less of a problem now than it was then.

I bought the camera new for under $1000. I don't think I can really say that I researched this purchase very carefully.

Although this camera got me a long way, I began to notice that other people were shooting better pictures of galaxies with the ZWO ASI 2600MC, launched in 2019. I agonized about whether to upgrade for a long time. What sold me in the end was that I saw a few YouTube videos in which people stated that deep-space objects required about a third the exposure time with this camera compared to the 294 model. I bit the bullet and bought one. I didn't buy new. I bought a refurbished one because at one point there was a product recall because of an oil leak problem. The refurbished one had the same two-year warranty as a new one so I took the plunge. The same Sony IMX571 chip is in other astronomy cameras, but the word on the street was that the ZWO model worked better with my image capture software, SharpCap Pro. Before anyone sues me for libel, I cannot say whether this would actually have been true, but I paid over $2000, and was afraid that this upgrade might not deliver the benefits anyway, so I was feeling a bit risk averse. I also had had a bad experience with a useless guide camera I bought from one of the other manufacturers, so I wasn't keen on taking a risk with them. Besides, by buying the refurbished model, I had saved the cost difference. So I stuck with ZWO. This time I had researched before I bought.

I did in fact get the desired benefits and did not have to sell the camera on at a loss. I really love this camera. It is shown in Fig. 2.24. I have a camera rotator, which is particularly useful when performing meridian flips, because I can turn the camera

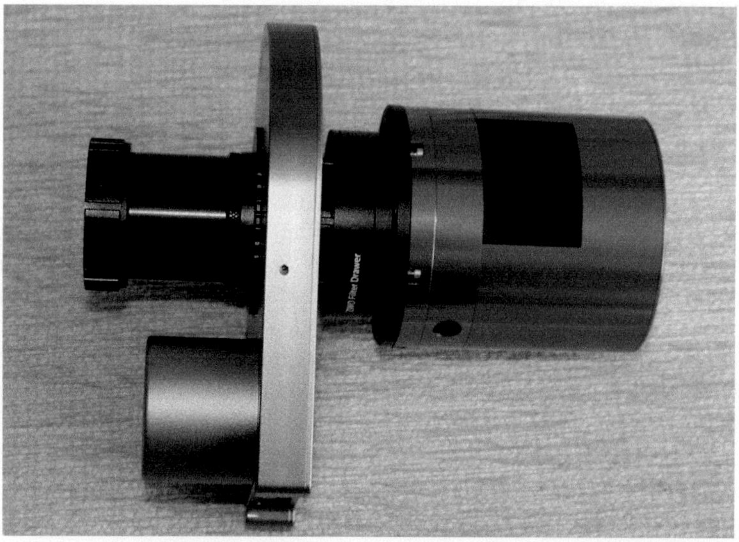

Fig. 2.24 My ZWO ASI2600MC camera. It requires 12 V DC power and communicates with my PC via a USB 3.0 lead. I use it with a ZWO 2-inch filter drawer, shown, and with a Pegasus Falcon camera rotator. (Image by the Author)

through 180° and the subsequent frames will stack without trouble. It is also useful if I want to fit an object on the chip. I occasionally use this camera to take pictures of the Moon, which barely fits on my chip. The galaxy M81 also benefits from having the camera rotated to fit it on the chip.

I don't always have the filter drawer there. It is useful for using a H-α filter. If I want to switch between that and white light, I put a UV-IR blocking filter in. This is because having no glass filter present will change the focal plane. The filter isn't necessary for UV: the camera has a built-in UV blocking filter.

It also has a built-in dew heater to keep the chip-protecting screen free of dew. I've never had a dew problem with it. The heater appears to be thermostatically controlled.

If I use this 16-bit camera at a gain of 100, it has a dynamic range of almost 14 stops. That's like being able to go from f/2 to f/256 and still get a usable image. It's astonishing to me. This means that the images taken can be stretched an awful lot without running out of tones.

Does it have any disadvantages? It's a bit heavy, but not massively so. This also means that the cooler can be a bit slow on warm nights. It could use a more powerful cooler; though users who go out to remote dark sites and rely on battery power may not appreciate this.

It doesn't have a very fast download speed, but this is an extremely minor irritant. You can't use it for planetary photography, but I would not want to: I have a separate camera for that.

It is also much less noise-prone than the ASI294MC.

I bought a used monochrome camera for applications where I really need monochrome. I experimented with it for H-α, but found no real benefit over my 2600 for this. I do occasionally take IR images of galaxies, and the mono camera is essential for that. And I can use the Baader-Bessell[3] range of UVBRI filters for astrometry. The one I have is a slightly older model, a ZWO ASI1600MM. It's not as good as the 2600 model, but I can't really justify the extra expense to myself. Its smaller chip means that I can get away with using 1.25″ filters, with only very slight vignetting at the edges. That saves me a lot of money. It also cools a lot better than the 2600. It can go to 45 °C (113 °F) below ambient temperature, and gets there quickly. Therefore for me, this camera satisfies a niche requirement. It is shown in Fig. 2.25.

Finally, in the background of Fig. 2.25, you can see my guide camera, a ZWO AIS120MM. You are better off with a mono camera for this application, and if I can expose for a second or less, which I usually can, I get much better guiding. Older CMOS cameras can't achieve exposures this short. It has a big enough chip that I can always find a star. You will not be surprised to hear that I bought this camera used.

[3] Michael S. Bessell, the filter expert, should not be confused with Friedrich Wilhelm Bessel, the eighteenth & nineteenth century astronomer & mathematician. Their surnames are spelled differently. Alas, even filter manufacturers sometimes get this wrong.

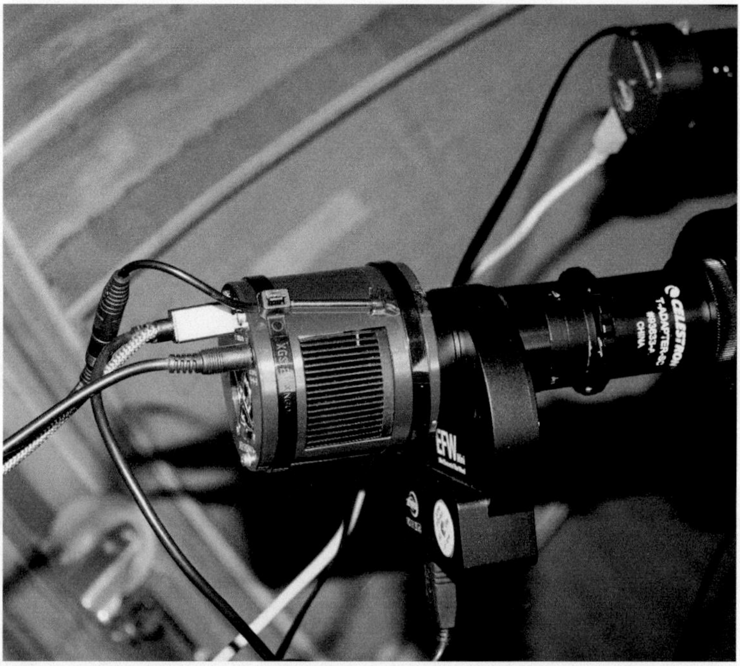

Fig. 2.25 My ZWO ASI1600MC camera in harness. I use it with a 5-wheel 1.25-inch filter. It requires 12 V DC power and communicates with my PC via a USB 3.0 lead. In the background, out of focus, you can see my ZWO ASI120MM guide camera. (Image by the Author)

Focus and Collimation

Focus

I do not recommend that you try to focus without help from an instrument. The most popular choice I see nowadays is a Bahtinov mask, correctly pronounced "**Bach**-tin-ov" with the CH as in Scottish "loch".

This device, two examples of which are shown in Fig. 2.26, needs to be sized for your equipment. They are available inexpensively online, though you can also make your own. Websites such as https://www.deepskywatch.com/Articles/make-bahtinov-mask.html offer templates. My method is to choose a bright star near my target and to focus on that, then slew the short distance to my target.

Since seeing in British skies is rarely good, I find the Bahtinov mask focusing tool in SharpCap, shown in Fig. 2.27, works well with my 73 mm refractor, but not so well with my 11-inch SCT. With the latter, I have learned through long practice to "eyeball" the focus with the Bahtinov mask. I might use the image in the focusing aid to look at the dotted lines it draws through the Bahtinov lines, but I don't take much notice of the numbers in the histogram graph, except with the refractor.

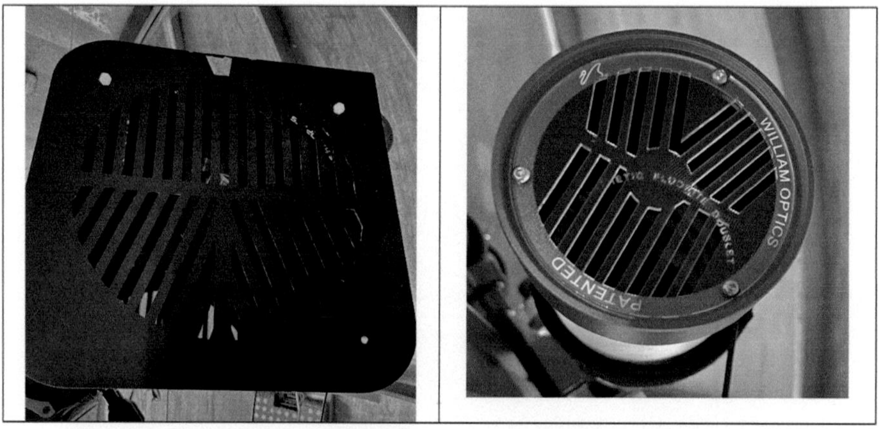

Fig. 2.26 Two examples of Bahtinov masks. (Images by the Author)

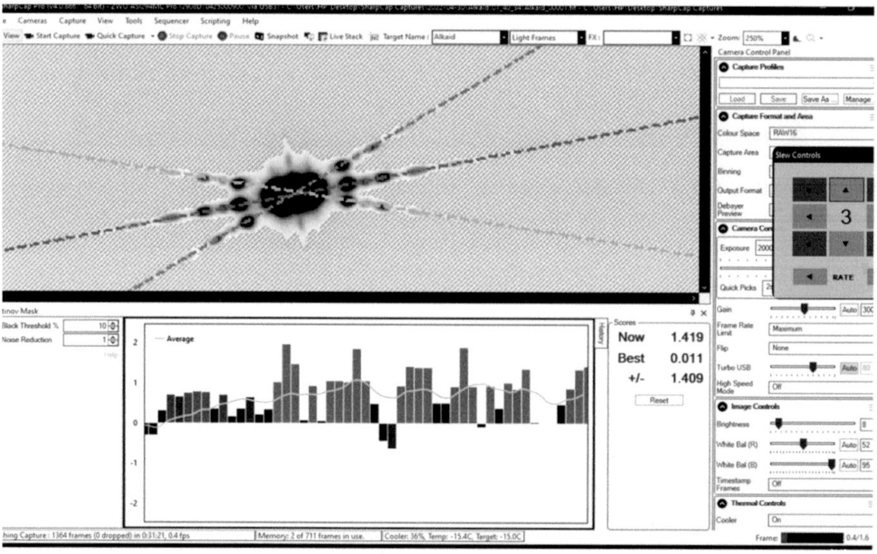

Fig. 2.27 The diffraction pattern produced by Bahtinov masks. In this image I have used the Bahtinov mask focusing tool in SharpCap. The middle line should cross the other two at the center of the star image. (Image by the Author)

I do the rough focusing with an exposure time around a quarter of a second, and then fine focus with an exposure time of 2 seconds.

If I refocus during a long run of data collection, I will use the Bahtinov mask on the brightest star in the frame showing my galaxy, and expose for 4 to 8 seconds as needed. I don't slew the telescope away from my target. I monitor sharpness using a measure in SharpCap called FWHM, Full Width Half Maximum, which is plotted for each frame if you have live stacking turned on.

Focus may drift if the telescope is still cooling. It also drifts somewhat with seeing conditions in my experience. In my maritime climate, temperature change is present but slow. In a Continental climate well to the south of the jet stream, the opposite is likely to be the case. Your seeing will be much better than mine, but your telescope is likely to experience more cooling.

If conditions allow, I will equilibrate my telescope thermally with the environment for at least two hours before starting to image. I do this by opening my observatory dome. At this point I switch on the dew heaters on the telescope and camera.

Image capture software like SharpCap and NINA can be programmed to refocus automatically every hour or so if you have a motorized focuser connected to the software.

Collimation

You only need to collimate telescopes with mirrors in them. Refractors do not need collimating.

Various tools are available to help with this. A device called a Cheshire Collimator is very cheap & helpful with Newtonians, though you really need a laser version to collimate at night.

Catadioptric telescopes are more difficult to collimate. You achieve rough collimation by throwing a bright star right out of focus. The secondary mirror produces a hole in the center of the out of focus star. You adjust the collimation screws very gently until the hole appears to be in the center (Fig. 2.28).

Many people find it rather difficult to tell when the hole is exactly centered. You can use tools to obtain a more accurate result.

Various tools are available to help with this. A device called a Cheshire Collimator is very cheap & helpful with Newtonians, though you really need a laser version to collimate at night.

In recent years, a tool called a tri-Bahtinov mask (Fig. 2.29) has become popular.

You set it in front to your corrector plate, or secondary mirror for a Newtonian. The three bumps on the inside ring must be aligned with the collimation screws on your secondary mirror, as shown in Fig. 2.29. I have installed Bob's Knobs (Fig. 2.30) on my Schmidt-Cassegrain telescope, replacing the collimation screws. These are inexpensive and easily available. Why would you bother? The main advantages are speed and safety. I don't want to be messing around in the dark with a screwdriver close to the expensive corrector plate.

The three individual slots are to place plastic screws in to hold the mask in place. They are adjustable. You then get an image on your camera chip or in your eyepiece like the one shown in Fig. 2.31.

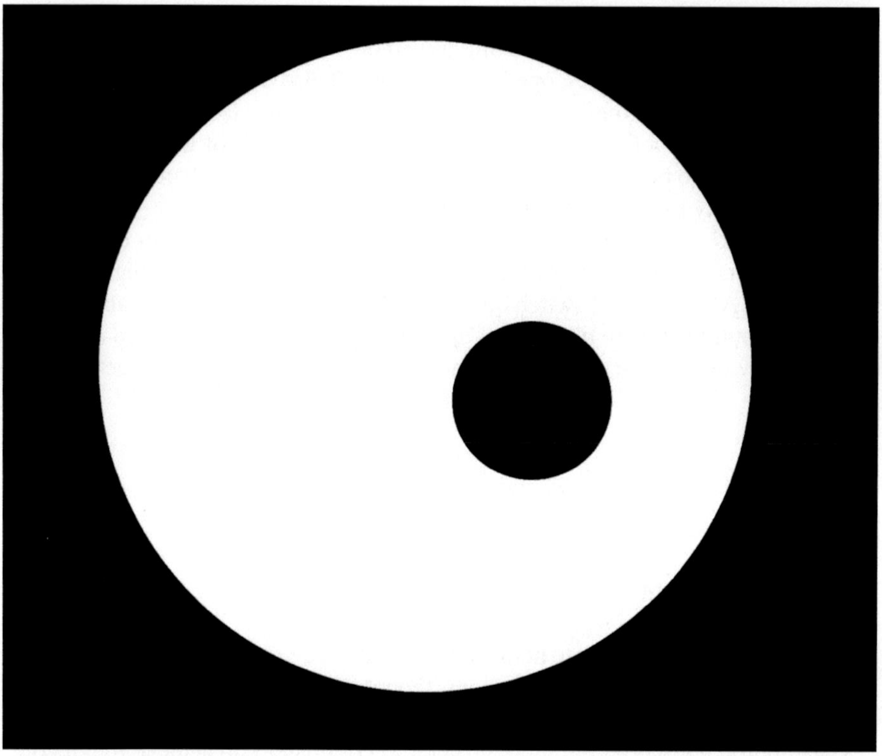

Fig. 2.28 A very out of focus, and out-of-collimation, star looks like this through a Schmidt-Cassegrain telescope, with an off-center hole in the middle. The aim is to collimate until the hole moves to the center. (Image by the Author)

The middle line in each triplet is supposed to be midway between the other two, as shown in Fig. 2.32.

The trickiest part of adjusting the three screws on the secondary mirror is knowing which screw affects which line. My way of dealing with this is to have a two-inch decorator's paintbrush and use it to obscure one set of lines as shown in Fig. 2.33. Why a paintbrush? It is inexpensive and the bristles are not likely to knock my scope out of line if I touch the mask in the dark.

The line shown as missing in Fig. 2.34 is the one that the collimating screw or knob *opposite* the paintbrush will alter.

You then adjust. It is vital to adjust collimation really gently. You are rarely far out of line. Plus, the very last thing you want to do is have a screw fall out in the dark.

I bought a software app called GoldFocus along with my mask. It will give you a quantitative readout of collimation – or lack thereof. At first, I found this to be essential, but now my eye is much more practiced, and I can accurately "eyeball" when the collimation is out.

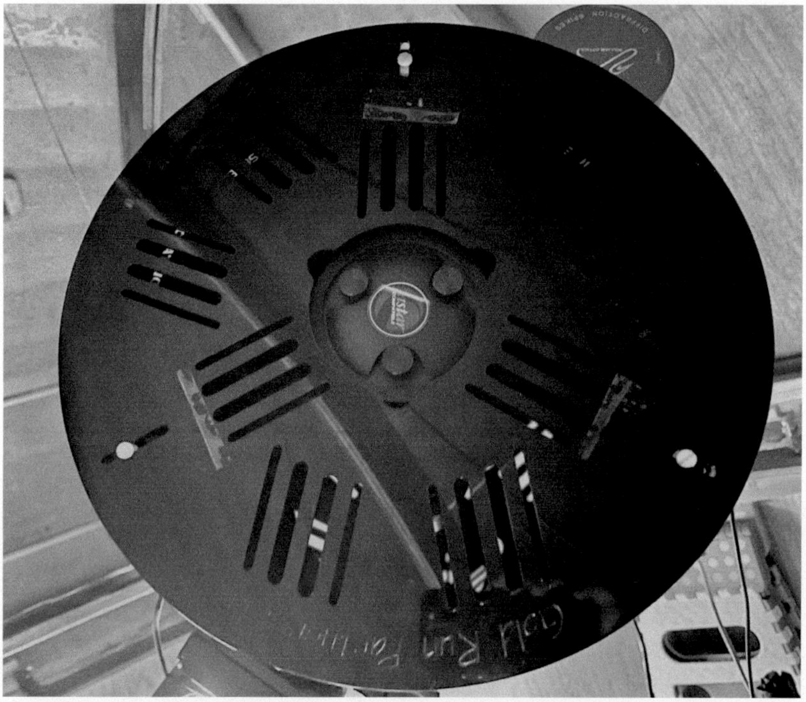

Fig. 2.29 The author's tri-Bahtinov mask. (Image by the Author)

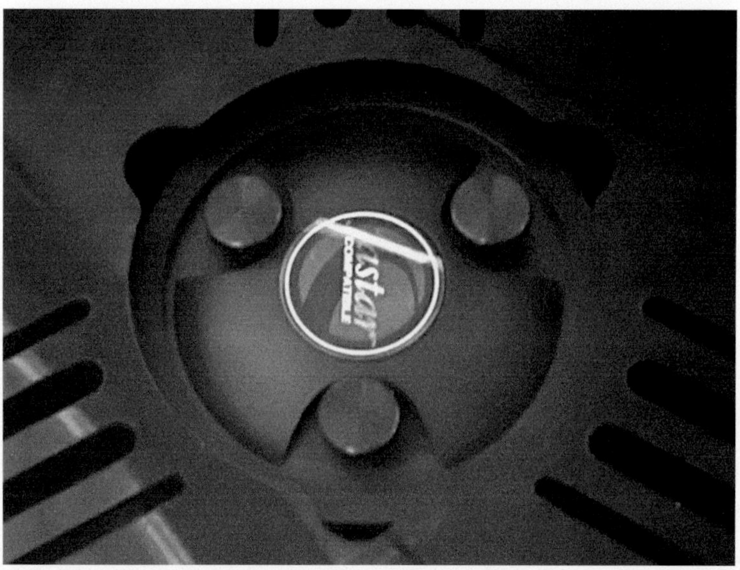

Fig. 2.30 Detail from Fig. 2.29 showing the Bob's Knobs that replace the collimation screws, and showing how the bumps on the inside of the mask are aligned with the collimating knobs or screws. (Image by the Author)

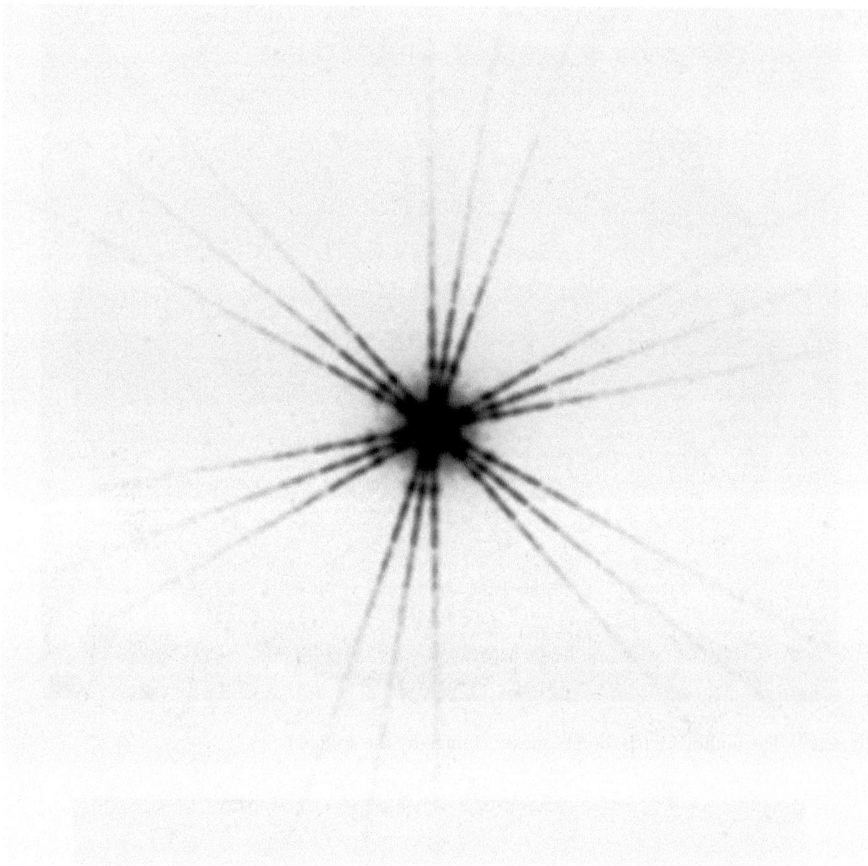

Fig. 2.31 A star looks like this through a tri-Bahtinov mask. (Image by the Author)

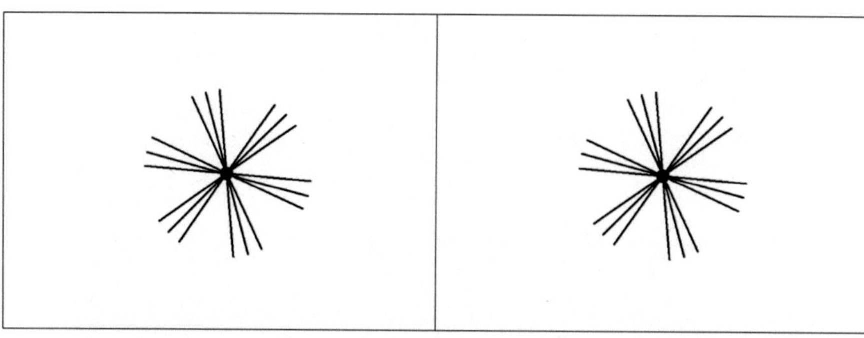

Fig. 2.32 A tri-Bahtinov mask image out of collimation (left) and in collimation (right). (Images by the Author)

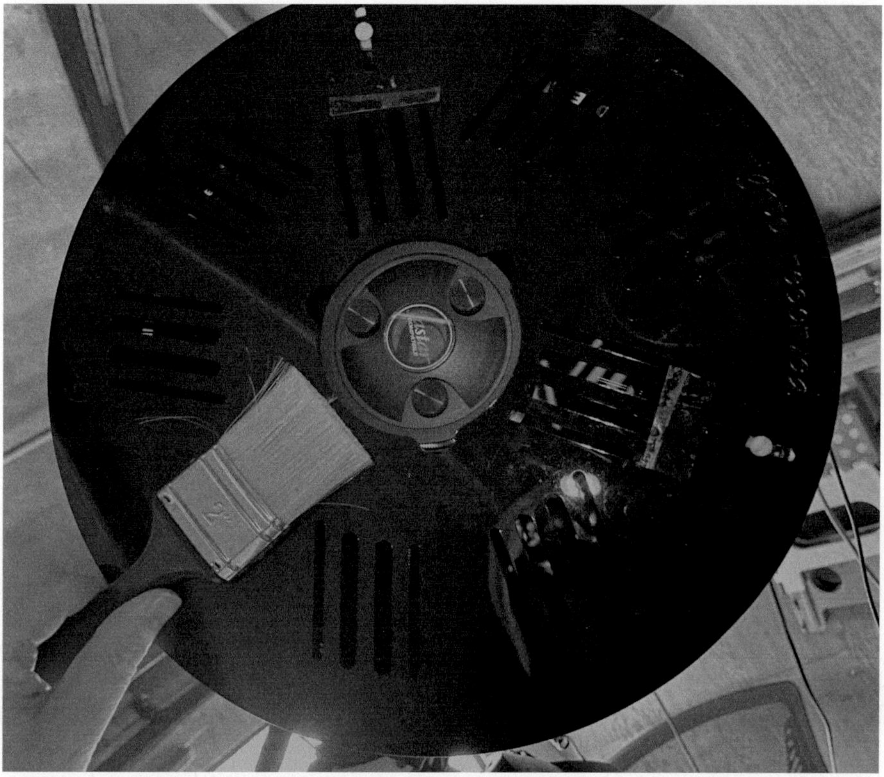

Fig. 2.33 Using a wide paintbrush to obscure a set of lines. The knob opposite the paintbrush will alter the missing line shown in Fig. 2.34. (Image by the Author)

Fig. 2.34 The paintbrush makes one line disappear from the tri-Bahtinov mask image. (Image by the Author)

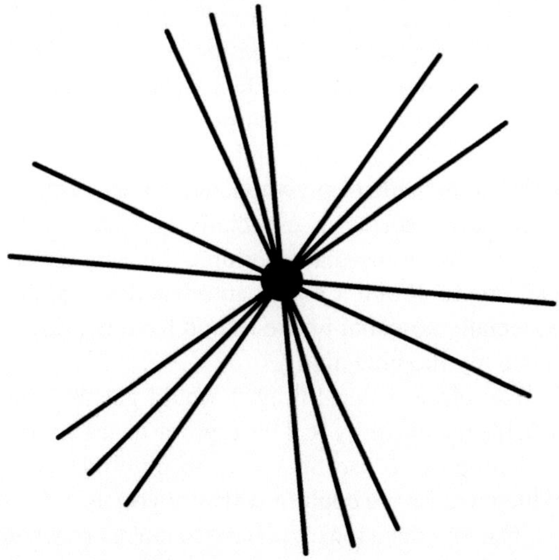

I sometimes see on the online forum the question, "Why are all three of my collimation lines equally out; and why can't I get then back into collimation?". The answer is that the questioner is also out of focus. You need to use the regular Bahtinov mask to achieve really good focus before you can use the tri-Bahtinov mask.

I do all this focusing and collimation through a camera. I turn the computer screen so that I can see it while I am doing the collimation. The thought of doing this through an eyepiece in the dark when you are alone is the stuff of nightmares.

You can quite easily collimate visually if you have a buddy with you, but I would not recommend it if you are alone.

Image Capture Software

With any digital camera, you need to send your images to a computer. On that computer, you need an app to capture, read and store the images. Even on my smartphone, there is an app to do this.

There are several ways to skin this particular cat. I have a desktop PC in my permanent observatory, and use a Windows app.

This is an obsolete way to work if you want portability. Then you should consider something like a ZWO ASAIR controller (https://www.zwoastro.com/product-category/asiair). This little box has a Raspberry PI computer inside it. It does rather lock you into using ZWO hardware, but it is very compact, and will control a lot of kit. It measures 100 × 70 × 26.5 mm (4 × 2.75 × 1.04 inches). You can even drive it from your smart phone. Competitors include Stellarmate (https://www.stellarmate.com) and Astroberry (https://www.astroberry.io).

Meanwhile, back at my desktop PC, viable deep sky software includes NINA and SharpCap pro. The former is freeware. The latter has a free version, but the annual fee for the Pro version is very modest and well worth it.

NINA is probably the easier to automate if you want to sleep while you gather data, although you can write scripts for SharpCap. But NINA is not good for Solar System work, whereas SharpCap is. That's where I started from, so I have somewhat invested in SharpCap knowledge. I also have a manually rotated observatory dome. It is made of plywood, and so heavy that I would probably have to spend big bucks to motorize it, especially to install a silent motor. I would really have to replace this home-made building, built by my friend Trevor Taylor, with a commercial one to obtain a viable automated setup. I don't feel any great need to do this, especially now that I have retired from my day job; and am happy to sit outside and nurse my imaging along.

Therefore, I don't think I would get the benefit of switching to NINA. People who use it swear by it. I have never heard a bad word said about it.

I propose to use SharpCap to highlight issues people might want to focus on, whichever image capture software you use.

The first decisions you have to make are whether to capture the full frame or part of it. For planetary work I typically capture 640 × 480 because calibration frames are not going to do much there. If worried, I might then shoot darks, but I never bother with flats. For deep sky stuff, I need flats, darks and biases; and it's much easier to do this for the full frame, so that's what I capture. I often crop at the final stage.

Next, what file format do you want? I recommend FITS files because you get a lot of text information for negligible storage cost; and because the postprocessing applications I will discuss in Chap. 3 use them.

10, 12 & 14-bit cameras have their images stored as 16-bit files. Storing as 8-bit throws information away, as we have already discussed.

I also typically expose 60-second frames. This decision is driven by living in a fickle climate where clouds are an occupational hazard. I would lose a lower percentage of frames if I were to shoot 30-second frames, but I would usually also run out of disk space. So I compromise on 60 seconds. Depending how good your guiding is, you could go as high as 5 minutes, but you might then need off-axis guiding rather than a separate guide scope.

With the gain at 100 for my ZWO ASI2600MC camera, I get a dynamic range of almost 14 stops. To shoot in H-α I need to set the gain to 300 to see anything.

All these settings are shown in Fig. 2.35. For ease of viewing in the book I have made negative images of the SharpCap screens. In practice, the screen is set in night mode, and looks very dark.

Since I have to do dome rotation manually, and since I actually like to sit & watch the heavens and chill, I don't program in automated sequences of capturing then stopping to dither or re-focus. I need to explore dithering, but the only time the kind of "walking noise" that dithering fixes is really a problem for me is when I shoot with a narrowband filter. You can certainly program such sequences into SharpCap and NINA.

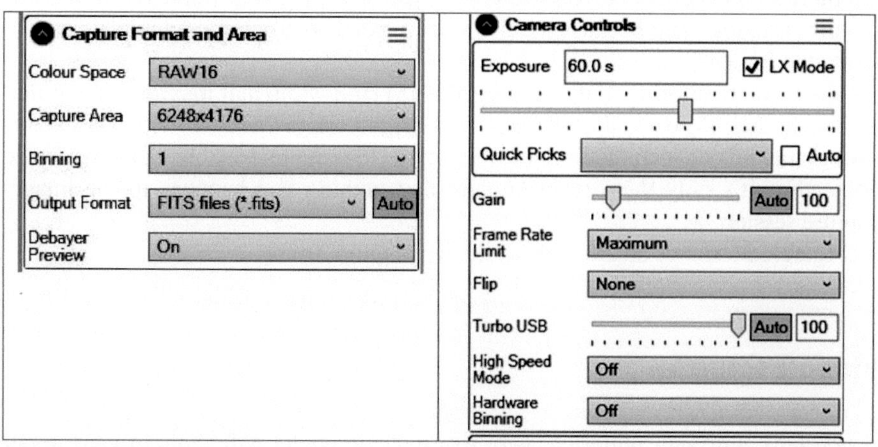

Fig. 2.35 Showing typical setup parameters that I use in SharpCap. I use full-frame FITS files in 16-bit mode. My usual recipe is to expose 60-second frames at a gain of 100 for as long as the weather, the dawn or the descent of the towards the horizon will allow me to. (Images by the Author)

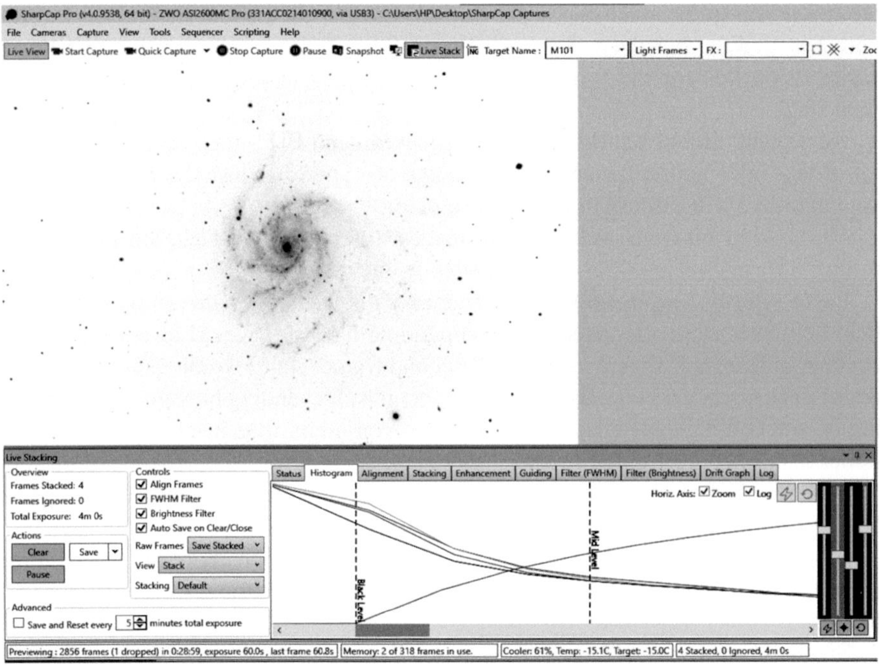

Fig. 2.36 The live stacking console in SharpCap. I adjust the histogram shown to be able to see the image, and monitor the FWHM tab for focus, and the Brightness tab for the effect of cloud. I have set cut-off values for these to stop stacking if the focus or the brightness is lost. (Image by the Author)

I do use the live stacking. I have tried to save and process live-stacked images, but there's no benefit once you know what you are doing. The reasons I do it are to monitor progress, and because the live stacking can be programmed not to save un-stackable frames.

The main monitoring tools I use are FWHM (full width half maximum) to monitor focus, and, if necessary, re-focus; and brightness to check on cloud. I use the histogram tools to keep the live stacking image looking reasonable, but I'm relaxed about this. It's only there to make sure that nothing bad happens to the image-capture session.

Figure 2.36 shows an example of the live capture frame.

Finding Galaxies

If you have any light pollution, very few galactic nuclei will show in your finder scope. You have to use other methods.

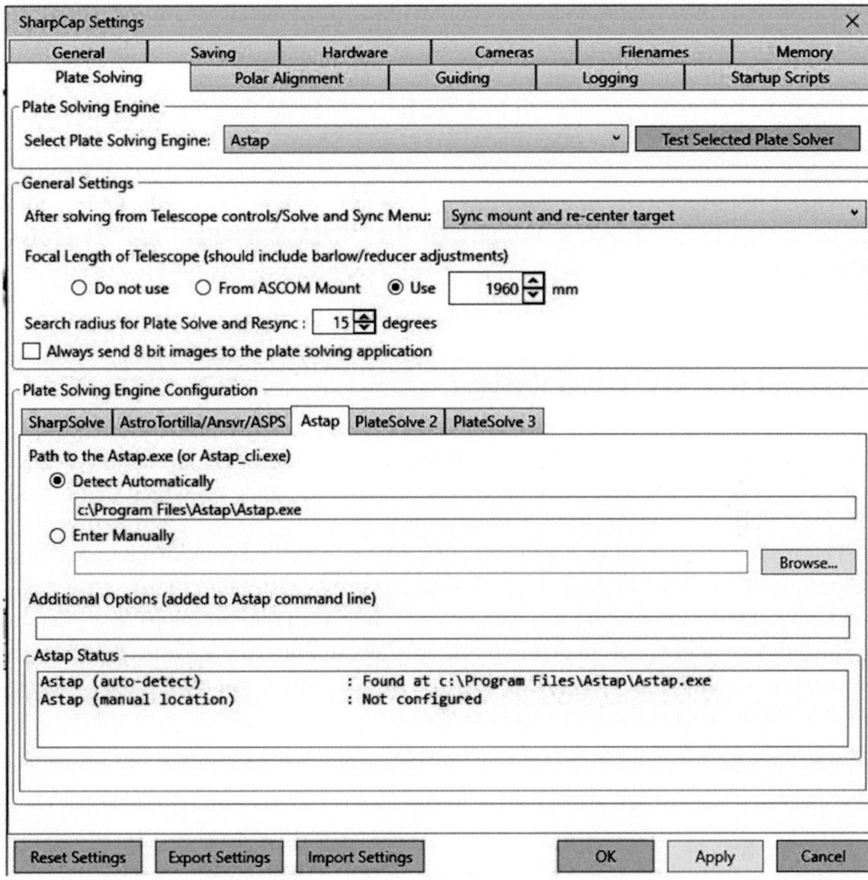

Fig. 2.37 Showing the plate solving setup in SharpCap. I have arbitrarily chosen to use the Astap database. (Image by the Author)

My first line of defense is that the software to control Celestron mounts from a PC, CPWI, can be programmed to have any number of stars in its model of the sky. I generally keep this number around 25–30. The GOTO is then very good; and most galaxies will be visible on my camera chip, if I expose for 4 seconds at a high gain. I then use the hand controller to slew the galaxy toward cross-hairs I can place in the live image in SharpCap.

The next step is a tool called plate solving. The trouble is that if the galaxy is not on your chip, you don't know where to find it. But if you download a star database, your computer will know where to find it.

I Googled a database for plate solving called Astap and downloaded it. The settings for SharpCap are shown in Fig. 2.37. Once I got this to work, I never seriously explored the alternatives. As they say, if it ain't broke, don't fix it.

In the Tools menu in SharpCap, I select Plate Solve & Resync. Sometimes I have to fiddle with the camera settings because 4 seconds and a gain of 400 don't always work. But once that's done, so long as I am within 15° of the target galaxy, SharpCap will send slewing commands to the mount software, and slew to the galaxy. This usually works first time, but occasionally takes a couple of attempts.

The plate solver knows what to look for because I told the GOTO on the mount control software CPWI to find this galaxy, and SharpCap & CPWI talk to one another.

When I first started, it could take me an hour to find the galaxy by hand. Luckily, the latest version of SharpCap has this facility. NINA and ASIair also have plate solving capability.

Where in the Sky Are Galaxies?

Cosmological theory posits that the universe is the same in all directions & at all distances from us. We have yet to discover the distance scale on which this is true. The local universe is certainly not evenly spaced.

The map in Fig. 2.38 shows a dot for each of 1200 galaxies of apparent blue magnitude of 14.4 or brighter. I got the data from a catalogue compiled by Allan Sandage & G. A. Tammann [36], some of which is available online [37].

It is immediately apparent that in Northern Declinations, the galaxies are strongly concentrated around the constellations Virgo, Leo, Coma Berenices, Canes Venatici & Ursa Major. This is the Virgo Cluster.

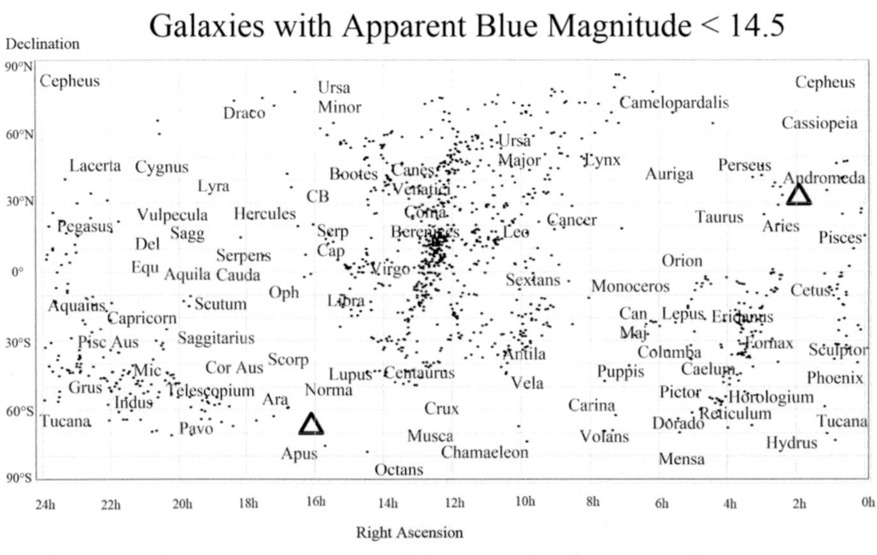

Fig. 2.38 Showing the distribution of bright galaxies in our sky. The two triangles are the Northern and Southern Triangulum constellations. (Image by the Author)

The picture at Southern declinations is more complex. There are three concentrations. One is a continuation of the Virgo Cluster as far as Centaurus. Another is loosely centered on Fornax. The third is around Grus, Microscopium, Indus, Telescopium and Pavo.

By knowing when the appropriate right ascension is on the meridian you can maximize your chances of finding galaxies. The Virgo cluster crosses the meridian in late March. The Fornax Cluster does so in mid-November. The third group crosses the Meridian at midnight from July to September. Southern hemisphere observers therefore have a good spread of observable galaxies across the Southern summer. If this is you, you are probably fortunate enough to live where it gets decently dark in the summer: not many people live south of 45°S.

The largest members of the Local Group of galaxies are in Andromeda and Triangulum; and are Northern autumn objects. The Large Magellanic cloud would be highest in Southern Skies in the Southern summer. I saw it in June: It was perfectly photographable then, so it's probably a good year-round target. The Small Magellanic Cloud is at its highest in about November.

Being Outside at Night

Long experience of star parties for the public tells me that, while you shouldn't have to remind people to dress very warmly, a minority seem to need to learn this the hard way before they take you seriously. Even in warm climates, it's colder at night than during the day. Where I live, in a temperate northern hemisphere maritime climate, I still need to wear a coat at night in July. In continental climates, the effect is even more marked. On cold winter nights, I may have a thick coat with zip and press studs, a thick cardigan, a dress with built in liner, woolly tights, pants[4], thick boots and warm gloves. If I need finger dexterity, I can wear disposable gloves under fingerless gloves. On the coldest nights I use fingerless gloves with a flap to keep my fingers warm. Here, a cold winter's night is at worst −5 °C (23 °F), but it will be windy. I dread to think what a cold night in Winnipeg might be. Presumably the good people of Manitoba have a better idea than I how to deal with this.

My own outfit of the night after a warm day in May is shown in Fig. 2.39.

If you travel to a remote site, have a means of calling for help if your car breaks down or whatever. Never forget that in mountainous areas the air becomes thinner, and poorly maintained cars may be more reluctant to start; and you won't normally notice the problem if you live at a low level.

They tell you not to wear dark clothes on the street at night, though no-one seems to take much notice. Dayglo jackets are a hindrance in my observatory; they spread light pollution.

[4] Trousers in the UK.

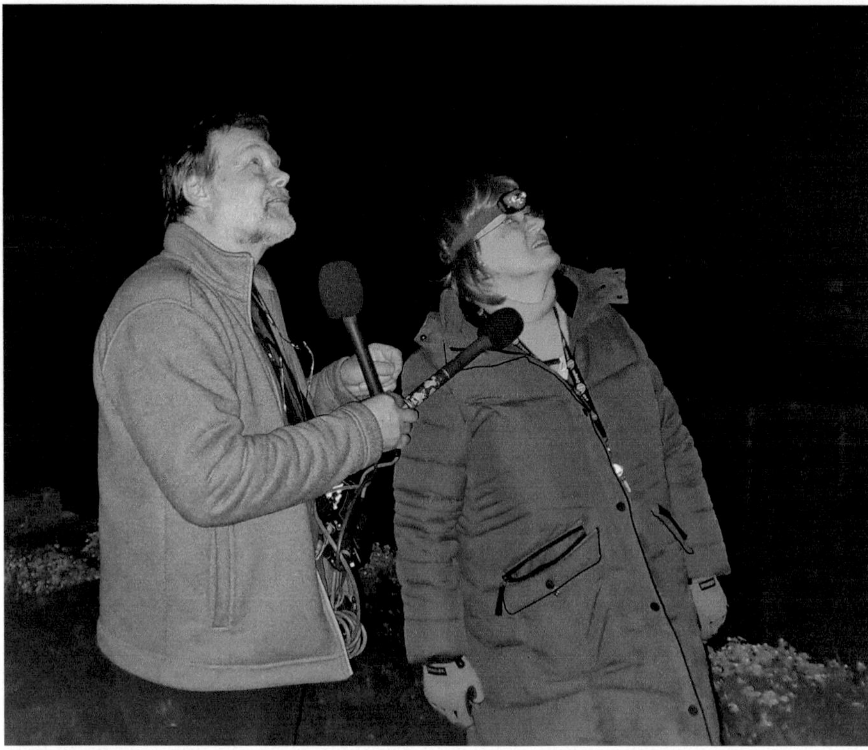

Fig. 2.39 The author, shown being interviewed by a BBC reporter about the supernova in M101 in May 2023. Notice the serious lack of glamor, and the warmth of the clothing. I bought that coat because it has four pockets, which are very useful while observing. (Image by Gregory Titley, used with permission)

Chapter 3
First Steps in Processing Your Captured Images

Do My Darks Fall Flat or Am I Just Biased?

I frequently see people asking on chat forums if they really need dark, flat and bias frames.

Let me put it this way: if you take a photo in daylight or even in an artificially lit room at night, the amount of light going into your camera and onto the chip means that you have a high signal-to-noise ratio. Therefore, the electronic noise in your image is insignificant.

If, on the other hand, you take a long exposure image at night, the electronic noise in the image is much more significant. Therefore, the answer is: yes, you do need to take these frames.

We do several things to fight back against spurious features in our images.

The first is to cool the camera chip if the camera has a cooler. I usually run mine at −15 °C. One of the few advantages of the maritime climate where I live is that we rarely get hot days. 80 °F (27 °C) is a really hot day for us. So a camera that can run at 35 °C below ambient will be able to reach −15 °C if the ambient temperature is +20 °C (68 °F) or less. We achieve this on all but the hottest handful of nights of the year. I appreciate that this is not likely to be the case in much of North America, South Africa or Australia. You may need to run your camera hotter.

That's all the more reason for taking *dark frames*. You need to subtract background noise and hot pixels from your image. Trust me, you are going to notice the difference in astrophotography. I have seen advertising claims that certain camera chips have such low noise that you don't need to take dark frames. If your camera maker claims that, dear reader, please proceed at your peril. Some of the pixels on your camera chip will become 'hot' over time. This means that the pixel behaves as if it were receiving the maximum amount of light even if it is receiving no light at all. I have never seen this not happen; and over the years I've owned over a dozen

J. Clark, *Photographing Galaxies from Light Polluted Skies*, The Patrick Moore Practical Astronomy Series, https://doi.org/10.1007/978-1-0716-4506-2_3

cameras for astronomy. None of them has been free of hot pixels. To take a dark frame, cover your telescope lens and take frames with the same exposure at the same chip temperature, and gain setting if the camera has one, as the actual pictures (called 'light frames') that you are taking. I usually take 100 such frames, though that is probably overkill. Twenty-five to fifty is probably plenty. A cloudy night is the perfect time to take such frames. You can't do anything else anyway. Since the lens cap is not a perfect light seal, I would not take dark frames in daylight.

From your dark frames, the processing software makes a 'master dark', which averages the signals of the dark frames. If you have a cooled camera where you can control the temperature, you can take your darks at a different time from your light frames; and you can re-use your darks. You only need to repeat them if you see extra hot pixels appearing. If you are using a DSLR or mirrorless camera, with no temperature control, you'll need to take some dark frames during every observing session.

Flat frames are taken to address two issues. The first is to correct the phenomenon of *vignetting*. The optical signal from your telescope is brightest in the middle of the chip, which is or should be, along your *optical axis*. The brightness falls off as you move from the center of the chip. This is shown in Fig. 3.1.

Flat Frames also help get rid of the dreaded *dust bunnies*, which are out of focus shadows cast by dust particles. Figure 3.2 shows a particularly nasty example. Dust gets everywhere. It is unavoidable. But your best defense is to buy camera chip cleaning kits, and clean your chip – or rather the window protecting it – regularly.

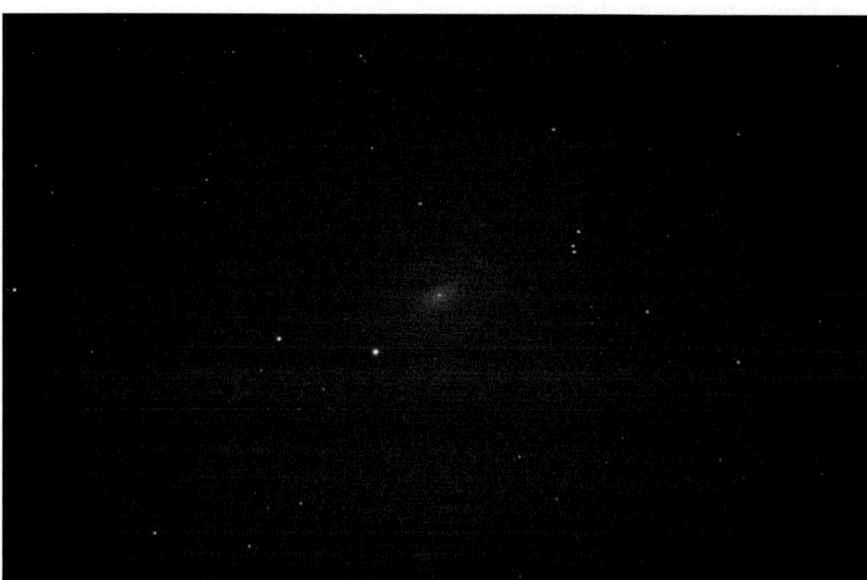

Fig. 3.1 This is a single frame, showing the effect of vignetting. The corners are quite a bit darker. The effect is less obvious across the rest of the image, but it is there. (Image by the Author)

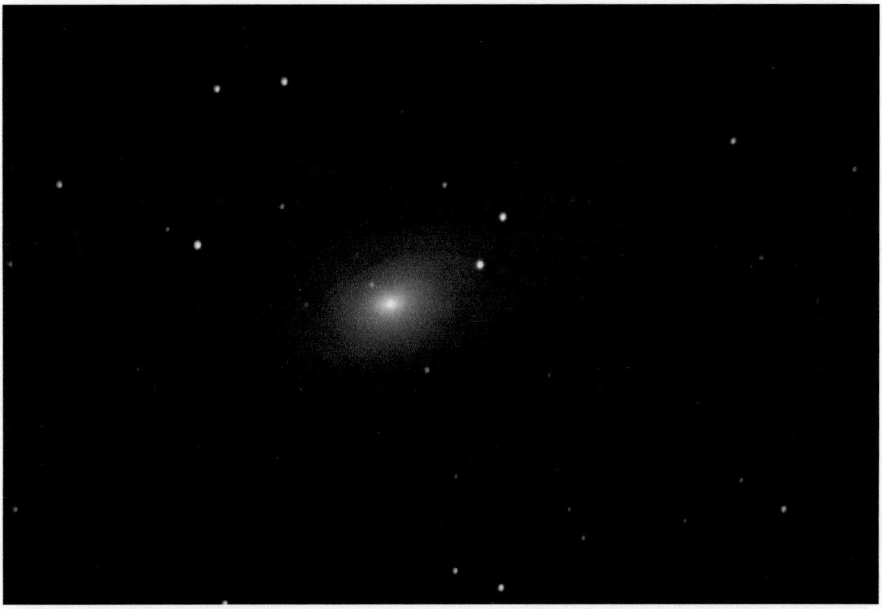

Fig. 3.2 This image is ruined by 'dust bunnies' – out of focus shadows cast by dust particles. (Image by the Author)

Any filters you use should also be kept clean. This must be done with great care. Scratches are obviously ruinous.

If I am running a campaign over several nights to obtain a particular image, I avoid removing the camera, if possible, because every time you change camera, and expose it to dust, you're going to need new flat frames to deal with the dust bunnies.

So how do you take flat frames? There are low-tech and hi-tech methods. I do this often enough that I have upgraded to a high-tech method.

I use a light-emitting diode (LED) panel that works with 12 V DC power (Fig. 3.3). I chose the Aurora flat field panel from Gerd Neumann Jr., in Germany. I chose it because it emits a continuous spectrum, unlike many low-cost LED panels. This enables me to use it with a narrowband filter, should I so choose.

The astute among you may notice a black mark in the middle of the panel. This happened when I dropped it once. It's a bit fragile, especially where the wire comes out, so I now hold it in place with small G-clamps. Of course, I have to be extra careful not to let these clamps touch the corrector plate of the telescope.

What exposure should you use for the flat frames? I set the exposure to about ¼ second, then check the automatic exposure box on Sharp Cap & let the camera determine the exposure. It usually suggests between ¼ and ½ second. If so, I round the exposure up to half a second per frame. I take 100 frames, but you can manage with fewer. Of course, your temperature, gain, if any, and everything else should be as near as possible to the conditions under which you shot the light frames.

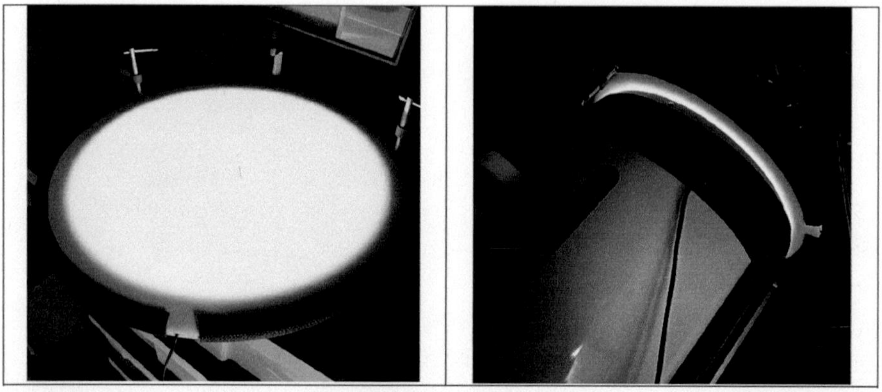

Fig. 3.3 My Aurora flat frame panel from Gerd Neumann, Jr. Left: the panel. Right: the panel mounted on my telescope. (Images by the Author)

While I am doing this job, I also shoot bias frames. Keep everything as it was for your flat frames, but turn your exposure to the shortest possible time. With one of my cameras this is $^{1}/_{16000}$ second. If it's $^{1}/_{1000}$ second or shorter, you will be fine. The idea is to get as close as possible to zero exposure, to record the image at zero brightness: there will still be a small electrical signal present.

You could shoot flat and dark flat frames, but this makes the processing messier, so I stick with flats and bias frames.

What's the low-tech way to shoot flats & bias frames? The lowest tech is to use the sky as your light source. In the United Kingdom it will usually be cloudy, and clouds are out of focus for a big scope. In sunnier climes, you may be using a blue sky. With a monochrome camera, this would be fine, but with a color camera, this may not give you quite the right flat frame. For slightly more sophisticated flats, cover the front of your telescope with tracing paper, a white or grey towel or even a white or grey T-shirt.

How does the software use the flat, dark & bias frames? This is a two-stage process, as illustrated in Fig. 3.4. The dark, flat and bias frames are each averaged to make master frames. The master frames are then used to calibrate each light frame.

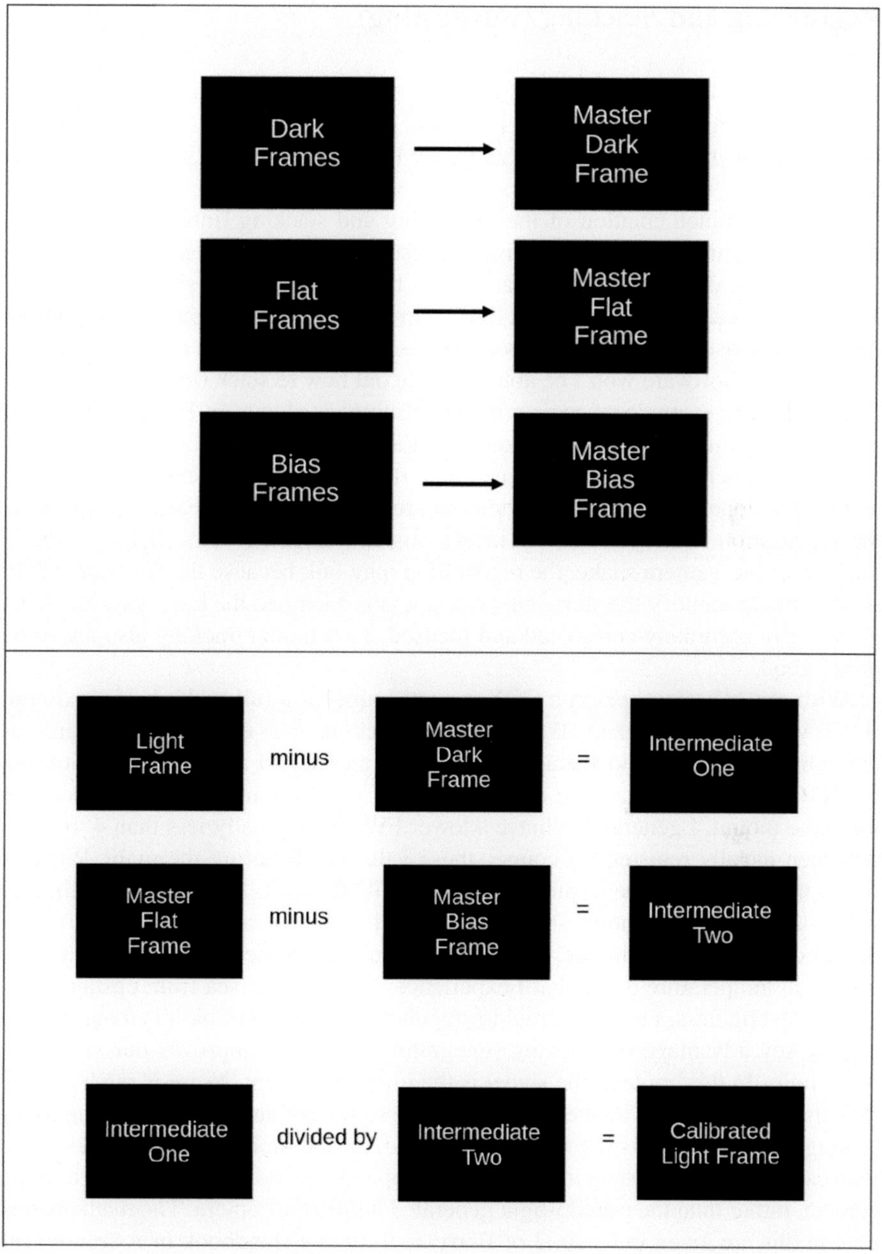

Fig. 3.4 Schematic diagram showing how the flat, bias & dark frames are used to calibrate each light frame. (Diagram by the Author)

Registering and Stacking (Integrating)

The principle of registering & stacking (integrating) is shown in Fig. 3.5. Confusingly, PixInsight calls stacking 'integrating'. In fairness to its creators, English is not their first language; and their English is a whole lot better than my Spanish.

In actual implementation of the registering and stacking (integrating) process, there are several sub-steps. We'll discuss these as we come to them.

There is no divinely-given guarantee that the registering and stacking (integrating) will succeed. Over the years, I have learned the hard way that the light frames that you, the user, supply to the processing software need to be of decent quality, otherwise the software won't be able to figure out how to stack (integrate) them.

An extreme example of registering two frames is shown in Fig. 3.6. This is a slightly frivolous example to demonstrate the principle. In practice, your frames need to be much more closely aligned than these two. I would expect registration software to cope with small translations and rotations of frames. For deep sky work, the registration usually works by identifying and aligning stars. If the image is unsharp or has camera shake, the registration may fail, because the software would not be able to identify the stars. This was a lesson I learned the hard way. You have to be really accurately collimated and focused. Your mount tracking also has to be pretty good.

With my 11″ telescope on a GCX mount, I aim for a full-width half-maximum (FWHM) value of less than 10 and an RMS tracking error of ½ to $^{2}/_{3}$ of an arcsecond. On nights with good seeing, I can achieve an FWHM around 7. I monitor the FWHM using the live stacking feature in SharpCap. With my 730 mm telescope on the same mount, I generally achieve a lower FWHM, typically less than 4. Frames will then usually register. Of course, these values drift during the night. Where I live, I think seeing is the dominant cause of FWHM drift. In places with climates where the night temperature drops a lot, loss of focus due to telescope contraction due to cooling may dominate. I suspect that only maritime climates have the low nocturnal temperature drops that I experience: I can see the sea from upstairs windows. Nevertheless, I am not afraid to stop and re-focus if I think it is required.

The key advantage of stacking (integrating) is that it improves our *signal-to-noise* ratio. In this context, the signal is the image you want. Noise is random stuff that also ends up in the image. You want more signal and less noise. Stacking (integrating) will only do this if you stack (integrate) the calibrated frames, because only then can the process of stacking (integrating) operate on the light that comes into the camera, rather than the pseudo-light generated within the camera. The mathematics behind this are given in Chap. 2 of Berry & Burnell's Handbook of Astronomical Image Processing [38]. These authors show that if you quadruple the exposure time, the noise halves. If you increase the exposure time a hundred-fold, you get a tenth of the noise. So there's a law of diminishing returns at work.

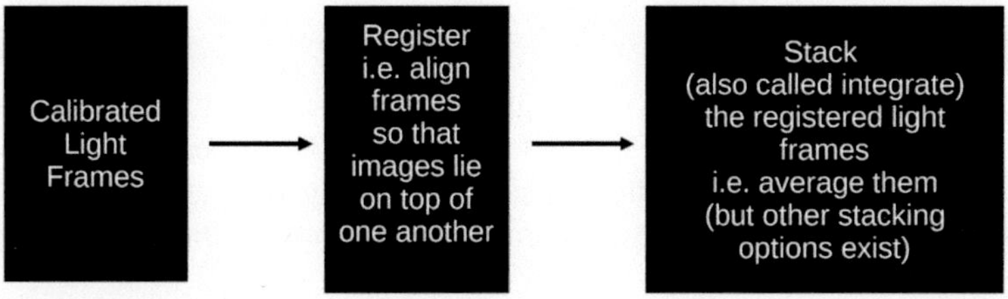

Fig. 3.5 Schematic diagram showing the principle of registering and stacking (integrating). (Diagram by the Author)

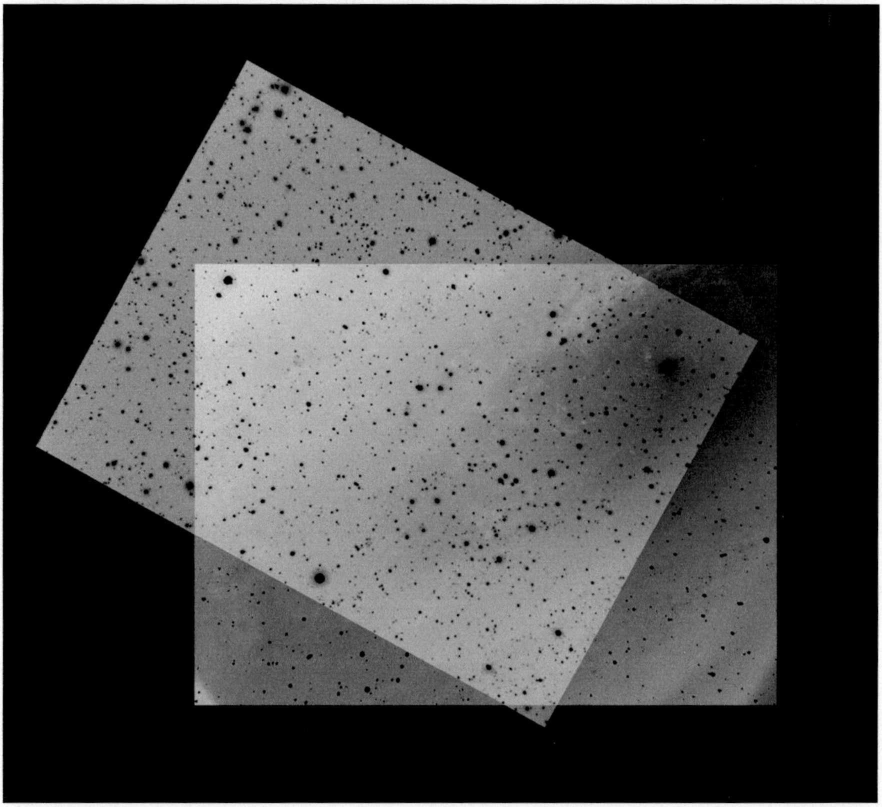

Fig. 3.6 An extreme example of two frames that have been registered, i.e. moved so that the images lie on top of one another. I would not seriously expect software to be able to register frames this severely mis-aligned. (Image by the Author)

Light Pollution

But here's a really key point: most light pollution behaves like random noise. It is light reflected back from random tiny smog and water particles in the air, which are so small that they float in air. Therefore, your strategy should be to expose for much longer, i.e. expose many more frames, to drown out the light pollution with signal.

Of course, this strategy won't work if your neighbor's light is shining straight into your scope. That's not random noise. But skyglow is.

I have a friend who images galaxies from a Bortle 8 sky in a city center without a filter. His results are excellent. He needs to expose for much longer than I do in my sky, which is Bortle 5 or 6, depending how humid the weather is.

I will now take you through a complete processing sequence for an image in a piece of free Software called Siril, available on Linux, MacOS and Windows. Then I will take you through the preprocessing of an image in a paid-for application called PixInsight.

Folders & Directories

When I learned computing back in the Middle Ages, groups of computer files were always stored in directories. Some 12-year-old whizz kid in Microsoft decided that that was a bad name, and started calling them folders. While this was supposed to reduce confusion among users, it seems to me to have had the opposite effect. There are now two names for the same thing. They are entirely synonymous: a folder is a directory and a directory is a folder.

Some computer programs use one name, some use the other.

Nice One, Siril

You download Siril from siril.org. I found it very easy to learn, at least at a basic level, which means that I can try more advanced techniques now if I want to because learning to walk before I could run wasn't very difficult. I'm going to assume that you have a one-shot color camera for now.

I began my deep-sky career with Siril. I would thoroughly recommend the latest version (1.2.1 at the time of writing) it to anyone who has to do astrophotography on a limited budget. I would also happily recommend it to anyone struggling to get started with image processing software. It will at least get you going before you lose heart. You won't have lost any money if you later upgrade to paid-for software.

Table 3.1 YouTube Videos to show you how to enable Developer Mode in Windows

Windows version	YouTube video
10	https://www.youtube.com/watch?v=zGjgZnJE9cE&ab_channel=TechMeSpot
11	https://www.youtube.com/watch?v=-2rXAzGgpJA&ab_channel=TechRelatedTips

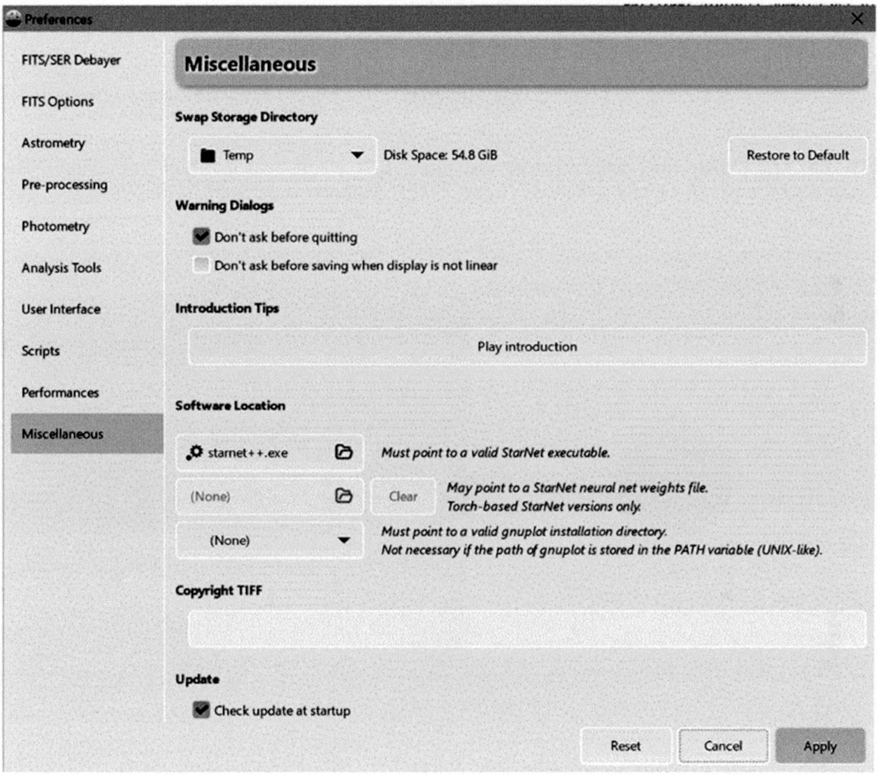

Fig. 3.7 The Siril Preferences menu, where you tell it the location of Starnet++. You access this from the button at the top right of the Siril console with three horizontal bars. (Image by the Author)

If you download Siril for Windows, there is one little twist. You need to enable Developer Mode in the operating system. By default, this is disabled, as a security feature for those less au fait with PCs. Table 3.1 shows you how to find a YouTube video for your version of Windows.

You will also have to download an executable called Starnet++, and store it on your PC somewhere safe: you can point Siril to the location. I did this following the instructions in https://www.youtube.com/watch?v=-SwL6vxgLbM&ab_channel=stefanastro.

My images of the Siril windows have been made into negatives for ease of reading in a book (Fig. 3.7).

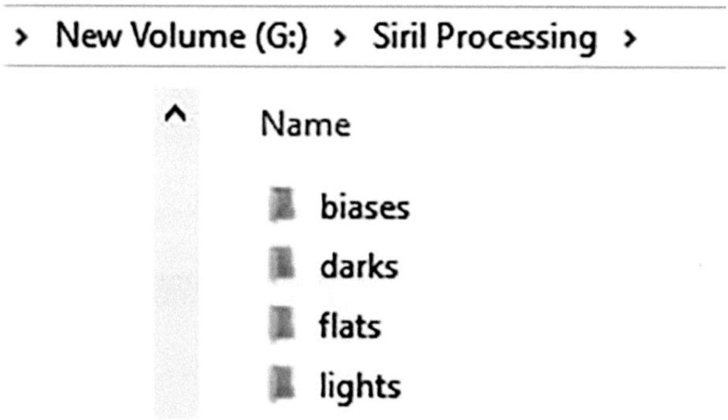

Fig. 3.8 The folders Siril needs to store your images. It is self-evident which goes where. The folder names must be exactly as shown. (Image by the Author)

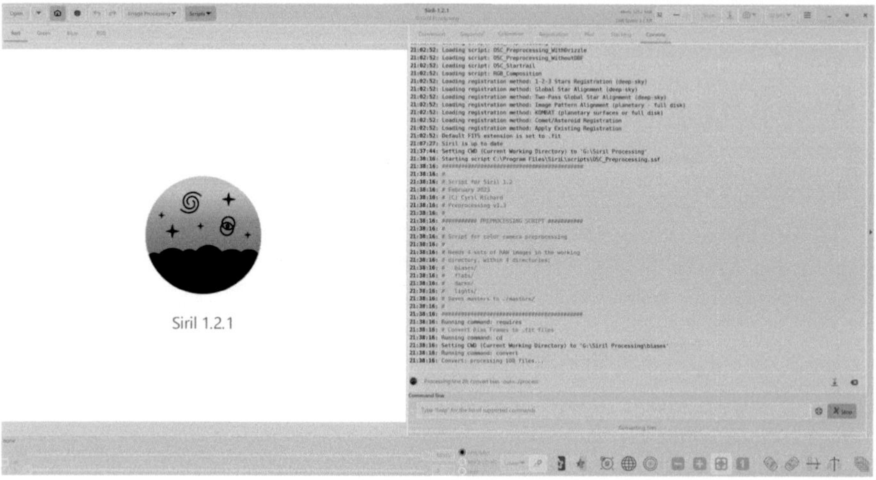

Fig. 3.9 The Siril Window just after you select the OSC_Proprocessing script. (Image by the Author)

Siril is a little bit rigid about the folder structure it requires. As long as you follow the structure shown, you'll be fine. You point the home icon the program at a folder called Siril Processing. Underneath this it you put your lights, flats, biases and darks folders. It is self-evident which frames go into which folders as shown in Fig. 3.8.

Once you have done this, you pick a script from the scripts menu along the top of the Siril window, and select OSC_preprocessing, if indeed you have a one-shot-color (OSC) camera (Fig. 3.9).

This script will take quite a while to run, and it will create master darks, flat and biases. It will use these to register and stack your light frames. It will place an output file in your Siril Processing home folder called something like result.fit. In the case I'm taking you through it called it result_12000s.fit. I made a copy of it in case of disaster. It also creates two folders called process and masters. The folder structure is now as shown in Fig. 3.10.

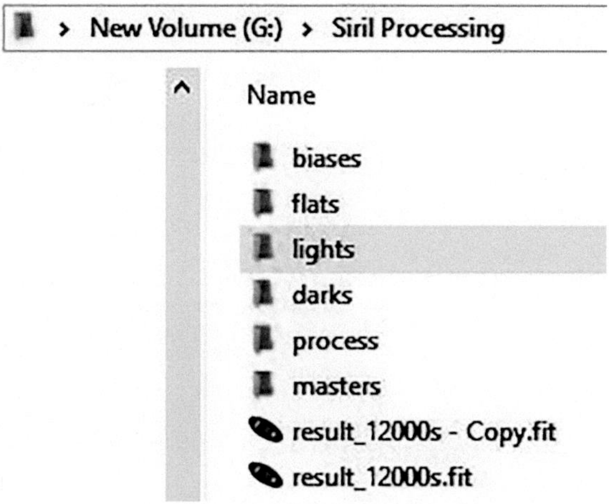

Fig. 3.10 The folder structure after running the OSC_Preprocessing script. (Image by the Author)

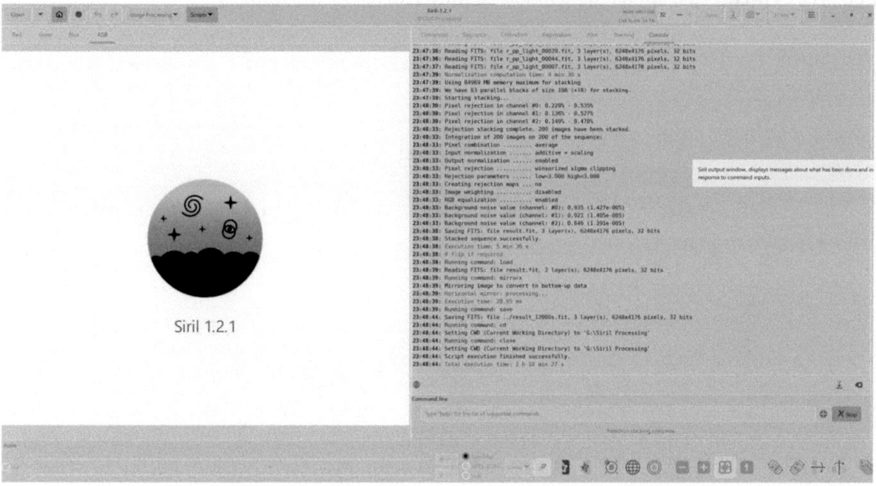

Fig. 3.11 The Siril Console after running the OSC_Preprocessor script. (Image by the Author)

Figure 3.11 shows the Siril console after the preprocessor script has run. You now need to load your result_XXXXX.fit file, where XXXXX is a text string, into Siril to postprocess it.

The next stage is to stretch the preprocessed image so that you can see stuff in it. I used the Generalized Hyperbolic stretch options, and picked the 'Modified arc-sinh' transform. ("Arcsinh" is either pronounced "arc-sign-ch" or "arc-shine". It is not the same as the arcsine function on your pocket calculator.) You click right in the left of the graph to set the symmetry point, then slide the stretch factor until you can see an image. Stretch to taste. By skillful overstretching, then placing the symmetry point to the right of the bell curve shown in Fig. 3.12, you can darken the background. This takes a bit of trial and error. There's always the undo arrow at the top.

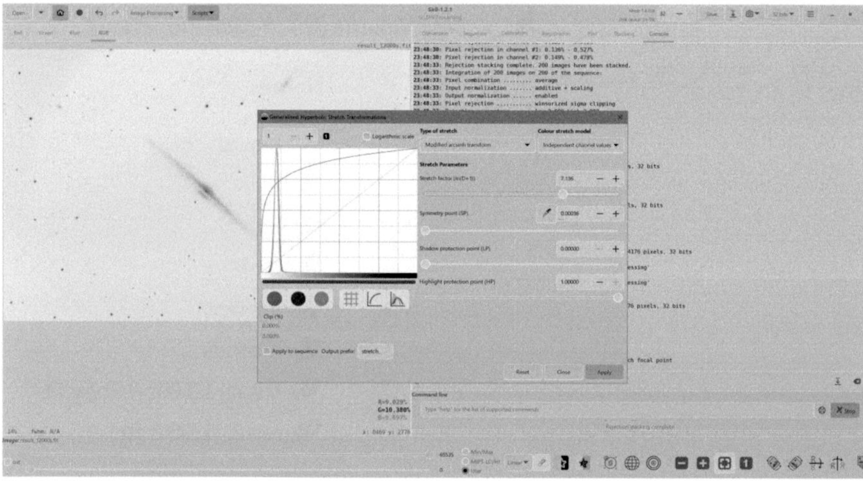

Fig. 3.12 The Arcsinh stretching window in Siril. (Image by the Author)

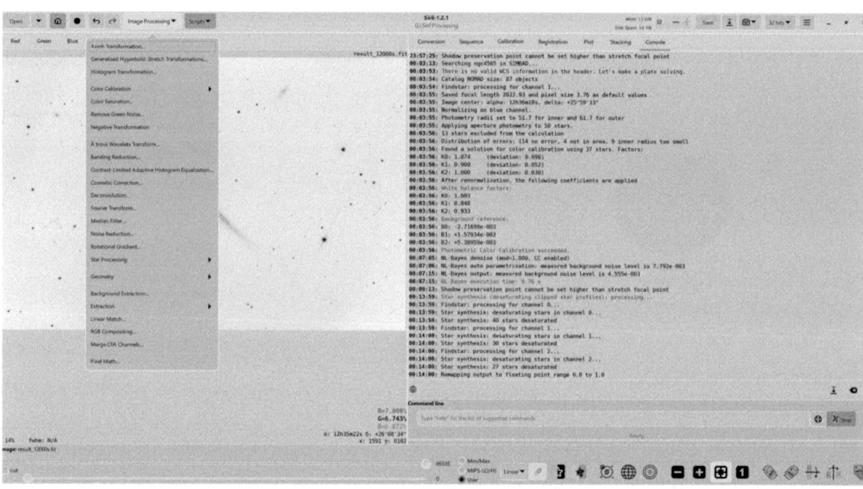

Fig. 3.13 The Siril Image Processing menu. (Image by the Author)

The Siril Image processing menu is shown in Fig. 3.13.

For your next trick you use the photometric color calibration, again from the image processing menu. You have to tell it what your target is – NGC4565 in this case – and it calibrates the colors from the surrounding stars, which it recognizes once it knows what the target is. It may offer more than one choice once you have told it to find the target. In this case it didn't. The menu is shown in Fig. 3.14.

I then chose to sharpen the image a little with the 'A trous wavelet transform' function. It only needed a very slight tweak, which I discovered by trial and error. This is shown in Fig. 3.15.

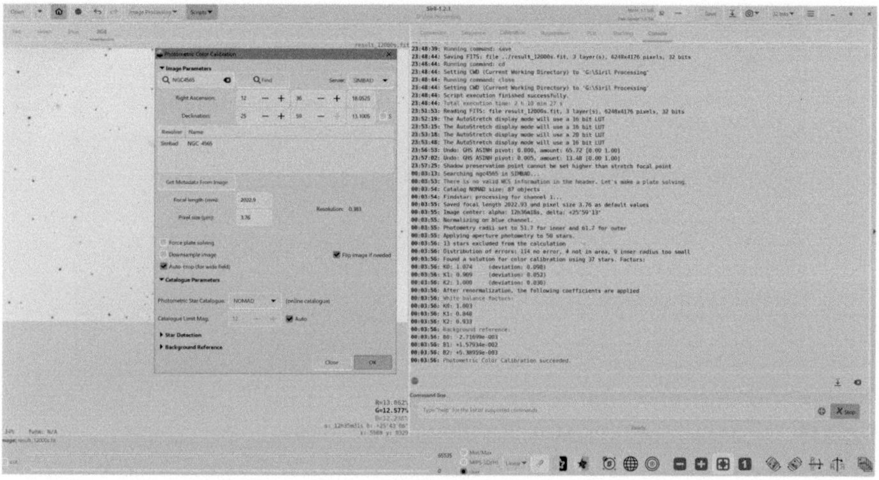

Fig. 3.14 The photometric color calibration menu in Siril, dawn down from the Image processing menu. (Image by the Author)

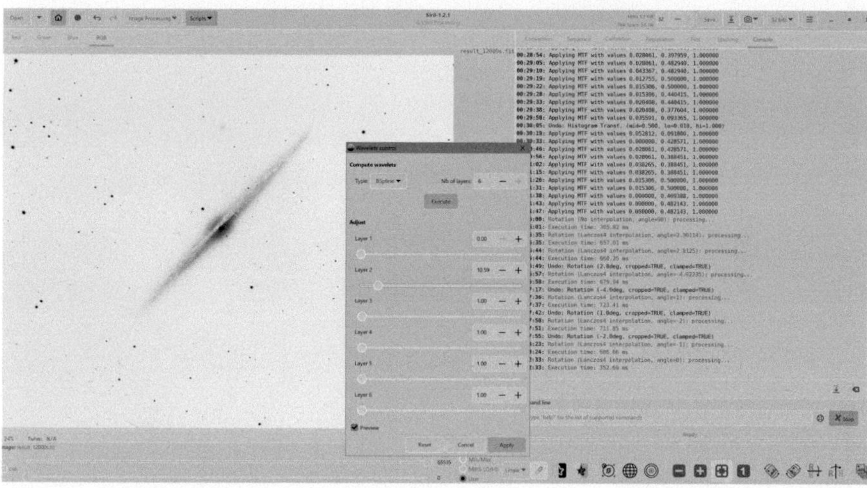

Fig. 3.15 Using the 'A trous wavelet transformation' function to sharpen the image slightly. (Image by the Author)

I next decided to remove the stars so that I could separately tone them down. To do this, I went into Image Processing → Star processing, and chose the StarNet Star Removal option (Fig. 3.16). The defaults worked fine. You then get one image with the stars in (Fig. 3.17) and one without (Fig. 3.18).

I loaded the stars into Siril and used the Histogram Stretch tool to tone them down. I could then have loaded the starless image and processed it further, but I didn't.

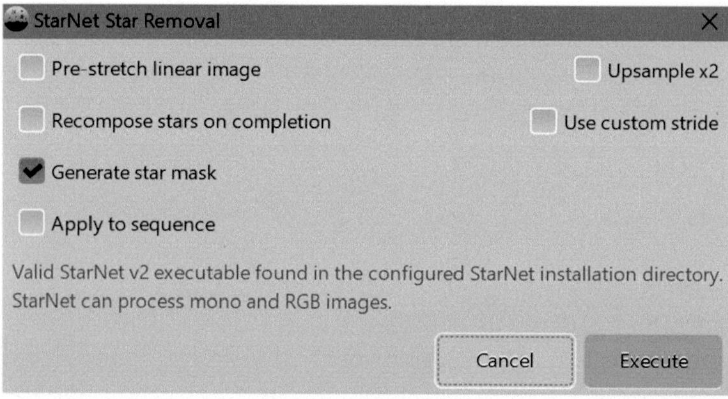

Fig. 3.16 The Starnet Star Removal option. (Image by the Author)

Fig. 3.17 The stars. (Image by the Author)

Once I had done that, I recombined the images using Pixel Math as shown in Fig. 3.19.

Finally, I end up with the image in Fig. 3.20. You could obtain a better image with PixInsight, but it is not a terrible image. I did take it into the free image

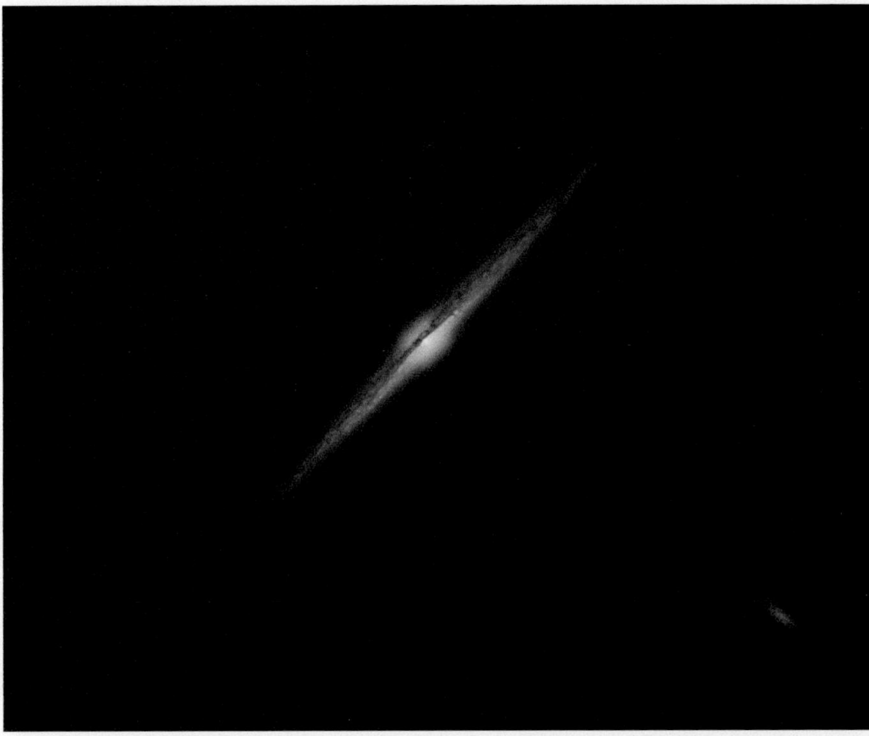

Fig. 3.18 The Starless image. (Image by the Author)

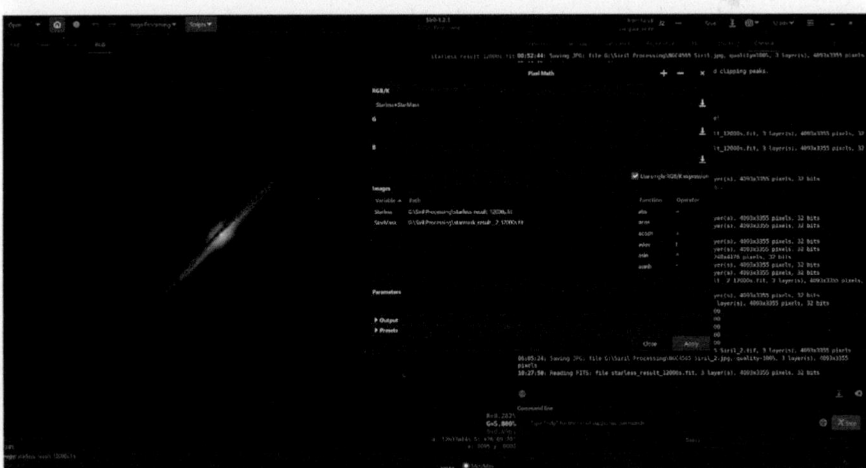

Fig. 3.19 The Pixel Math menu, showing how I used it to add the images. (Image by the Author)

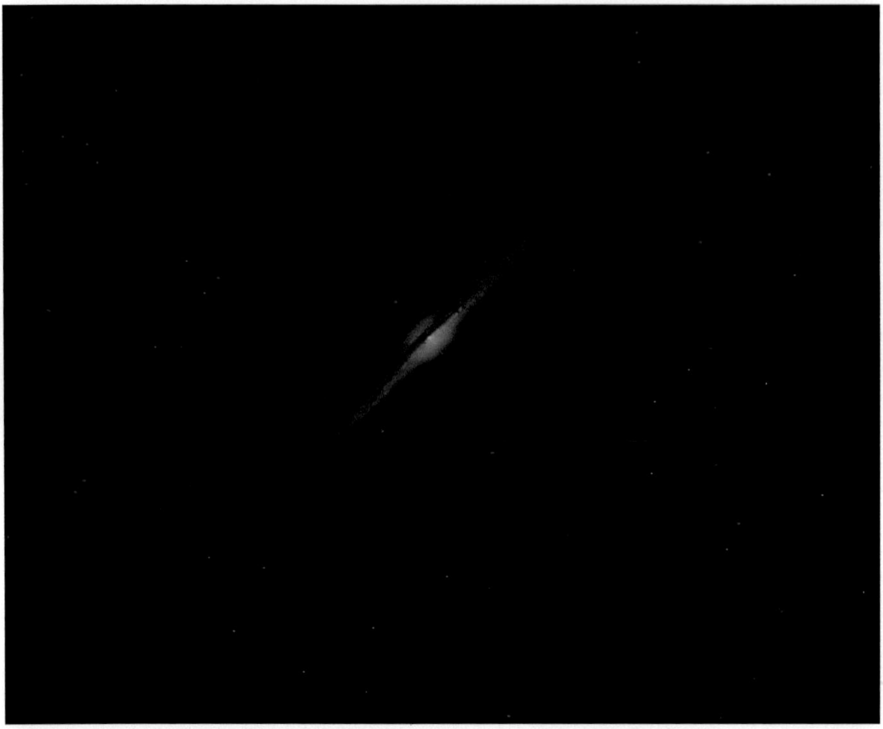

Fig. 3.20 NGC 4565, the Needle Galaxy, processed in Siril. (Image by the Author)

processing application Gimp to lower the black point, something with which Siril seems to struggle.

Buying PixInsight

The web site is pixinsight.com. They offer a generous trial period of 45 days.

I have to say that I spent a long time getting nowhere with PixInsight. This is not a rare experience. This book should help you to get started.

I think I would try a few processes at a time to get used to it. You can always drag and drop an image processed elsewhere, for example one stacked in Siril or Deep Sky Stacker, make a duplicate to protect the original, and play around with the stretching and masking tools.

In the end I took advantage of the fact that I have retired from my day job, and spent a week solidly trying to get this program to work. At first, I stacked in Siril, and stretched and used masks in PixInsight, but eventually I took the plunge and used the Weighted Batch Preprocessor to stack in PixInsight.

By following the tips I give in the rest of this chapter, you should get the Weighted Batch Preprocessor to work as easily as the stacking in Siril.

There is, however, one prerequisite: you need sharp frames to process. You may need to do some work on your focusing and collimation. I certainly did. For a long time, I tried to stack fuzzy images. It never ended well.

Calibrating, Registering and Stacking (Integrating) in PixInsight

In this section I'm going to go through the routine stuff I do before getting on to those parts of processing that require a bit more skill. I might be able to do it all with a script: perhaps I should try.

The built-in script I'm going to use is Weighted Batch Preprocessing, found from the PixInsight menu as shown in Fig. 3.21. Everyone I know seems to call it WBPP, so that is what I propose to do.

The version I am describing is WBPP version 2.7.0. As with any software, things may change.

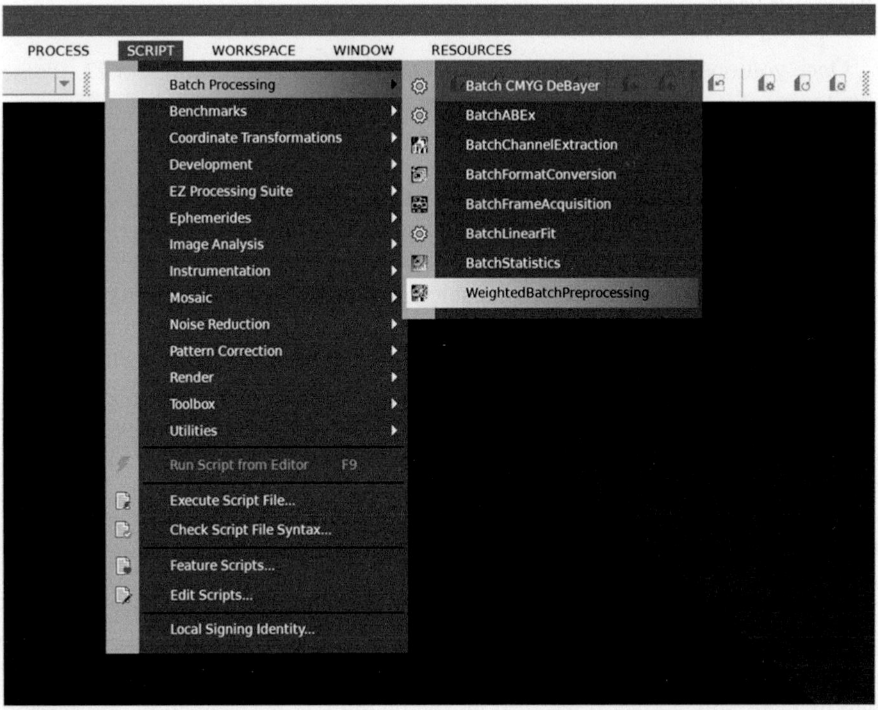

Fig. 3.21 Where to find the Weighted Batch Preprocessing (WBPP) script in PixInsight. (Image by the Author)

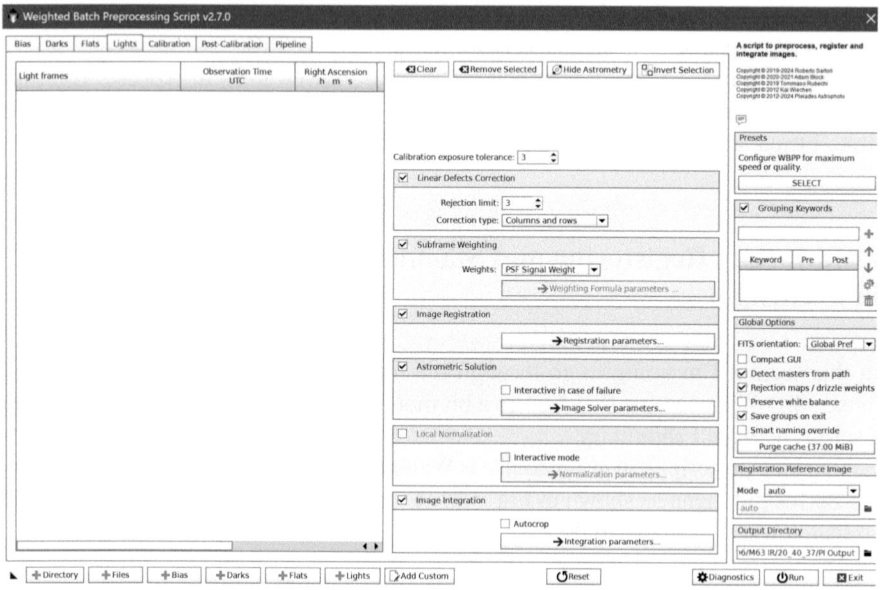

Fig. 3.22 The opening menu of the Weighted Batch Preprocessing (WBPP) script in PixInsight. (Image by the Author)

Once you open WBPP, you see a menu like the one shown in Fig. 3.22. Please don't panic if it takes a few seconds to appear. That's normal.

WBPP may open the tab where you were when you last closed it. It generally remembers previous settings. If so, just find the Lights tab on the top left.

To load your frames from your camera – the actual images of the galaxy – press the +Lights along the bottom. All the plus signs along the buttons at the bottom left are green; and they all work similarly. A Windows-Explorer-like window opens whenever you press one of them, to enable you to select the files of that type (light, dark, flat or bias) that you are trying to load. The process is the same for all of them (Fig. 3.23). If you want to select all the images in the folder, you can simply press CTRL-A at this point.

It is often quicker to do this, and then use the Remove Selected button towards the top in Fig. 3.22 to remove frames previously noted to be unwanted. In my experience, WBPP does not do a bad job of rejecting frames all by itself.

What if I have several folders of files, and the sequences all have the same list of filenames? Don't worry. WBPP can tell them apart anyway. Repeatedly press +Lights to add files from other folders.

Figure 3.24 shows the information transferred as the header part of the FITS files to WBPP. SharpCap has picked up the time information from the computer, the pixel information from the camera, and the focal length information I manually fed into its Plate Solving tab. It has also picked up the RA & Declination information from my mount, to which it is coupled via Celestron's CPWI software to drive the

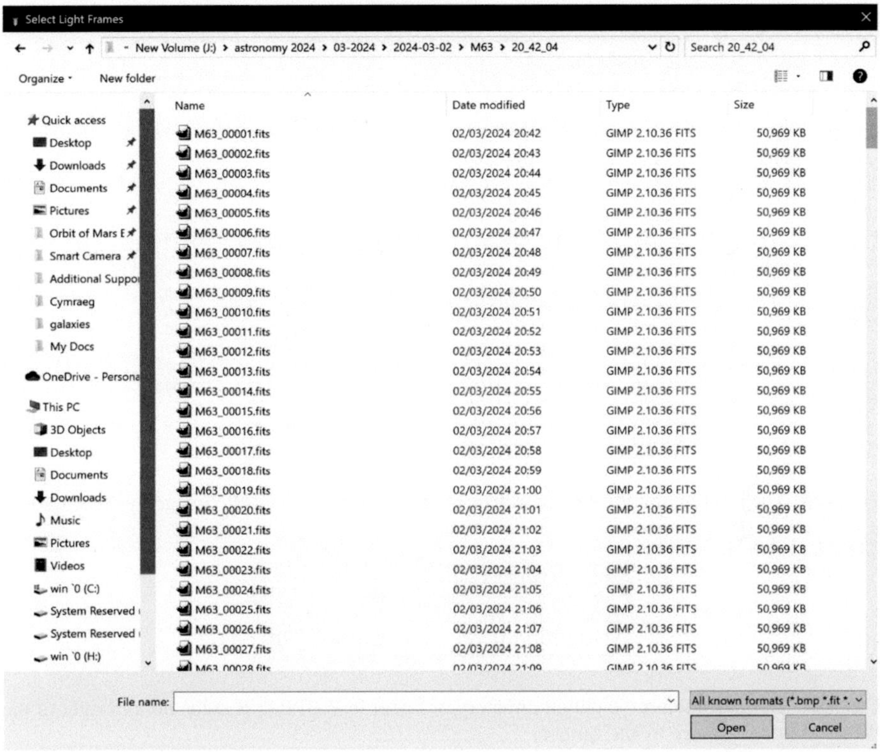

Fig. 3.23 The window to add light frames to the Weighted Batch Preprocessing (WBPP) script in PixInsight. (Image by the Author)

mount. It also picks up other information, such as exposure time and filter, from the FITS header. You can see in Fig. 3.24 that it knows that my frames were exposed for 60 s with no filter.

What happens if some frames were exposed for 60 s and others for, say, 120 s? WBPP will sort the frames by exposure times in the FITS header, and weight their contributions accordingly. That's why it is called a "weighted" batch preprocessor. You do have to warn it, though. In Fig. 3.22, you can see that I set the Calibration exposure tolerance to 3 seconds. You'd need to set this to a number greater than the difference in exposures, say 63 s. You also have to set this in a second place on the Post-Calibration tab. You'd need separate dark and flat frames for the two cases. We'll discuss how to do that later.

In the chapter on using remote telescopes (Chap. 8) I'll show you how to process separate luminosity, red, green & blue (LRGB) images. With my own telescope, I normally use a one-shot color camera. The main reason for this is that in a changeable, cloudy climate, I can't really plan a set of L, R, G & B frames.

You go through very similar processes to add your bias, flat & dark frames. If you forget to add these frames, WBPP will issue a warning when you press the Run button.

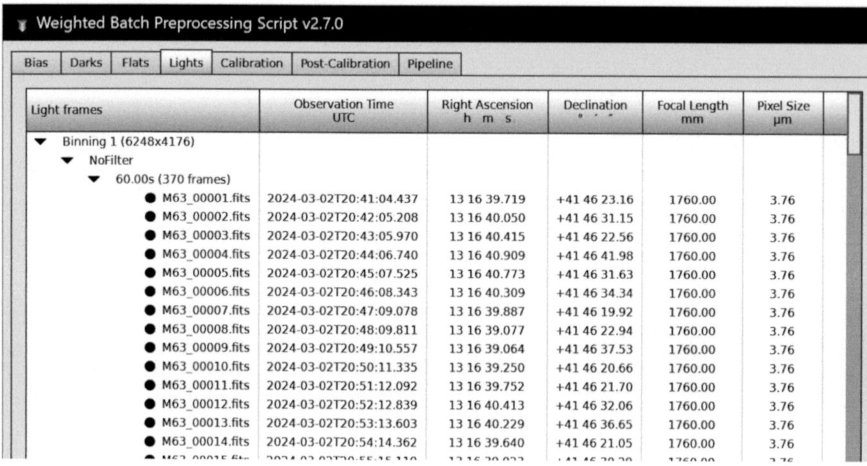

Light frames	Observation Time UTC	Right Ascension h m s	Declination ° ' "	Focal Length mm	Pixel Size μm
M63_00001.fits	2024-03-02T20:41:04.437	13 16 39.719	+41 46 23.16	1760.00	3.76
M63_00002.fits	2024-03-02T20:42:05.208	13 16 40.050	+41 46 31.15	1760.00	3.76
M63_00003.fits	2024-03-02T20:43:05.970	13 16 40.415	+41 46 22.56	1760.00	3.76
M63_00004.fits	2024-03-02T20:44:06.740	13 16 40.909	+41 46 41.98	1760.00	3.76
M63_00005.fits	2024-03-02T20:45:07.525	13 16 40.773	+41 46 31.63	1760.00	3.76
M63_00006.fits	2024-03-02T20:46:08.343	13 16 40.309	+41 46 34.34	1760.00	3.76
M63_00007.fits	2024-03-02T20:47:09.078	13 16 39.887	+41 46 19.92	1760.00	3.76
M63_00008.fits	2024-03-02T20:48:09.811	13 16 39.077	+41 46 22.94	1760.00	3.76
M63_00009.fits	2024-03-02T20:49:10.557	13 16 39.064	+41 46 37.53	1760.00	3.76
M63_00010.fits	2024-03-02T20:50:11.335	13 16 39.250	+41 46 20.66	1760.00	3.76
M63_00011.fits	2024-03-02T20:51:12.092	13 16 39.752	+41 46 21.70	1760.00	3.76
M63_00012.fits	2024-03-02T20:52:12.839	13 16 40.413	+41 46 32.06	1760.00	3.76
M63_00013.fits	2024-03-02T20:53:13.603	13 16 40.229	+41 46 36.65	1760.00	3.76
M63_00014.fits	2024-03-02T20:54:14.362	13 16 39.640	+41 46 21.05	1760.00	3.76

Fig. 3.24 The information transferred as part of the FITS files to WBPP. (Image by the Author)

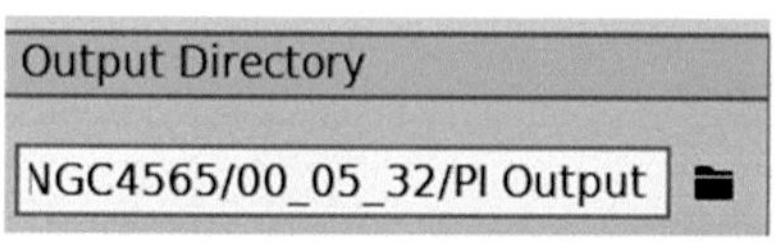

Fig. 3.25 The place to set the output directory or folder in WBPP. It is at the bottom right of the WBPP window. (Image by the Author)

Running WBPP is a fairly automatic process. After an observing session you can load it up, command it to run and go to bed. But there are a few tips and tricks where you may wish to challenge the default values.

The first one, which I frequently forget, is to set the output directory. This is a little box tucked away in a corner of the WBPP window (Fig. 3.25).

Just above this section of the window is the opportunity to change the reference image for registration. I leave this in the "auto" mode.

In the Lights tab, you can set the astrometric solution parameters. This is worth doing, as I have experienced cases where the information in the FITS files, which ultimately comes from my mount control software CPWI, is not always correct. The sub-window for this is shown in Fig. 3.26. The position needs to be fairly accurate; and the time should be a reasonable average of the times at which the frames were captured. It is obviously a bit excessive to quote this to the nearest second if you have exposed for six hours over six weeks because the weather was cloudy. This question is asked because the databases take account of the proper motions of stars.

You could force the values, but I have never needed to. The only thing that I have found to cause a problem is a vastly incorrect focal length for the telescope: the image solver will then fail. It tolerates small errors in focal length, but not big ones.

The next sub-window in the Lights tab that is worth a look is the Image Integration (i.e. stacking) section. I mostly don't argue with the defaults, but Fig. 3.27 shows

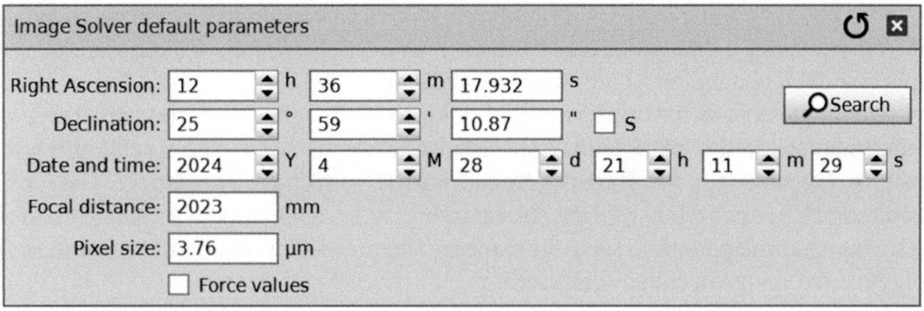

Fig. 3.26 The image solver parameters. You should search for the object in the database using the button shown, and check that the date is reasonable. (Image by the Author)

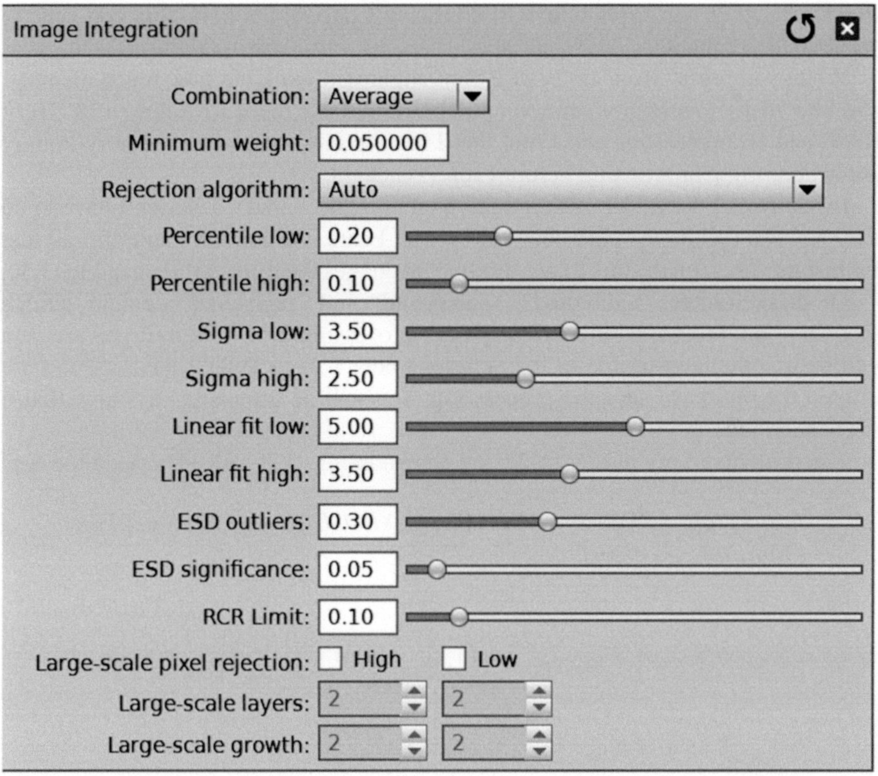

Fig. 3.27 The Image integration (stacking) parameters that I use. (Image by the Author)

that I have non-default values for "Sigma low" & "Sigma high". The defaults are 4.00 and 2.00 respectively. The integration (stacking) will then sometimes fail. So I have loosened these values to 3.50 and 2.50 respectively. Integration (stacking) then usually works for me.

Moving over to the Post-Calibration tab, there's a decision that you have to make if you are using a one-shot color camera: it's called "Channels Configuration". In this context a channel is a color channel: red, green, or blue.

The process of debayering is to add color to an image. Terrestrial cameras and smart phones do this by default. Astronomical cameras don't. The so-called Bayer pattern, invented by a Mr. Bayer of Kodak, varied from camera to camera. The most common Bayer pattern is RGGB, shown in Fig. 3.28. Your camera's manual should tell you what the pattern is for your camera. The process of adding the color to displays of the image is called 'debayering'.

Your options are to create master frames of the integrated (stacked) image as a single image or as separate images for each color channel. These options are shown in Fig. 3.29.

Why might you want to do this? An example is to add a hydrogen-alpha channel to your image. I will explain how to do this in Chap. 6, but to do this you need to separate the channels.

When you press "Run" on WBPP, PixInsight will work out how much memory, and how many processors your computer has, and use them all. I have 128 GB of RAM and 16 processors; and I find that I have to close any other memory-hungry applications that are running before I run PixInsight.

I usually call the folders where I store output "PI Output". You don't have to do this. You can call them "pink unicorns" for all I care – or for that matter for all that PixInsight cares. Inside this folder, PixInsight creates the folders shown in Fig. 3.30.

Of these folders, "calibrated", "debayered" and "registered" contain scratch files: they can be deleted immediately after processing. But wait until you are sure you have something usable in the "master folder". After WBPP has run, you are offered the option to save the process log. You will be astonished to know that it saves it in "logs".

The contents of my master folder are shown in Fig. 3.31. WBPP placed the file

$$masterLight\_BIN-1\_6248x4176\_EXPOSURE-60.00s\_FILTER$$
$$-NoFilter\_RGB.xisf$$

Red	Green
Green	Blue

Fig. 3.28 The RGGB Bayer pattern in a color digital camera is achieved by putting tiny filters in front of each pixel. It should be noted that, although most cameras have one red, one blue & two green filters repeatedly across the CMOS or CCD chip, not all do. (Image by the Author)

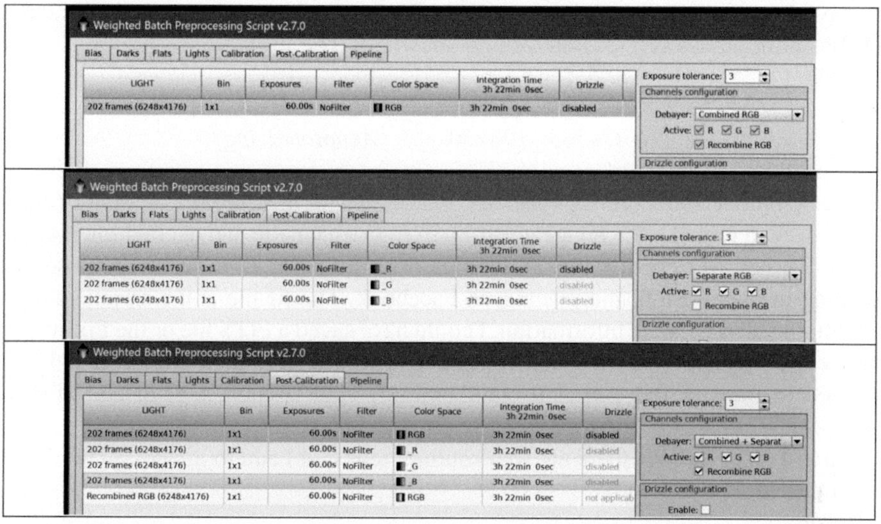

Fig. 3.29 The options for Channels configuration. (Images by the Author)

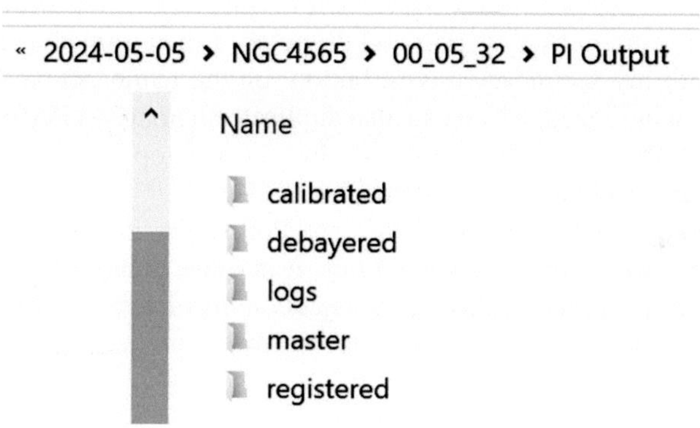

Fig. 3.30 The folders created by WBPP. (Image by the Author)

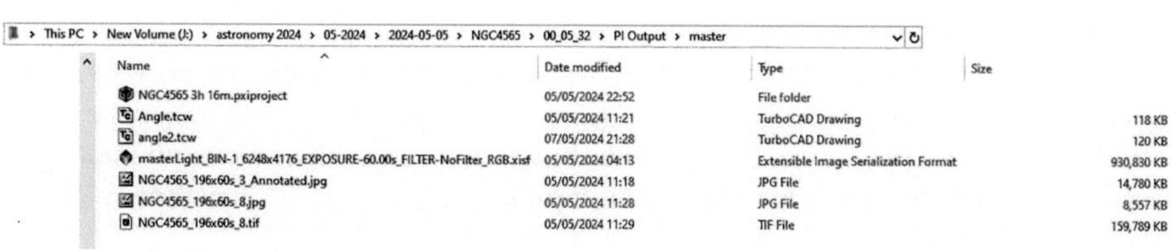

Fig. 3.31 The contents of my “master folder”. (Image by the Author)

in the folder, but all the others are ones I added later. The .pxiproject file, a record of the entire processing session, can go anywhere. The jpegs and tif files are ones I created. I fed the file

$$NGC4565\_196x60s\_3\_Annotated.jpg$$

to the computer-aided drafting package TurboCAD to measure the amount of rotation I needed to make the image north-up in right ascension and declination coordinates. TurboCAD created the TCW files. The jpeg file is for such purposes as parking in manuscripts and placing on Facebook. The tif file is the serious output containing all the image information. I could have saved a FITS file or the PixInsight native file format ".xisf" if I had wanted to. I would have done the latter if I had wanted to transfer the output to another PixInsight session, e.g. to add H-α to an image.

After I close WBPP, there are four routine steps that I add to images after WBPP has finished. I load the file

$$masterLight\_BIN-1\_6248x4176\_EXPOSURE$$
$$-60.00s\_FILTER-NoFilter\_RGB.xisf$$

into PixInsight, usually by dragging and dropping. I then immediately duplicate it by clicking with my secondary mouse button[1] on the name tab on the left of my image, as shown in Fig. 3.32. You can also duplicate an image in PixInsight by dragging the name tab to the left onto the PixInsight desktop. Having duplicated the image, I close the original before I damage it.

When you add a master light frame into PixInsight, it also adds a couple of images showing what WBPP screened out. It is quite good at removing satellite trails, and you can see this by making the rejection image active and clicking on one of the two the radioactive symbols on the menu. You won't need the rejection images so you can close them.

Next, I give the duplicated image a shorter, less clumsy name. I usually suffix the identifier with a _2 and then whenever I make new images, the identifiers go up: _3, _4, etc. I do this regularly so that I can go back if I don't like a processing result. Not that I ever mess up, you understand. Goodness me, no.

Now, above all, do not permanently stretch the image yet.

First, I do a plate solve with Script → Image Analysis → ImageSolver. This script wants to know the RA, Dec & Date of your image, as well as the telescope focal length and camera pixel size. If you use a Barlow lens or focal reducer, you need the focal length after allowing for these. It will pick up these parameters from WBPP if

[1] Most authors will describe this operation as "right click". However, like 10% of the population I am left-handed, and reverse the order of my mouse buttons. Therefore, for me a right click is a left click. That's a great way to confuse a girl, so I use "primary click" and "secondary click".

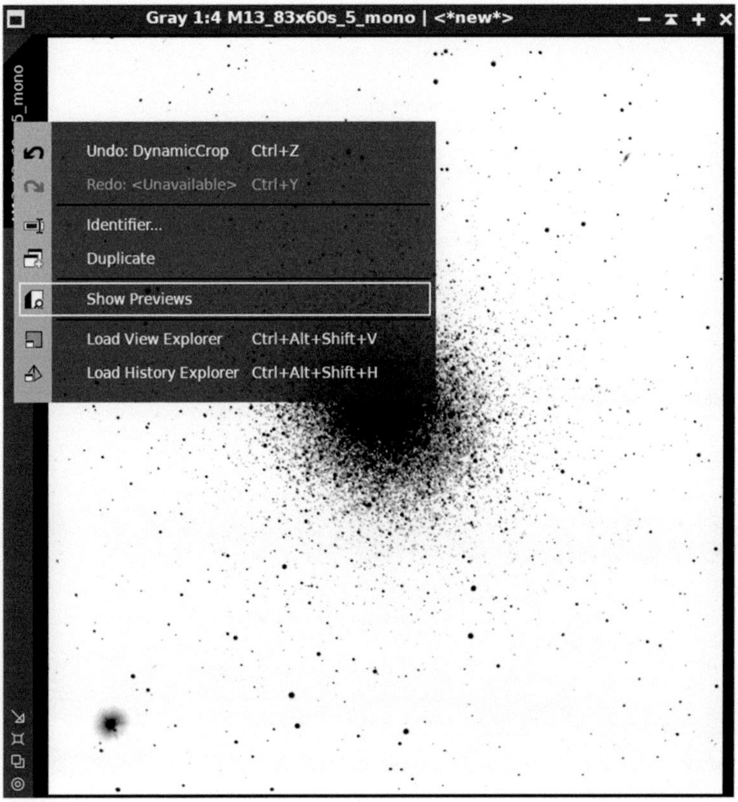

Fig. 3.32 The menu to duplicate an image or change its name (identifier). (Image by the Author)

the Image solve in WBPP worked. If not, you will need to add them. But always check the parameters in case they are wrong.

I do this ready for the next step,

Process → ColorCalibration → SpectrophotometricColorCalibration,

which requires a plate solved image. The menu is shown in Fig. 3.33.

This option will not work unless you have downloaded the GAIA database of stars. You have to log in to the downloads section of their website pixinsight.com and download the Gaia DR3/SP database. There are three choices. If you have the space, download the full set.

I override the automatic limit magnitude, which goes down to mag. 12, and put in a stupidly high magnitude. This means SPCC will use as many stars as it can find.

You can use ALT-N to add a preview region away from the target object and then click the Region of Interest to add to the window in Fig. 3.33.

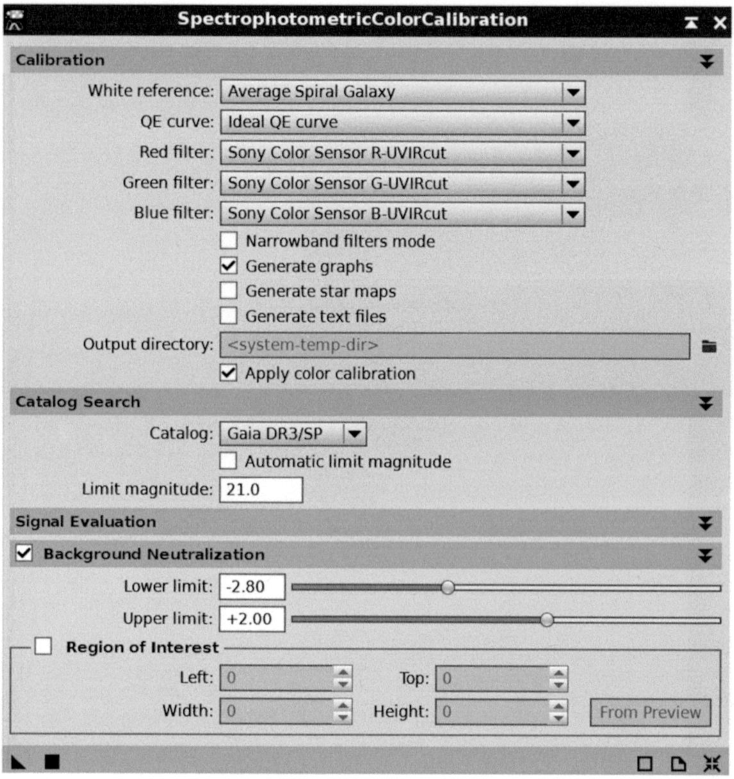

Fig. 3.33 The Spectrophotometric Color Calibration window. (Image by the Author)

The drop-down menu where it says "Average Spiral galaxy" in Fig. 3.33 has a large number of options such as "Elliptical Galaxy" and galaxies given their Hubble classification (Fig. 1.14).

I don't change the QE curve option. I do make sure that the red, green and blue filters correspond to my camera, which contains a Sony IMX571 chip. I check "Generate Graphs" and "Apply Color Calibration". (Why else would I do this if I did not want to apply it?)

The graphs it generates can be saved in the logs folder by clicking on the pdf icon that appears on them. I usually do this: it doesn't cost much storage space.

Now I am ready to go, and drag the little right-angle triangle at the bottom of the window shown in Fig. 3.33 onto my target image. This is actually a blue triangle.

This takes a few minutes, but by the end, using the radioactive button for a temporary stretch will show you that the colors now look more sensible.

Now to my third step. I use the RC-Astro BlurXTerminator tool, Process → Deconvolution → BlurXterminator, shown in Fig. 3.34. This is a paid-for add-in, so you may not have it. I thoroughly recommend it. I use the defaults and drag the little right-angle triangle at the bottom left onto my image. It takes a few minutes to run.

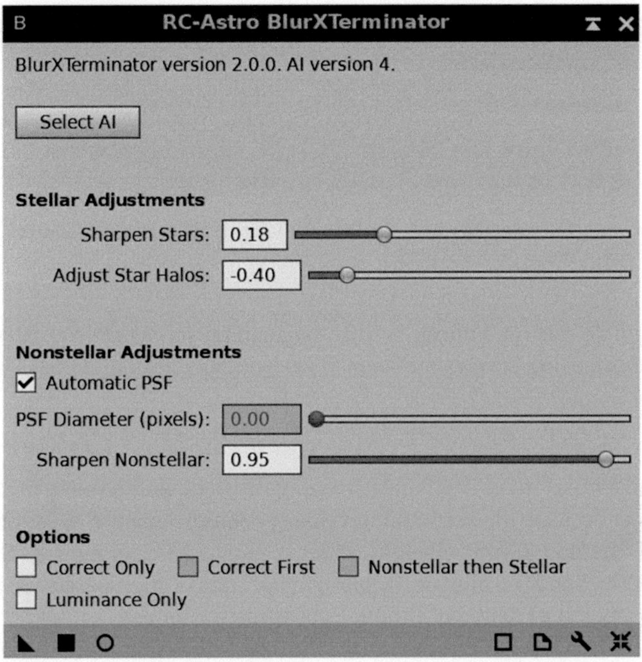

Fig. 3.34 The RC-Astro BlurXTerminator window. This is a paid-for add-in to PixInsight. (Image by the Author)

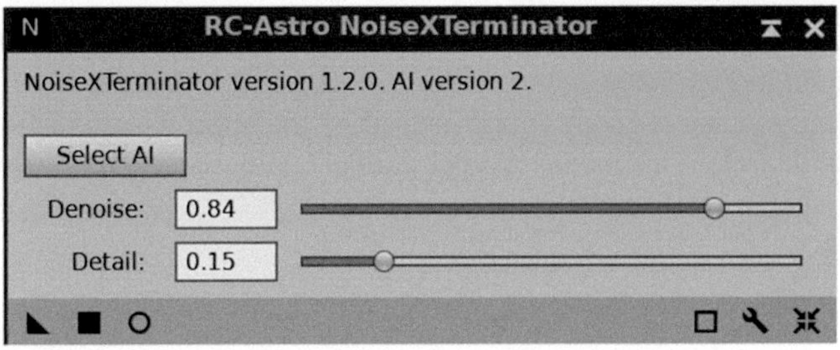

Fig. 3.35 The RC-Astro NoiseXTerminator window. This is a paid-for add-in to PixInsight. (Image by the Author)

This should be done before the next step, which is noise reduction. The order of these operations matters.

My fourth step is to use Process → NoiseReduction → NoiseXterminator to open the window shown in Fig. 3.35. This again is a paid-for add-in to Pixinsight so you may not have it. I use the defaults, and drag the right-angle triangle at the bottom left onto my target image. This doesn't always give enough noise reduction, but there are options to override the defaults.

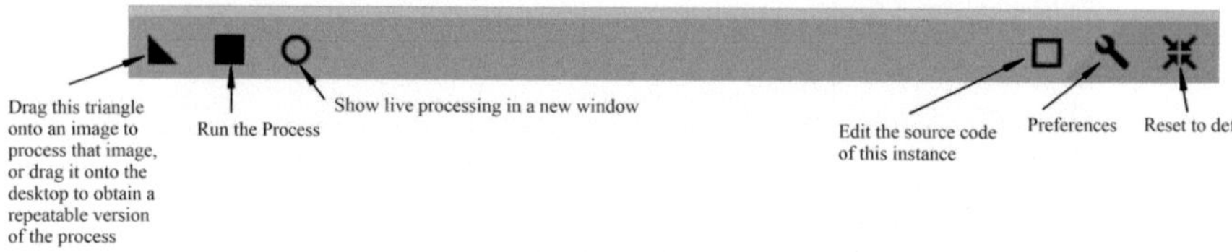

Fig. 3.36 Showing the meanings of typical icons along the bootoms of windows for scripts & tools in PixInsight. (Image by the Author)

Having gotten this far, I duplicate the image, to preserve the original, then stretch the image, often using several methods to see which is best, and then I use

$$Process \rightarrow MaskGeneration \rightarrow StarXTerminator,$$

a paid-for add-in, to make a separate image with stars, and remove the stars in my image. This is because the stars and the galaxy require separate stretching. There is an included alternative in PixInsight, Process → MaskGeneration → StarMask. It isn't as good as StarXTerminator. More manual intervention is required to remove the stars. Neither is perfect, but I think I got value for money.

Symbols Used in PixInsight Windows

PixInsight is not always consistent about these symbols, but it usually is. The main ones are shown in Fig. 3.36.

Sometimes you see a green check instead of the solid square. Sometimes you also see a solid circle. This means "Apply Global". You may need to use this instead of the solid square.

Chapter 4
The Phantom Galaxy: An Exercise in Disaster Recovery

Early Efforts

The Phantom Galaxy, M74 or NGC 628, has its opposition about a week before the end of October. It is then accessibly high in the sky for observers in the English-speaking parts of both hemispheres.

It is often said that it is the faintest Messier Catalogue object except for M101. I'm afraid I don't know who started this myth. It took me thirty seconds with Google to find someone who repeated it. I don't want to embarrass the unfortunate person by naming them. But I beg to differ. The Shapley-Ames Catalog [39] lists the apparent blue magnitudes as shown in Table 4.1.

M101 is 1.59 magnitudes brighter, or about 4.3 times as bright. In fact, there are over a dozen galaxies which are fainter than M74 and are in the Messier Catalog. Nevertheless, it is faint and therefore tricky to photograph from an urban or suburban environment. That makes it an interesting challenge.

It is a face-on galaxy with two spirals. They are a bit distorted but clearly identifiable.

My first attempt in November 2021 was beyond disastrous. I nearly wrote that I'm too ashamed to show it. But that's not the point. The point is to show you how I found my way from dire to decent, since many readers will start in the same state of ignorance that I did. The first image is shown in Fig. 4.1. By October 2022 I was getting a bit better, and produced the image in Figs. 4.2, 4.3, and 4.4. Figure 4.1 shows a live stack, using the arrangement shown in Fig. 4.3. Figure 4.4 shows the image stacked in Siril. I don't think this is a good place to go through my stacking technique, because it wasn't the best.

The final effort is not very sharp, is way too green, has feeble colors, is full of dust bunnies, and shows amp glow on the lower left. Yuck!

A little later that month, I bit the bullet of upgrading to a ZWO ASI2600MC camera. This was far from cheap. It cost me over $2000 for a refurbished version.

J. Clark, *Photographing Galaxies from Light Polluted Skies*, The Patrick Moore Practical Astronomy Series, https://doi.org/10.1007/978-1-0716-4506-2_4

Table 4.1 The magnitudes of M77 and M101

Galaxy	Apparent blue magnitude
M74 or NGC 628	9.77
M101 or NGC 5457	8.18
Difference	1.59

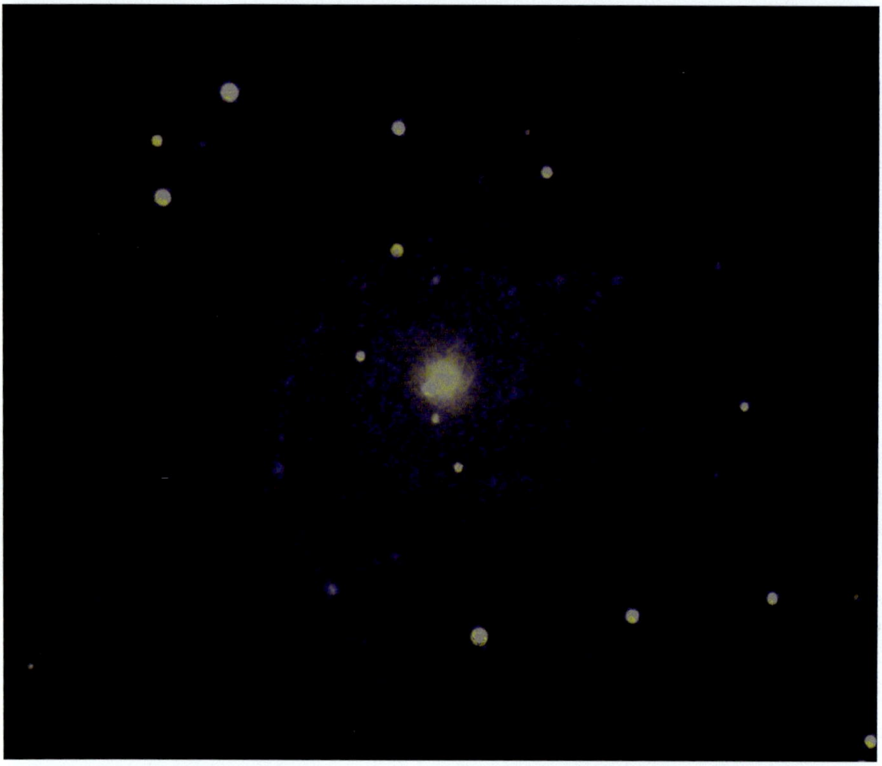

Fig. 4.1 My first attempt at M74. 2021-11-27. 30 × 120 s, Celestron C11 telescope, f/6.3 focal reducer, ZWO ASI294MC camera. Stacked with Deep Sky Stacker. (Image by the author)

But it was immediately apparent that this was a huge improvement, so I quickly sold the ZWO ASI294MC to help pay for the new instrument.

A preliminary live stack result (Fig. 4.5) shows that a three-hour live stack produced a much nicer looking picture, which I fed to Topaz Denoise AI. The colors came from the adjuster bars on the live stack. In other words, I did not use any tools to try to get good color balance. The stars are big and blobby, are saturated to white, and show chromatic aberration. I don't think my focus and collimation were great.

So I still had a long way to go.

I was busy elsewhere, and did not revisit M74 until the next opposition in 2023.

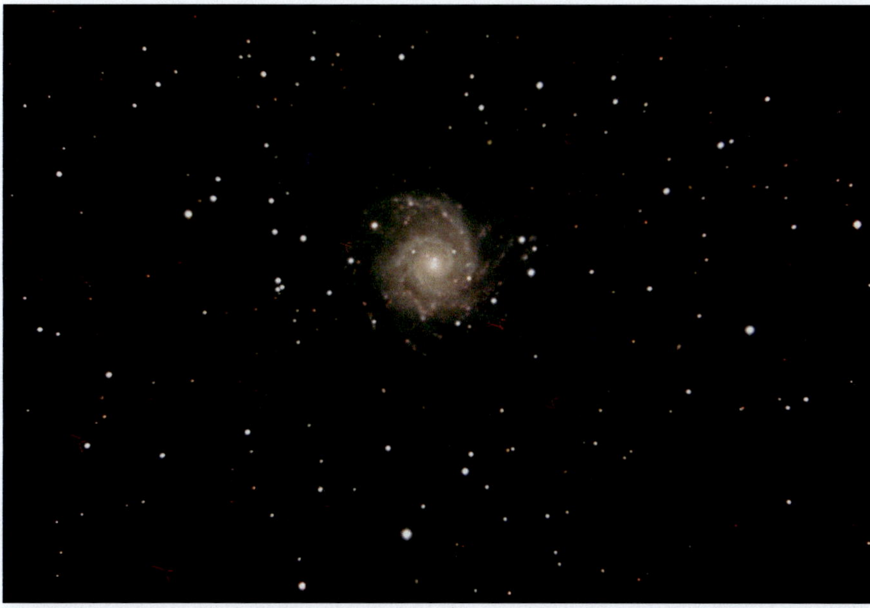

Fig. 4.2 My first recognizable attempt at M74. 2022-10-01. 102 × 120 s, Celestron C11 telescope, f/6.3 focal reducer, ZWO ASI294MC camera. Live stack with no dark or flat frames. Live stack from SharpCap. (Image by the author)

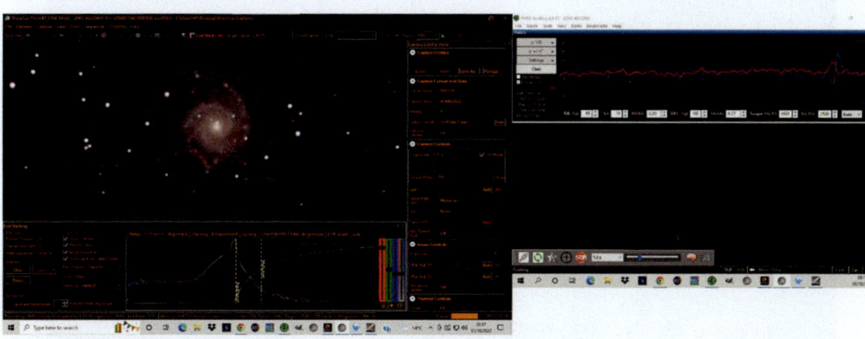

Fig. 4.3 Showing the live stacking set-up in SharpCap 4.0. The guiding with PHD2 is shown at the right. (Image by the author)

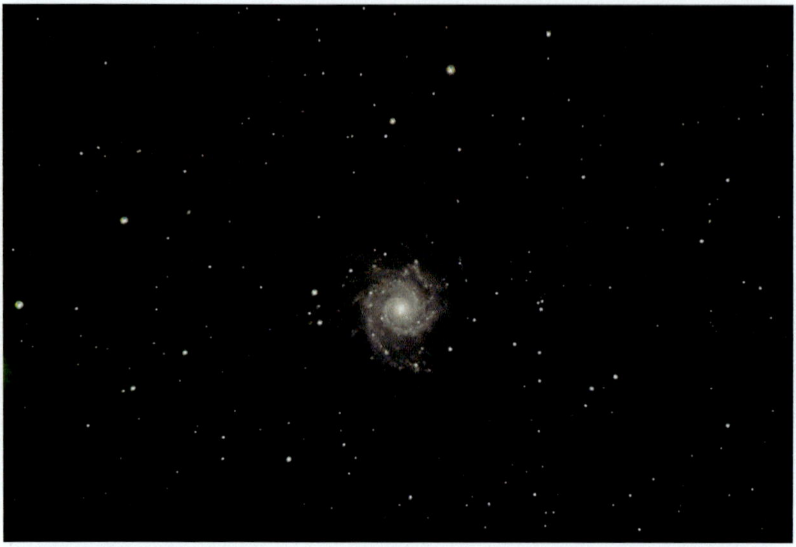

Fig. 4.4 My first recognizable attempt at M74, tidied up a little. 2022-10-01. 102 × 120 s, Celestron C11 telescope, f/6.3 focal reducer, ZWO ASI294MC camera, stacked in Siril. (Image by the author)

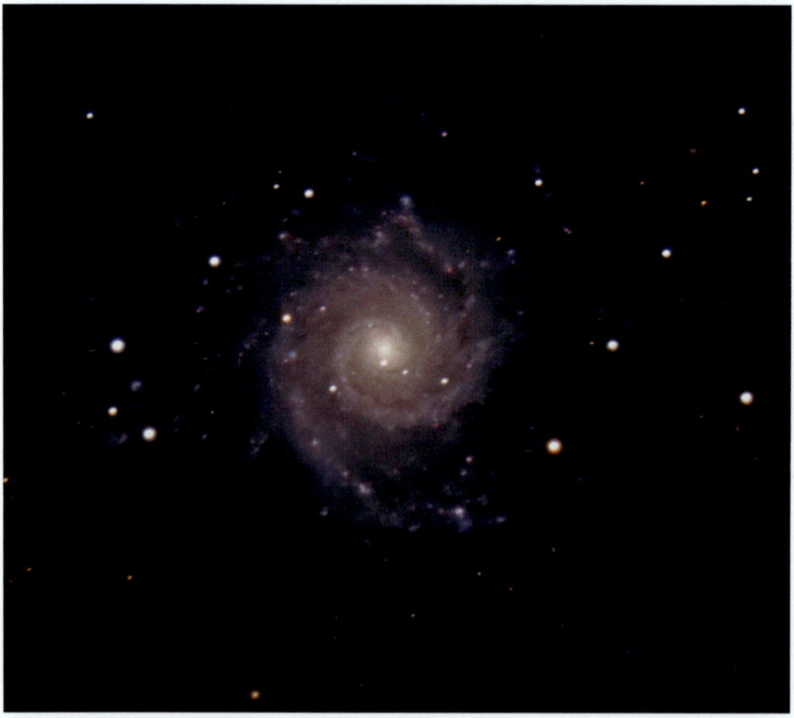

Fig. 4.5 My next attempt at M74. 2022-10-18. 90 × 120 s, Celestron C11 telescope, f/6.3 focal reducer, ZWO ASI2600MC camera. Live stack with no dark or flat frames. Live stack from SharpCap 4.0. (Image by the author)

The Green Phantom Wheel

The 2023 British weather more than lived up to its terrible reputation. In the end I collected just shy of twelve hours of data, but that took six weeks, with no other astrophotography on the go at the same time. I lost two clear nights while visiting family, but even so, that's grim. The weather is like certain politicians I could name. Just when you think they could not get any more awful, they do. So does cloudy weather. I have many tricks up my sleeve for telescopes behaving badly, but alas I have none for seeing galaxies through cloud.

I started on September 23 2023. I thought that I was doing everything right except for living in a cloud-less desert, but just look at this image in Fig. 4.6.

This has been stacked, with darks, flats & biases, and stretched. The galaxy itself looks kinda reasonable, though it could benefit from even more stretching, but it's in the middle of a horrible green wheel-like feature of similar brightness.

I tried all sorts of tricks with masks made by PixInsight's Game Script and Range masks, and its background removal tools, to subtract or divide the data away – please see Fig. 4.7.

Nothing doing. I was caught on the horns of a dilemma. The unwanted feature was presumably due to an unwanted reflection in the optics somewhere. Do I discard three hours of otherwise good data and take the telescope and camera apart to

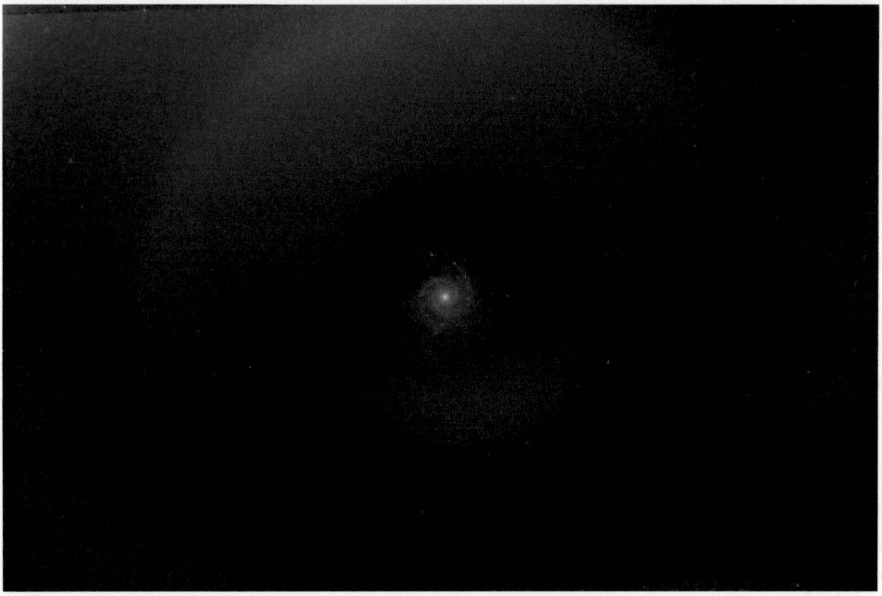

Fig. 4.6 My opening gambit for the 2023 M74 season. Celestron C11 telescope, f/6.3 focal reducer, ZWO ASI2600MC camera, 150 × 60s, processed in PixInsight 1.8.9-2. This image is what's left after the stars have been removed. (Image by the author)

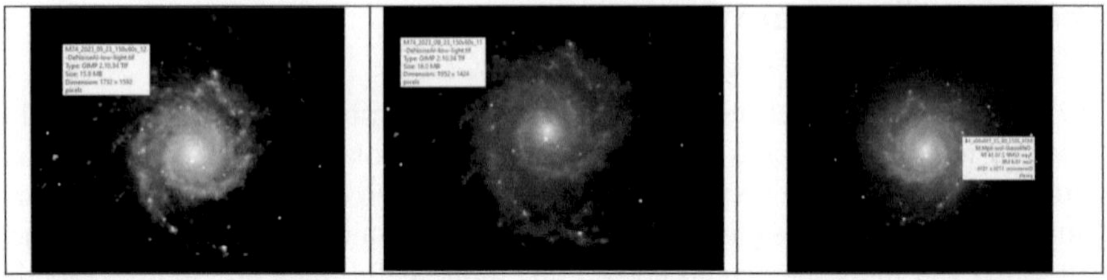

Fig. 4.7 Attempts to mask my way out of the fault shown in Fig. 4.6. The rightmost image has some potential, but I can't tell what is signal and what is unwanted noise around the edge of the galaxy. (Images by the author)

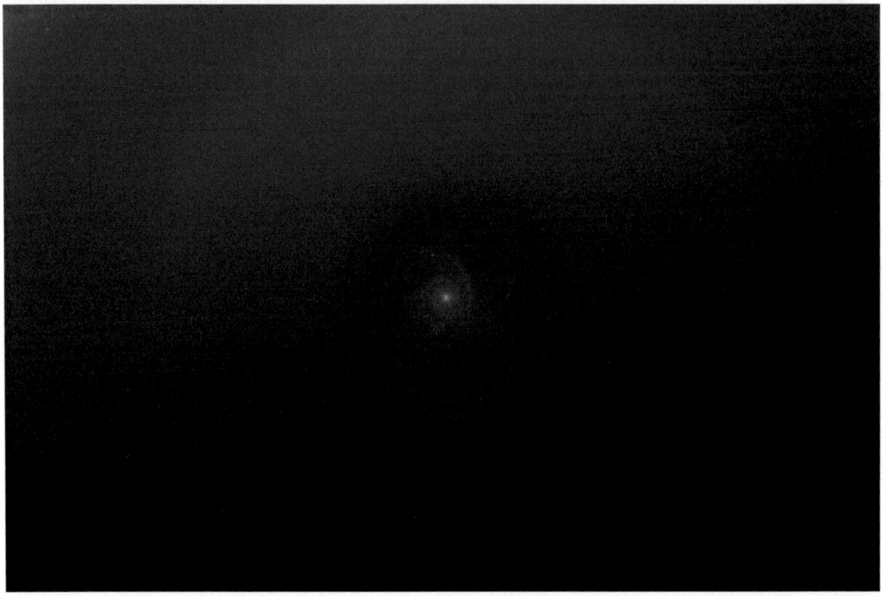

Fig. 4.8 Combined first & second nights for the 2023 M74 season. Celestron C11 telescope, f/6.3 focal reducer, ZWO ASI2600MC camera, 341 × 60s, processed in PixInsight 1.8.9-2. This image is what's left after the stars have been removed. (Image by the author)

trace the source of the damage, or do I assume that I will get very few clear nights and hope I think of a way to remove the ghastly green "phantom wheel"?

I re-shot my flats to see if the problem went away. It didn't.

So I decided to soldier on with the telescope as it is, and collect more data, while I tried to think of a trick to get round my little problem. There might have some random source of light pollution on the first night – someone might have left a light on without drawing their curtains.

Nothing doing, unfortunately. Figure 4.8 shows what's left after I had removed the stars.

This is just as bad. The problem has not gone away.

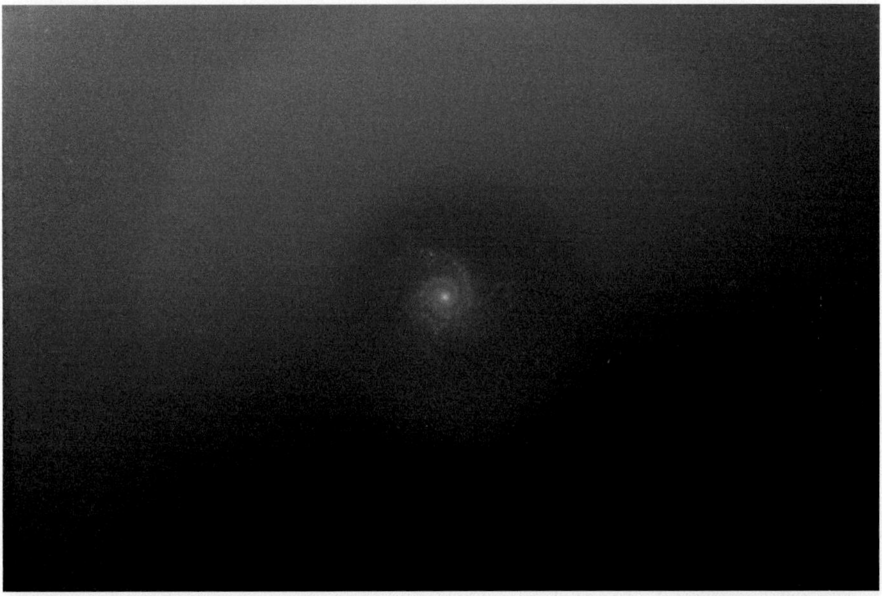

Fig. 4.9 Combined first & second nights for the 2023 M74 season, after using the GraXpert option to stretch the image as well as suppress the background. Celestron C11 telescope, f/6.3 focal reducer, ZWO ASI2600MC camera, 341 × 60s, processed in PixInsight 1.8.9-2. This image is what's left after the stars have been removed. (Image by the author)

I did try a new piece of software that was written as an activity for the Covid pandemic. It's free, and is called GraXpert, from https://www.graxpert.com. Its purpose is to remove background gradients. It's used a bit like the DynamicBackgroundExtraction tool in PixInsight, and it got good reviews on YouTube, so I thought I'd give it a go. It helped, but not enough. In fairness, it's not really designed to get me out of this mess. I have tried it on lesser problems, and liked it.

The signal-to-noise ratio (SNR) is better, but still not good enough. This left me feeling really frustrated. There's a good image in Fig. 4.9 if only I could pull it out.

One possible culprit was that at the time, we had a lot of high cloud above the intermittent low cloud, as shown in Fig. 4.10.

The only remedy for the weather was to wait. And wait. I had other fish to fry: I went to meet my new granddaughter for the first time. Since they live 700 miles away in Berlin, that was a major expedition, and a very welcome distraction from my astronomical woes.

The next suitable night, October 22 2023, had no high clouds that I could detect, but the gap in the lower clouds only enable me to "bag" 99 minutes' worth of 60 s frames. I figured I'd wait for the next clear night. This came ten days later. Why, oh why, did I ever get into this hobby? And why was I born in a cloudy country?

As I sulked and stewed over this, I had a brainwave. Why don't I shoot some frames of an area of sky with no galaxies, or at least no nearby ones, or well-known

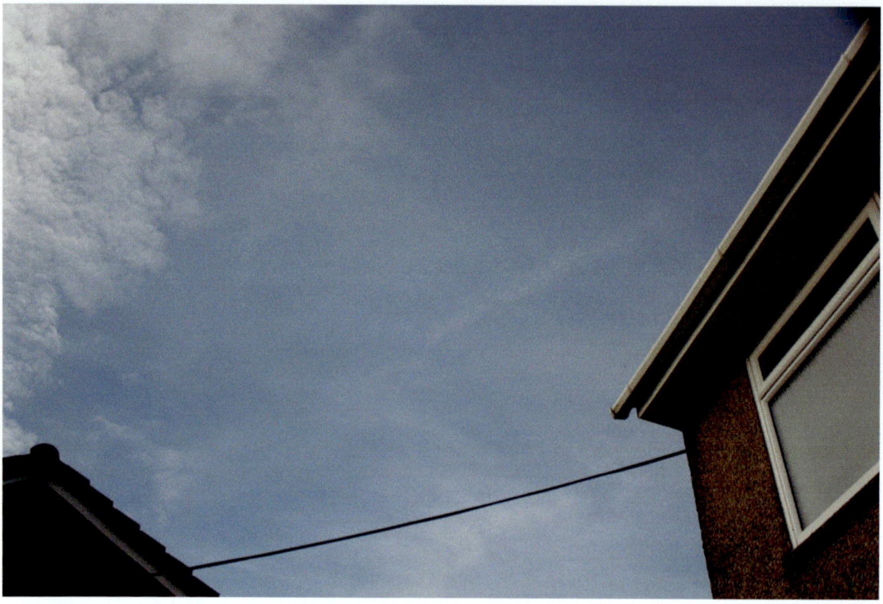

Fig. 4.10 You could easily see the milky, very high cloud during the day, but it was not very visible at night. (Image by the author)

nebulae, and subtract the resultant image of my green "phantom wheel"? That's gotta be worth a try.

As it happed, three days later, now on November 6th, there was a short clear spell around 4 a.m. So I found a bit of blank sky not too far from the track my telescope had been following, and managed to expose 90 × 60 s frames. The following night, in defiance of the best weather forecasts, was beautifully clear. I did not pass up this opportunity.

First, I'll show you my null shot (Fig. 4.11).

This shot found the dreaded green "phantom wheel" all right. It seems to have spotted a couple of faint nebulae away from the area of interest. I'm not going to lose any sleep about them. More concerning is that when I zoom in on the image, it's noisy. In particular, it's streaky (Fig. 4.12).

How about putting it through Topaz Denoise AI? Figure 4.13 shows the whole image, and Fig. 4.14 shows what happened when I zoomed in. This image is less noisy, but still streaky. Near the nucleus of M74, this doesn't much matter, but unfortunately the streakiness shows up in the outskirts of M74 after I have subtracted the Topaz-Denoised null image (Fig. 4.13) Oh dear, that's no good. Oh, well. If at first you don't succeed, sulk.

But think as well as sulk. It wasn't long before an idea came my way. There had to be an app that would blur my null image, just slightly, but enough to kill off the streaky noise in Fig. 4.15.

Fig. 4.11 My first shot of an area of blank sky. 90 × 60 s. Same equipment as before in same configuration. (Image by the author)

Fig. 4.12 My null image is streaky. (Image by the author)

Fig. 4.13 The null image put through Topaz Denoise AI. (Image by the author)

Fig. 4.14 When zoomed in, the Topaz-Denoised-version is better, but not perfect. (Image by the author)

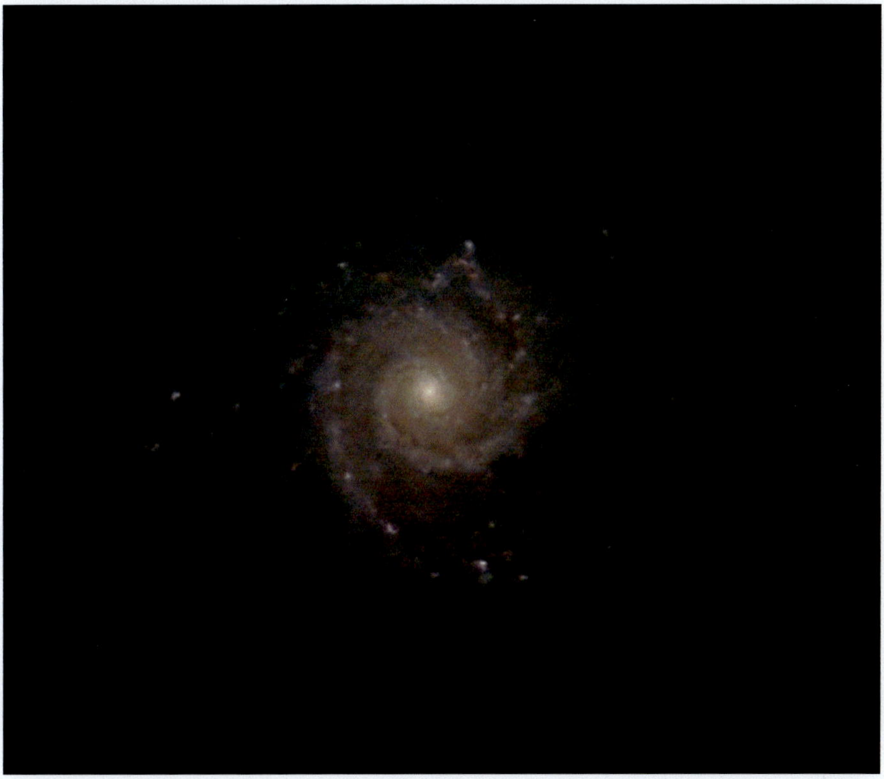

Fig. 4.15 The noise from the null image shows up in this image. 286 × 60 s, otherwise the same conditions as in Fig. 4.6. Horizontal streaky noise is clearly visible, even near the galaxy nucleus. (Image by the author)

I tried Gimp. For those who don't know, it's a Photoshop competitor, only it's free. The Blur tool is illustrated in Figs. 4.16 and 4.17. I chose Gaussian blur and set the blur size to three times the default, not for any good reason. It was just a first guess. How did I do? Have a look at Fig. 4.18.

This image now looks a lot better. I don't much like the colors, but at least I have got shot of the noise and the green "phantom wheel" of doom.

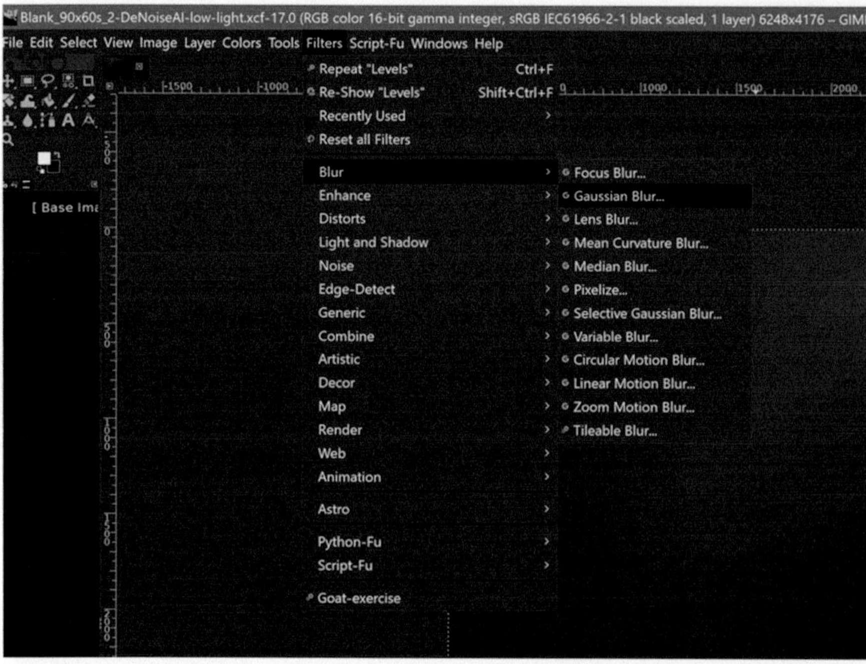

Fig. 4.16 How to find the 'Blur' tool in Gimp. (Image by the author)

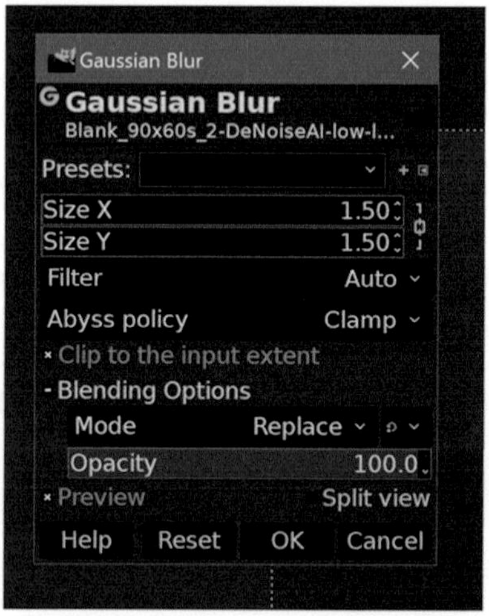

Fig. 4.17 The 'Blur' tool in Gimp opens this window. I trebled the default blur size offered. (Image by the author)

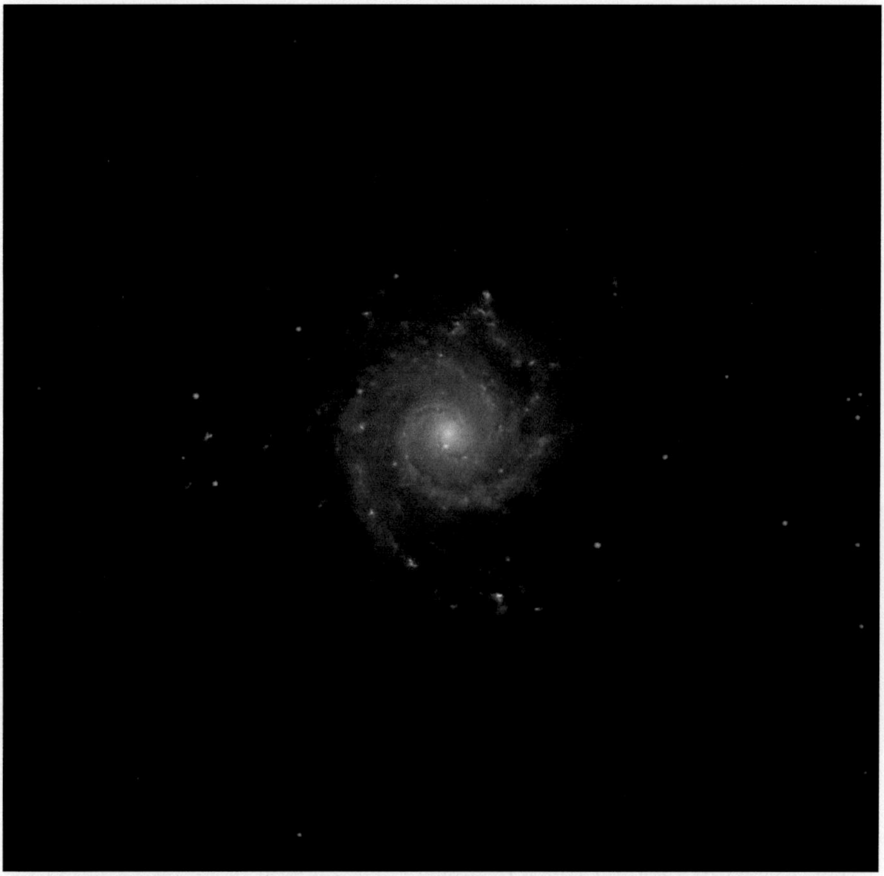

Fig. 4.18 The final image of the galaxy, using the version of the null image blurred by Gimp. I stacked all the usable raw images I had, giving me 709 × 60 s worth of exposure. Image by the author

How I Processed M74 in 2023

I ran my frames through the Blink tool in PixInsight, and rejected a few, either because they showed camera shake or because there was too much light reflected back from this cloud. Also, if you start imaging before the end of astronomical twilight, you will end up throwing away frames: it's a waste of time shooting faint galaxies in twilight.

Then I preprocessed the remaining frames using PixInsight's WBPP tool described in Chap. 3. This gave me a master image in PixInsight's xisf format. Same process, different day.

As always, I protected this master file by duplicating it the minute I brought it back into PixInsight for postprocessing. The duplicate is shown in Fig. 4.19.

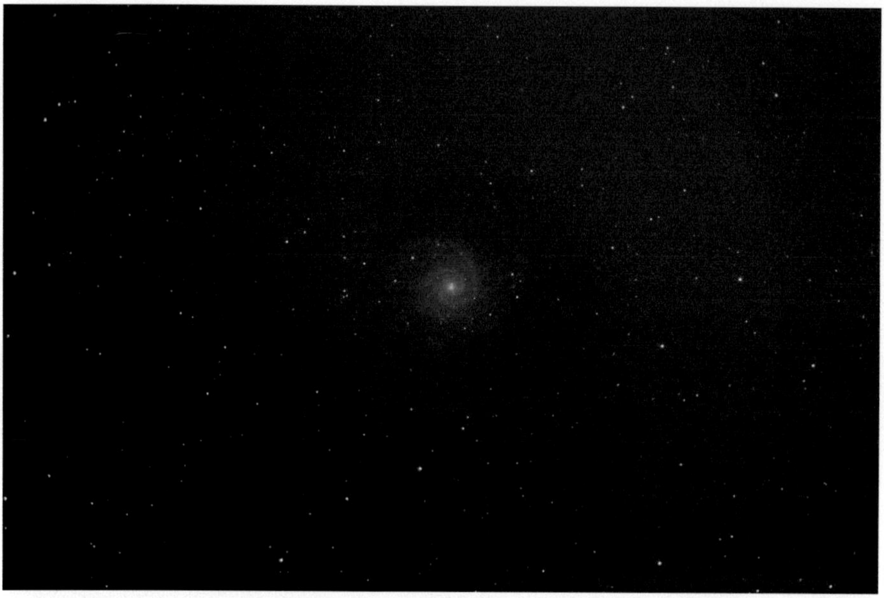

Fig. 4.19 My 709 × 60 s image with minimal preprocessing. (Image by the author)

In fact by the time it gets there I have given it the following postprocessing.

1. Use the Image Solver Script. The next stage requires this. The caption of Fig. 4.20 shows the menu path.
2. Perform Spectrophotometric Color Calibration (Fig. 4.21). It gives the graph in Fig. 4.22 showing how well it did.
3. Use RC-Astro's add-on BlurXTerminator. The caption of Fig. 4.23 shows the menu path. This takes a while. You can make coffee or watch a YouTube video while you are waiting.
4. Use RC-Astro's add-on NoiseXTerminator. The caption of Fig. 4.24 shows the menu path. This runs pretty fast.

After these jobs have been done, I use a script called EZ Soft Stretch to begin stretching the image. This is illustrated in Fig. 4.25.

There's a twiddly bit to this script. It stopped working after PixInsight was upgraded from version 1.8.9-1. It was only ever an add-on to PixInsight. However, a kind person whose username is Mike1485 debugged it, and posted the debugged version here: https://pixinsight.com/forum/index.php?threads/ez_softstretch-resurrected.21314/. You have to download and install EZ_SoftStretch.js and EZ_Common.js. These are not my intellectual property, though they have been made freely available. Please don't be unkind to me if they stop working in a future release of PixInsight. I show the location of these files on my Windows 10 PC in Fig. 4.26.

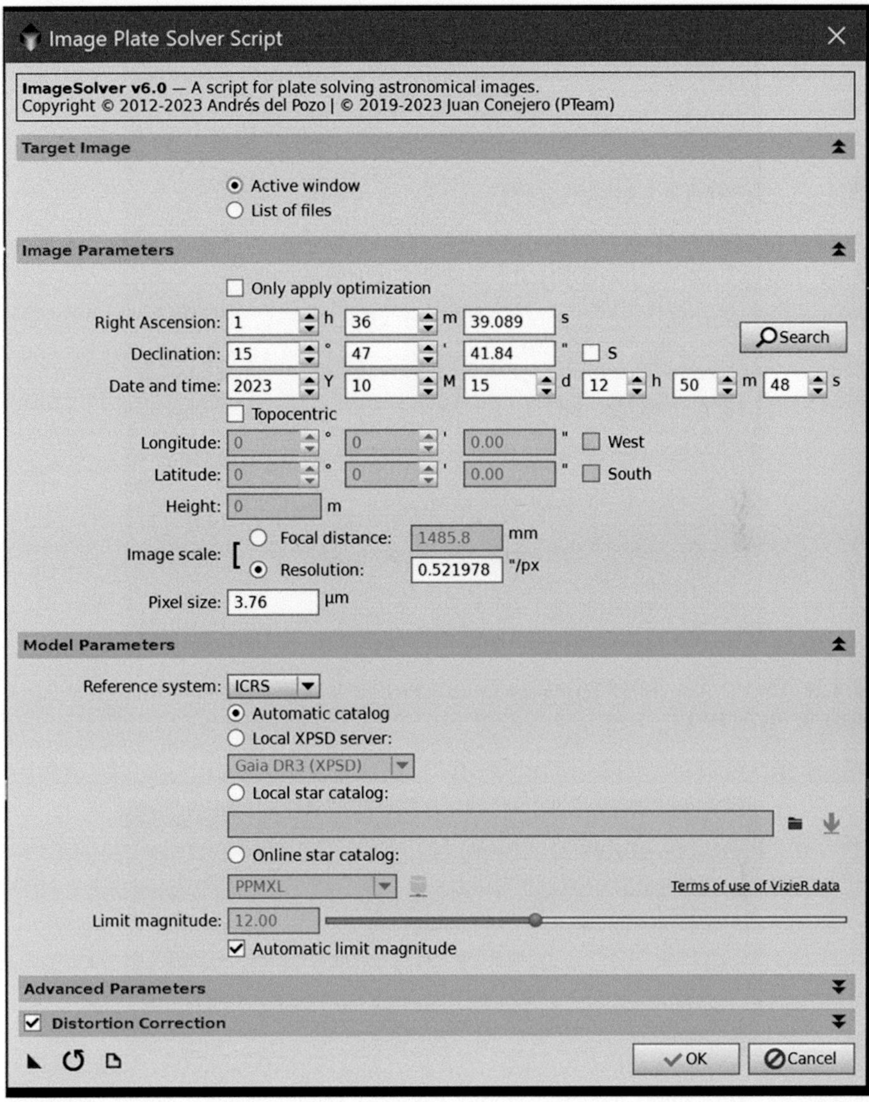

Fig. 4.20 The Image Solver Script. Script →Image Analysis → ImageSolver. (Image by the author)

Now I have the phantom wheel shown in Fig. 4.27. You can also see it in the preview in Fig. 4.25.

Now to the actual processing. I duplicated the Image shown in Fig. 4.27 so that I could remove the stars from it.

Star Removal is done with the add-on RC-Astro StarXTerminator, Fig. 4.28.

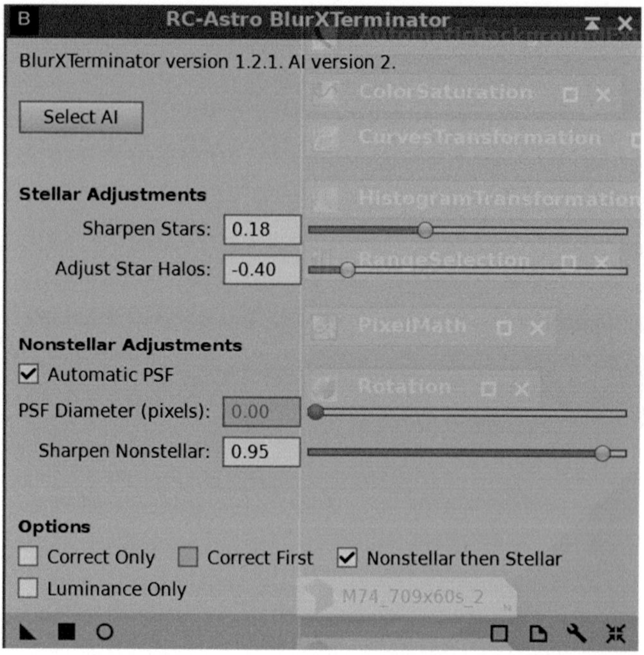

Fig. 4.21 The RC-Astro BlurXTerminator Process. Process → Deconvolution → BlurXTerminator. (Image by the author)

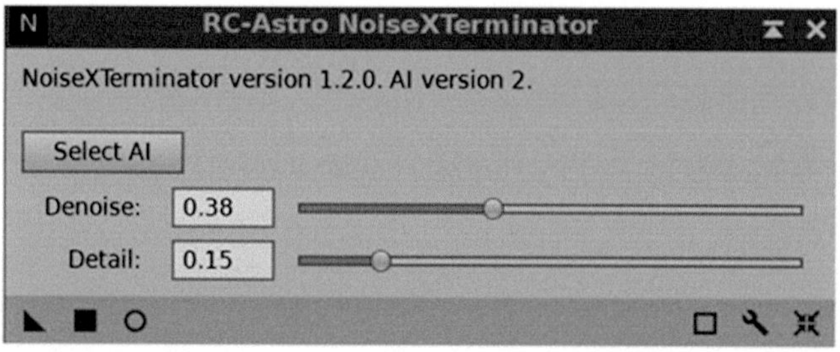

Fig. 4.22 RC-Astro NoiseXTerminator Process. Process → NoiseReduction → NoiseXTerminator. (Image by the author)

The output from RC Star Xterminator is that your original diagram has no stars, and there's a new image containing the stars. Notice that I had to check a box in Fig. 4.28 to obtain the separate star image.

My blank image of the wheel of death is in Fig. 4.13.

My next move is to use PixelMath to subtract this from the starless image in Fig. 4.29. The expression I used is shown in Fig. 4.30. I experimented with the number 1.00 shown in the box. I tried different values, but found no advantage to

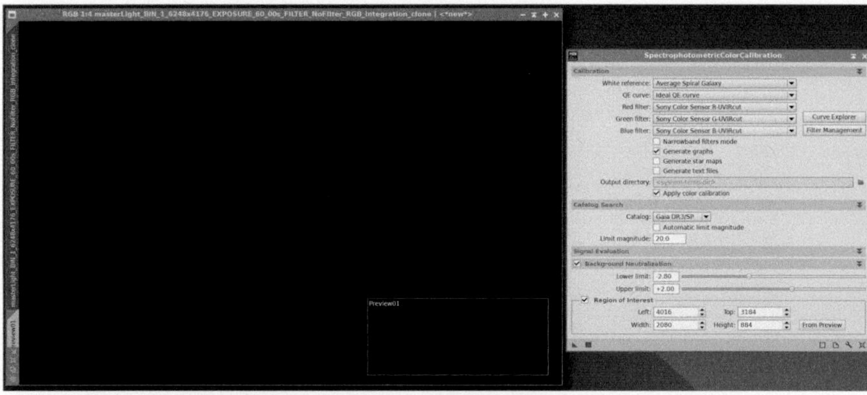

Fig. 4.23 Spectrophotometric color calibration has to work on the unstretched image. As you can see I gave this app a preview area well away from the galaxy. The menu path is Process → ColorCalibration → SpectrophotometricColorCalibration. (Image by the author)

doing this. Note that I have checked the boxes "Use a single RGB/K Expression" and "Create new Image".

The resulting image is shown in Fig. 4.31. I have pretty well removed the unwanted part of the image. The next step is to stretch the image further. I made one stretch without masking, but found that any further stretching brought out unwanted background fogging around the edges.

I did my stretching with the histogram transformation tool, shown in Fig. 4.32. My experience is that it's best to do this gently in stages.

Since doing this always brings a certain amount of unwanted background light into the image, I made some masks to simplify my task. As is usual in my experience, this involved some trial and error. In the end the best thing for this particular image proved to be a binary mask made with the GAME script and then to use the Range Mask tool to blur its edges.

The method of using the GAME script is described in Fig. 4.33. The Range Selection tool is shown in Fig. 4.34. The final mask is shown in Fig. 4.35.

There's a particular way to apply the mask to the image. Arrange mask and image as shown in Fig. 4.36. The mask has to be to the right of the image. You then drag the identifier tab of the mask onto the left window bar of the target image as shown. The result is shown in Fig. 4.37: edits now only apply to the non-red areas. You can also invert the mask to cover the galaxy and just edit the background. This is done by Mask → Invert Mask. The result is shown in Fig. 4.38. You would then use the Histogram Transformation tool as shown in Fig. 4.39.

When you have finished with the masks, use Mask → Remove mask. The identifier tag then ceases to be brown, and goes back to grey, to indicate that the mask is now inoperative.

I now have the image shown in Fig. 4.40. Before recombining this with the stars, I'm going to save it as a 32-bit TIFF file, and feed that to Topaz Denoise AI. This doesn't always help, but in this case it did. You should always do this to the image

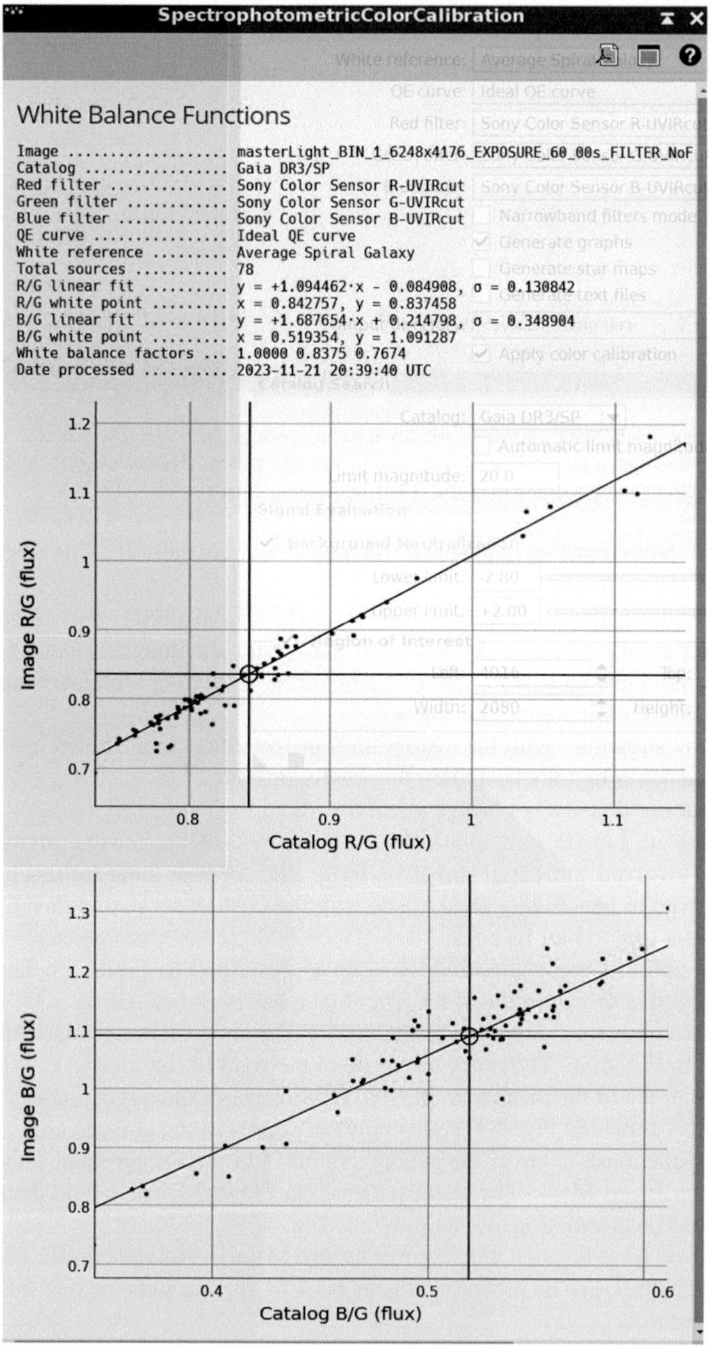

Fig. 4.24 The graph showing how well the spectrophotometric color calibration worked. (Image by the author)

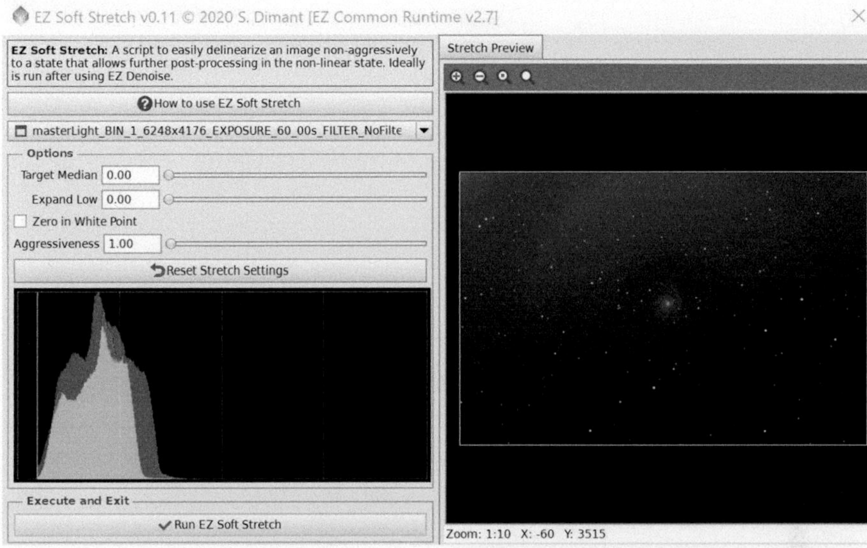

Fig. 4.25 The script EZ Soft Stretch produces this window. (Image by the author)

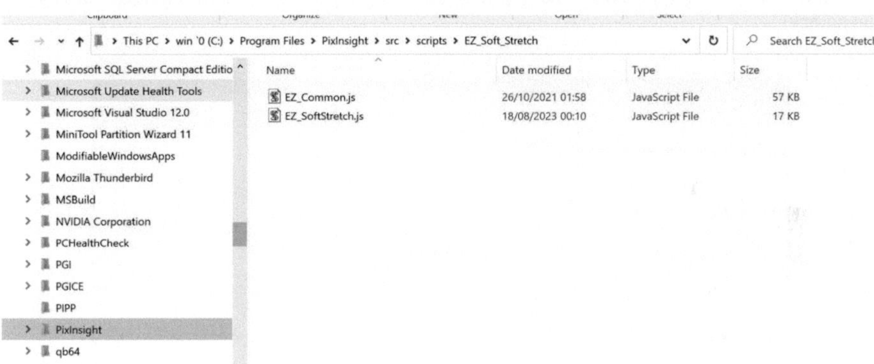

Fig. 4.26 The location of the debugged version of EZ Soft Stretch on my Windows 10 PC. I obtained this from https://pixinsight.com/forum/index.php?threads/ez_softstretch-resurrected.21314/. (Image by the author)

without the stars, because Topaz Denise AI will mangle up your stars. Figure 4.41 shows the subtle noise removal. Don't go crazy with this. You don't always benefit from this process. A light touch is best.

Next, I duplicated the image of the stars in Fig. 4.29, and did two things to it. First, I applied the Color Saturation tool to enhance the star colors (Fig. 4.42). This is purely for aesthetic reasons. There's no "scientific" justification for doing this. Finally, I tend to like to tone the stars down. They aren't the focus of the image, and serve as something of a distraction. This is also an aesthetic choice, not a scientific one. The Histogram Transformation tool is used rather as in Fig. 4.39.

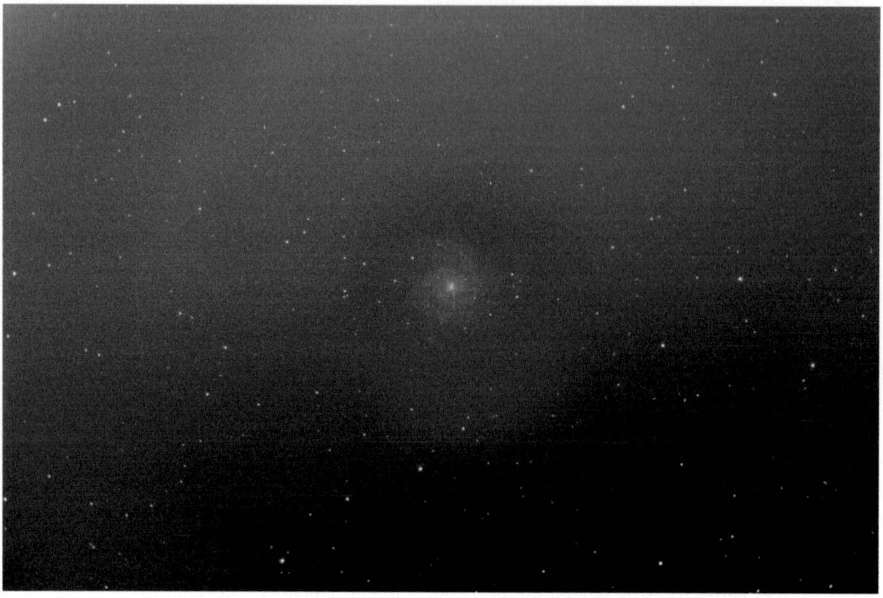

Fig. 4.27 The EZ-ily stretched image with 709 × 60 s worth of stacked frames. (Image by the author)

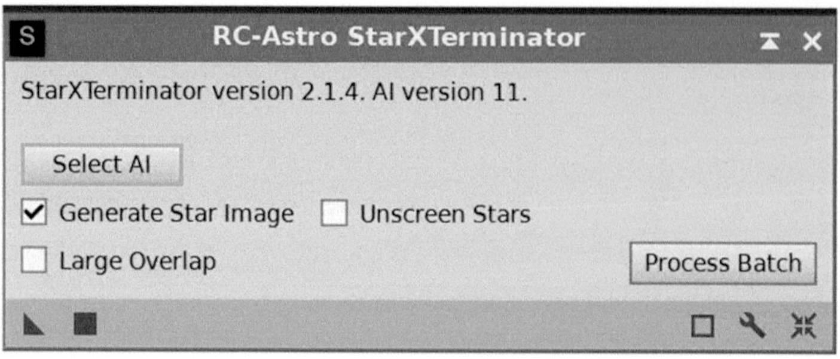

Fig. 4.28 the app RC StarXterminator found at Process → ObjectRecognition → StarXTerminator. You drag the bottom-left triangle onto the image from which you want to remove stars. (Image by the author)

I now have a star image I can add to my stretched, denoised image. This is done with PixelMath, with the expression shown in Fig. 4.43. Not everyone likes this simple way to add the stars in. There is a certain amount of YouTube discussion about using more sophisticated methods, which, alas aren't very easy to follow from the videos I have watched. In any case, I am not dissatisfied with the results from my method. If I were photographing nebulae, I might think differently.

At this stage I also use Process → Geometry → Rotation if I want to, in order to reorient the image. Do this before you crop, not after. Then I use Process →

Fig. 4.29 My image has been turned into two images, one without the stars, and one with them. (Images by the author)

Geometry → DynamicCrop to crop the image. There's a slightly irritating trick in the latter. You have to click a green check mark in the bottom left of the tool to do the crop after you have used your mouse to select the frame of the cropped image. I've shown this in Fig. 4.44.

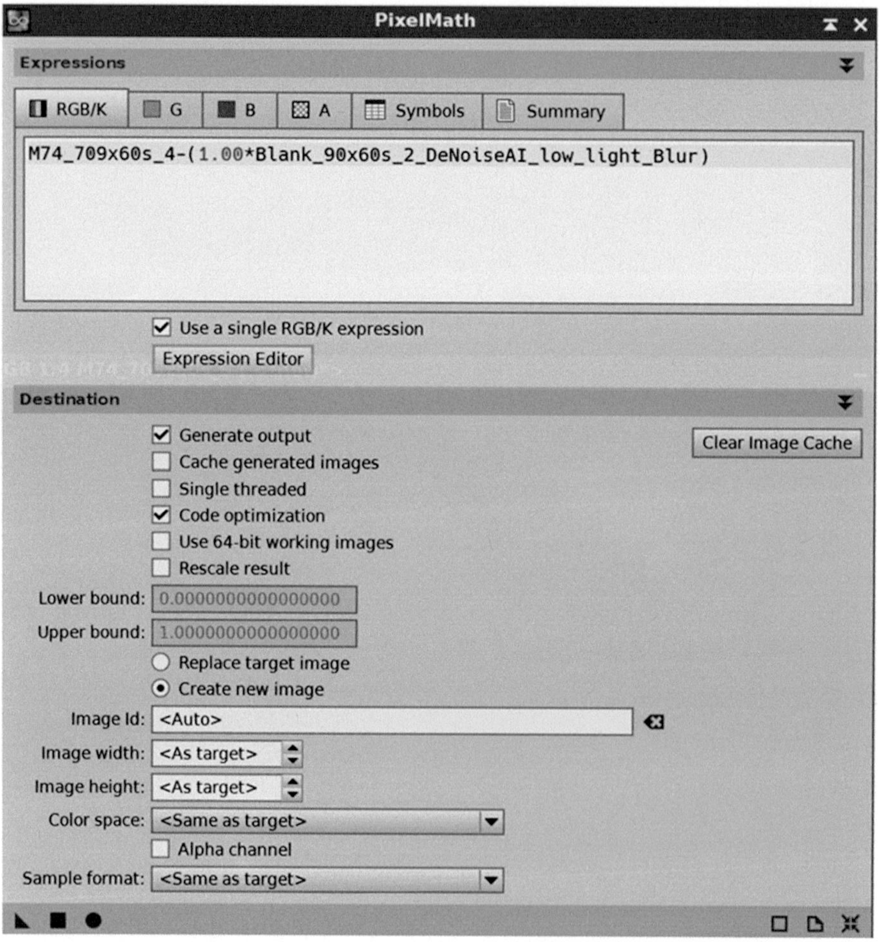

Fig. 4.30 Using the PixelMath app to subtract of the "Wheel of Death". Notice that I checked the box to generate a new image. Process → PixelMath → PixelMath. (Image by the author)

You can also do the rotation with this tool. There's no good reason why I don't.

My rotated and cropped image with stars is shown in Fig. 4.45. I'm happy with the level of detail and general look of this image, but I am far from happy with the color. The galaxy has a green cast that I do not like. (I do not like green casts on galaxies. I do not like them, Sam-I-am. Sorry.) This has arisen from my subtraction of the "wheel of death".

So I gave up on trying to preserve the colors from the spectrophotometric color calibration, and played around with the Assisted Color Calibration tool (Fig. 4.46) and the color saturation tool (Fig. 4.42) and ended up with the image shown in Fig. 4.47.

Fig. 4.31 My image of M74 with the “Phantom Wheel” subtracted from it. I have pretty well removed the problem. The next step is to stretch the image further. (Image by the author)

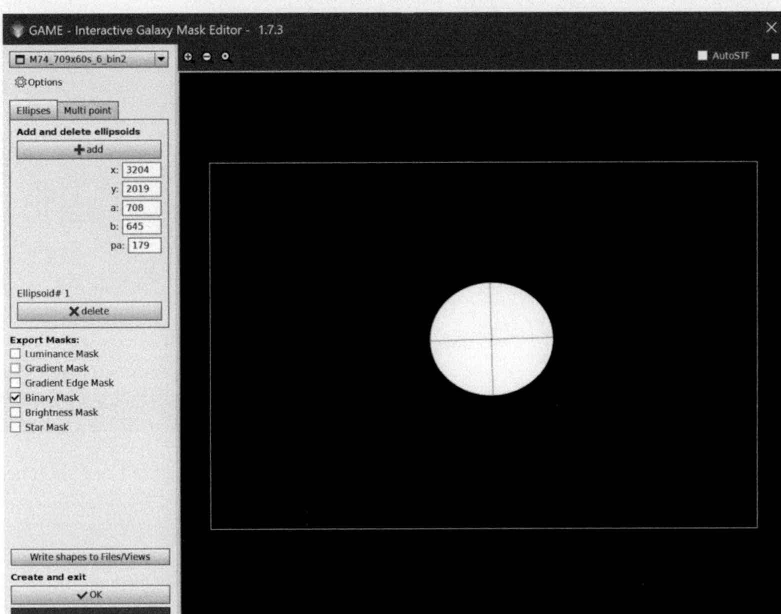

Fig. 4.32 The GAME script in the present instance. You add an ellipse, stretch it to the shape you want, which in this case was fairly circular, check the “Binary Mask” box and click OK. It generates you a black-and-white mask shown, with sharp edges. Script → Utilities → Game. (Image by the author)

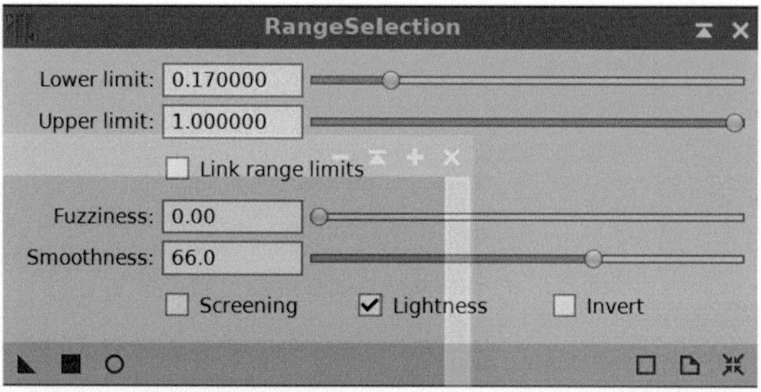

Fig. 4.33 The Range Selection tool showing the settings I used. Process → MaskGeneration → RangeSelection. (Image by the author)

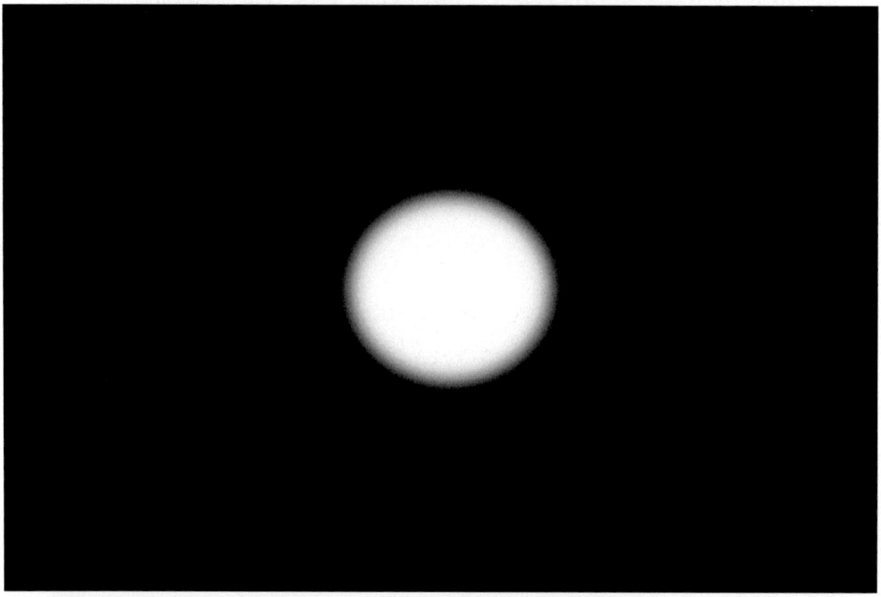

Fig. 4.34 The mask generated to enable me to suppress unwanted background as I stretched my galaxy image. (Image by the author)

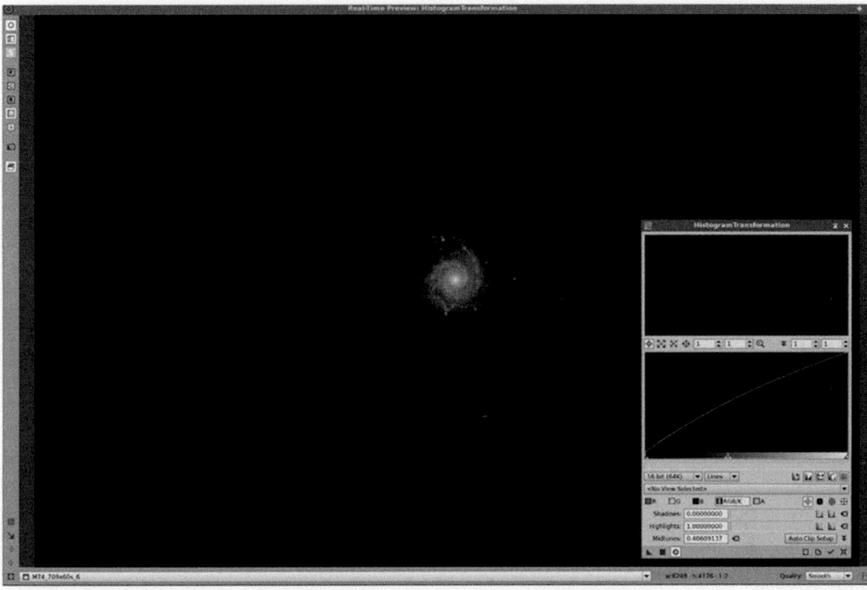

Fig. 4.35 Stretching with the Histogram Transformation tool. Clicking the round button at the bottom left of the tool window makes a 'live' image of the effect of your stretching appear. Pressing the square button at the bottom left makes your changes permanent, bit not so permanent that you can't undo them. Pressing the shape that looks a bit like an X at the bottom right resets the tool. Process → IntensityTransformations → HistogramTransformation. (Image by the author)

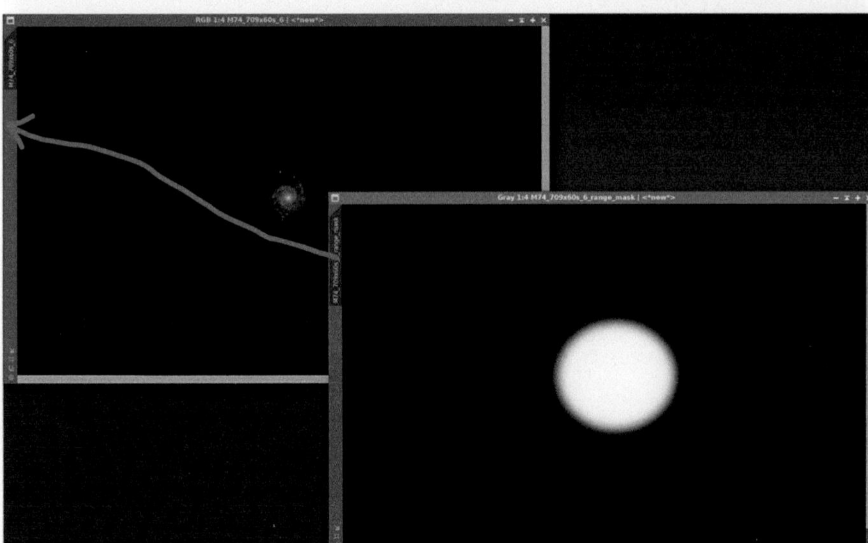

Fig. 4.36 The process for adding a mask to an image. (Image by the author)

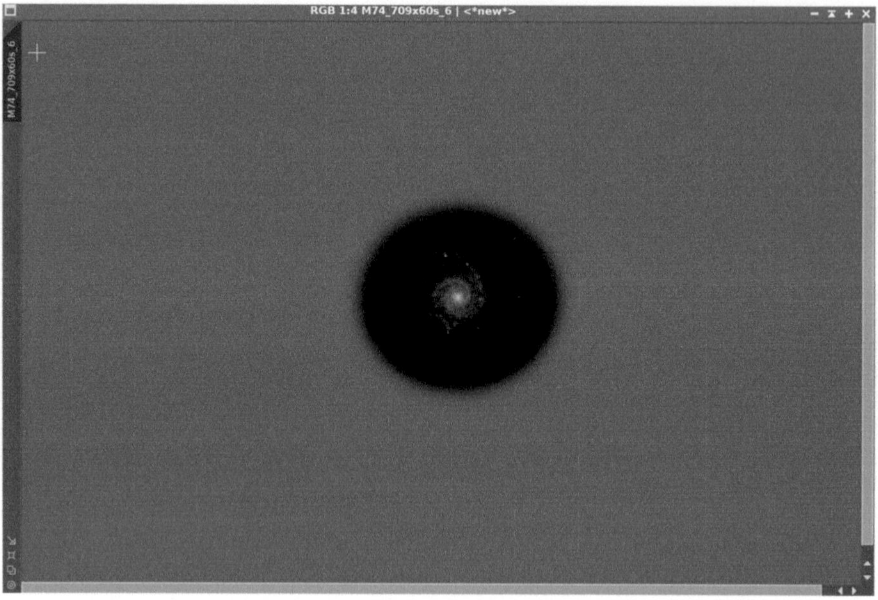

Fig. 4.37 The image as masked. You can now only edit the parts that aren't red. Note that the identifier tab has turned brown. (Image by the author)

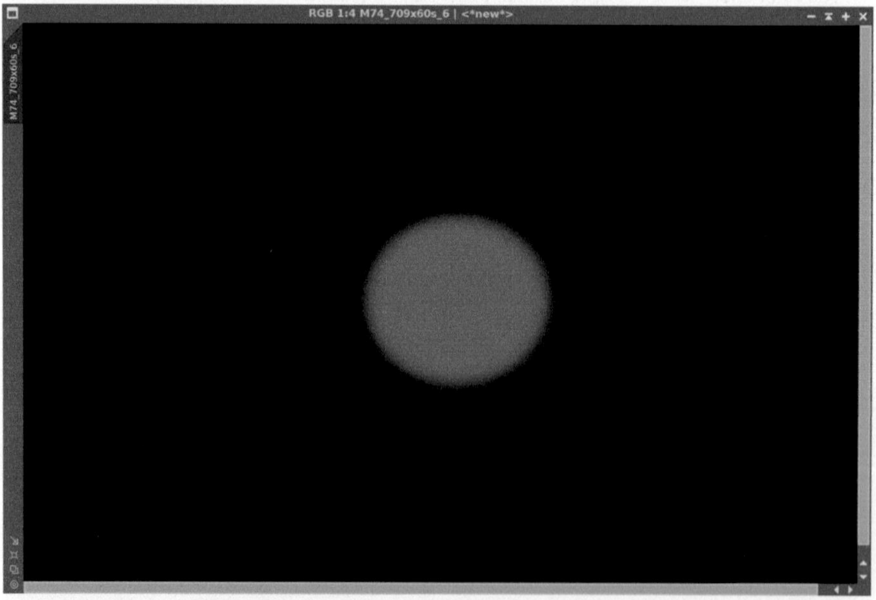

Fig. 4.38 Inverting the mask with Mask → Invert Mask. (Image by the author)

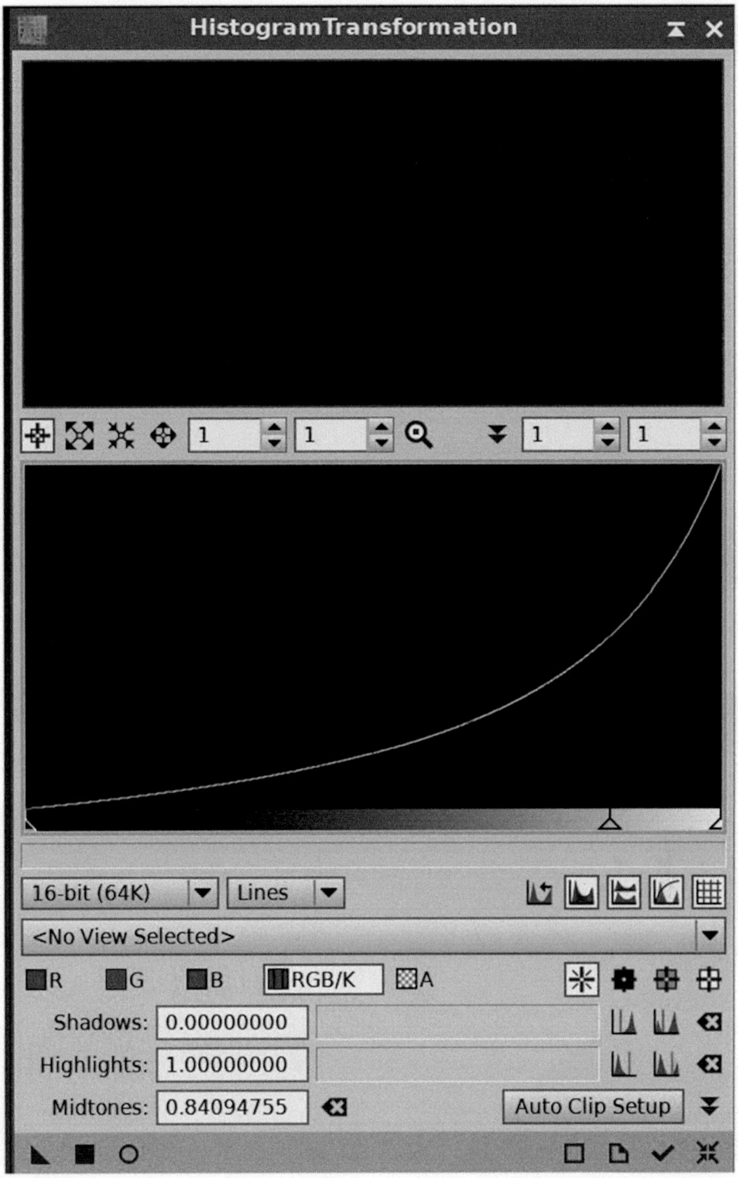

Fig. 4.39 Using the Histogram Transformation tool to suppress an area of the image. This is also best done gently in stages. Process → IntensityTransformations → HistogramTransformation. (Image by the author)

This image looks respectable enough, but I am not satisfied by not knowing whether the colors bear any resemblance to reality, as opposed to my imagination.

To do better, I will need to track down the root cause of the "wheel of death".

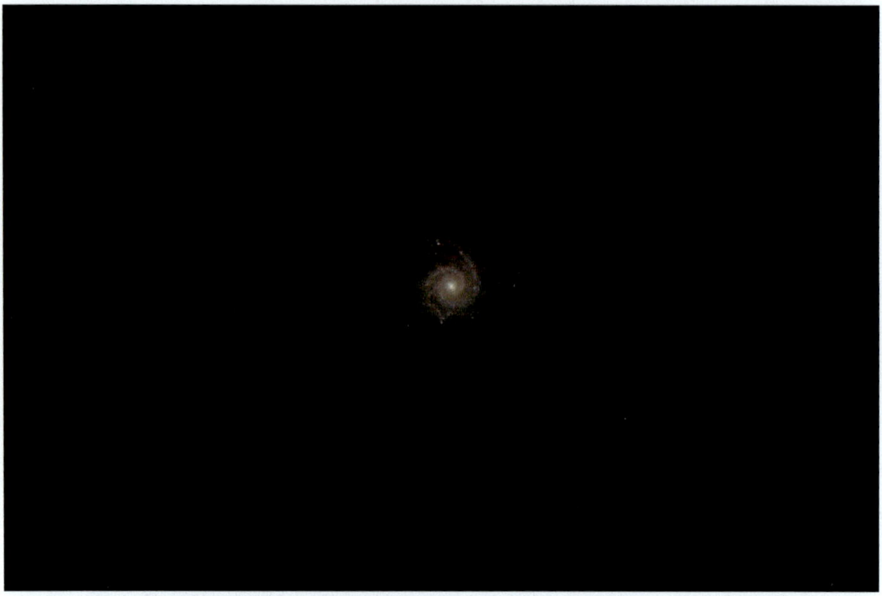

Fig. 4.40 The stretched image. (Image by the author)

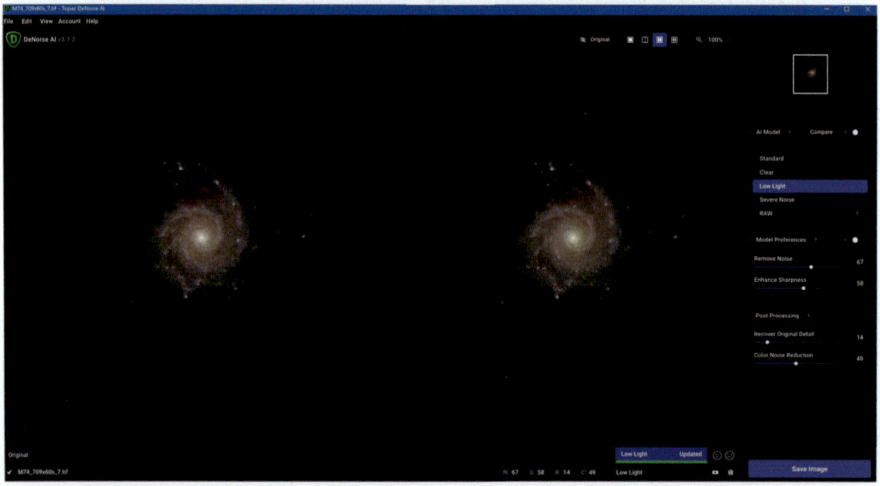

Fig. 4.41 Topaz Denoise AI in action. I usually find that the Low Light option gives a better result than the Severe Noise option. I always experiment with the noise removal and sharpness enhancing sliders. When done, you save, close all images and you are done. (Image by the author)

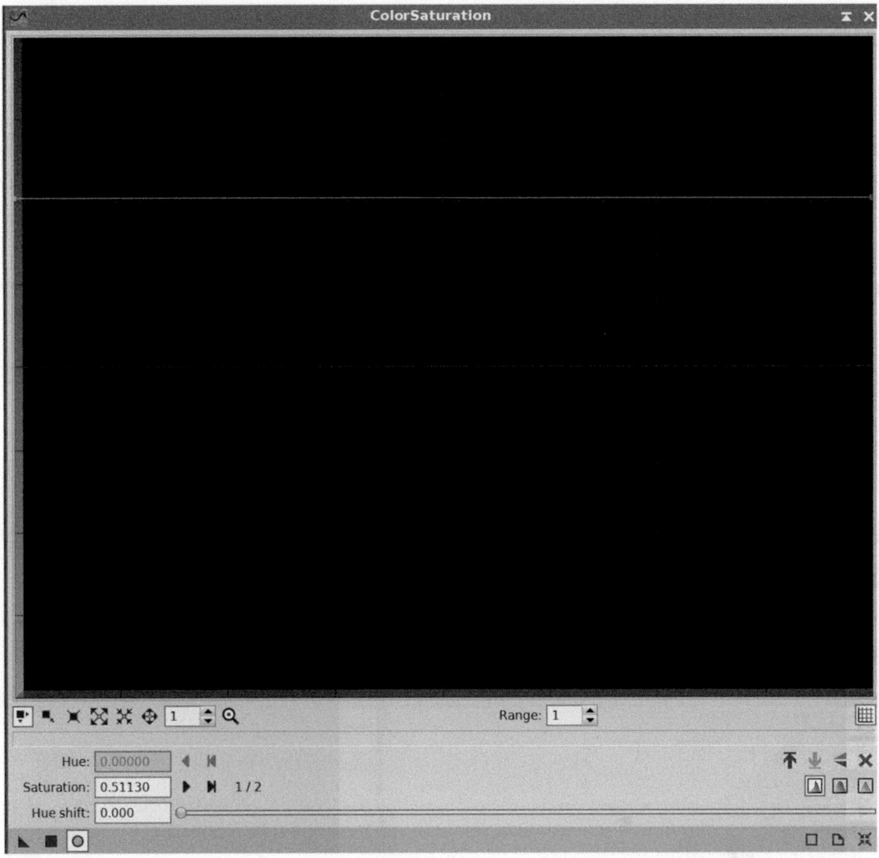

Fig. 4.42 with this tool, you click the circle button to the bottom left to make a live view of your actions. In the purple on the left side, there's a triangle you can grab with your mouse, and drag upwards, to increase the color saturation uniformly. This is a bit fiddly to do. Then when you are happy, click the square button, and close the tool. Process → IntensityTransformations → ColorSaturation. (Image by the author)

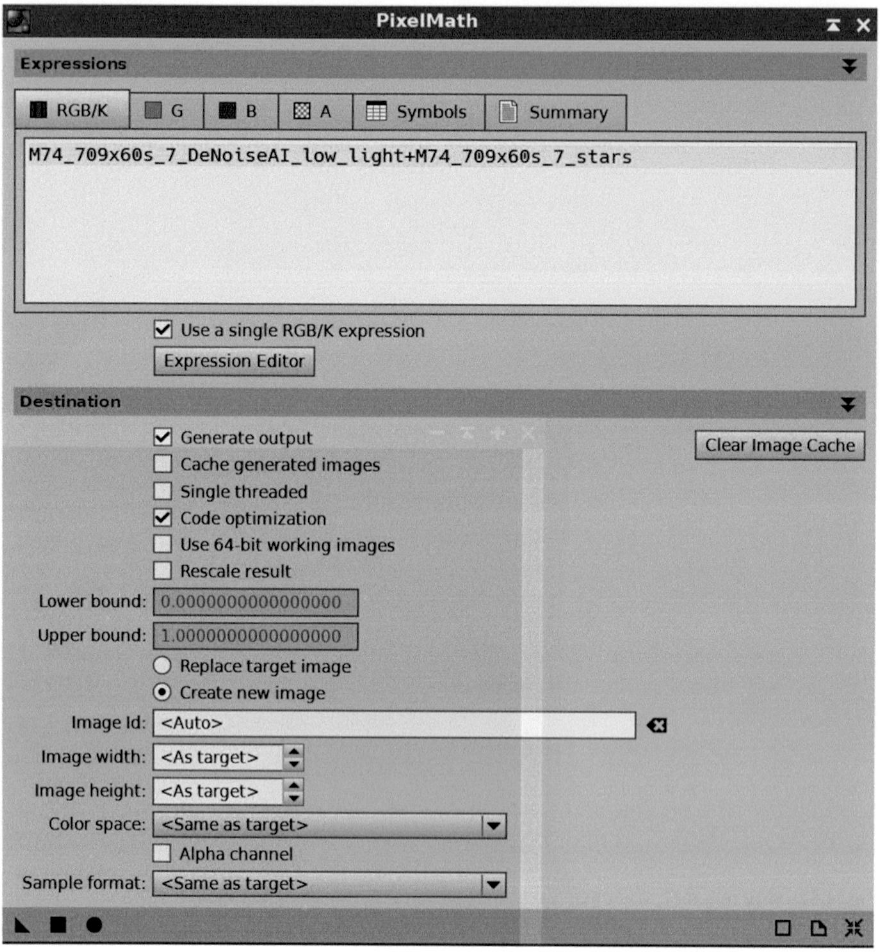

Fig. 4.43 Using PixelMath to add the stars back in. Process → PixelMath → PixelMath. (Image by the author)

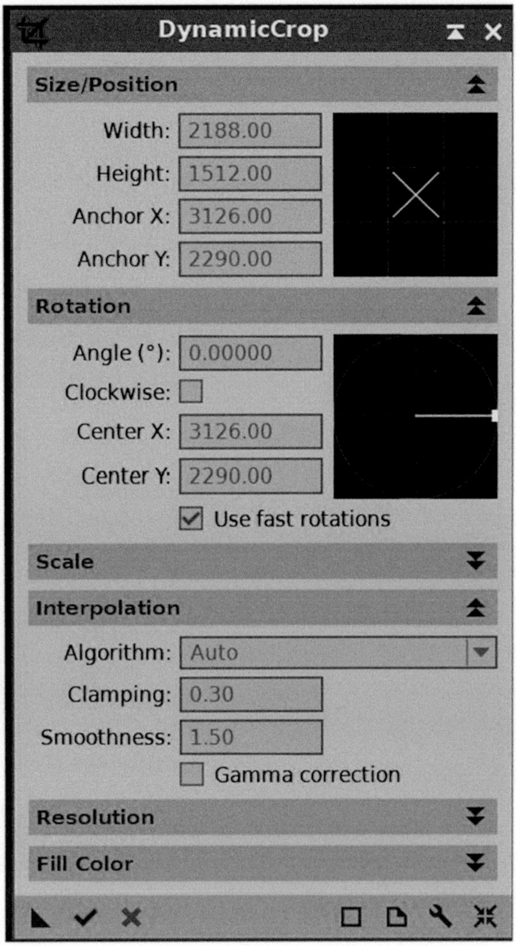

Fig. 4.44 The Dynamic Crop tool has a green check mark, not a square shape, which you have to click to do the cropping. Process → Geometry → DynamicCrop. (Image by the author)

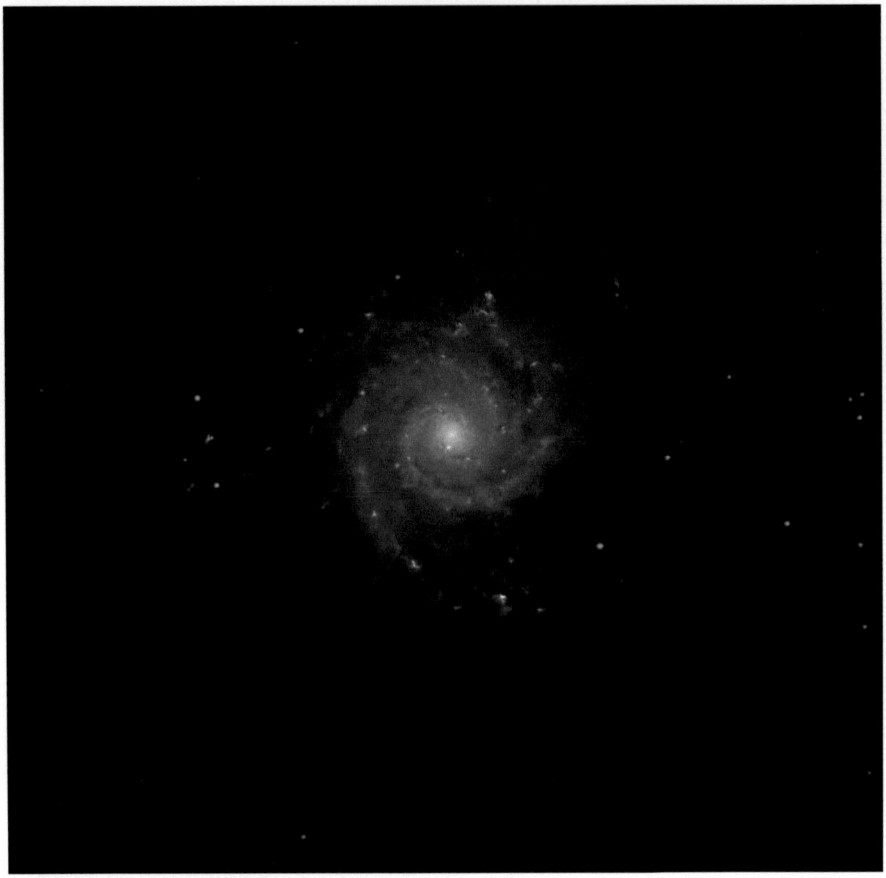

Fig. 4.45 My rotated and cropped image with stars. (Image by the author)

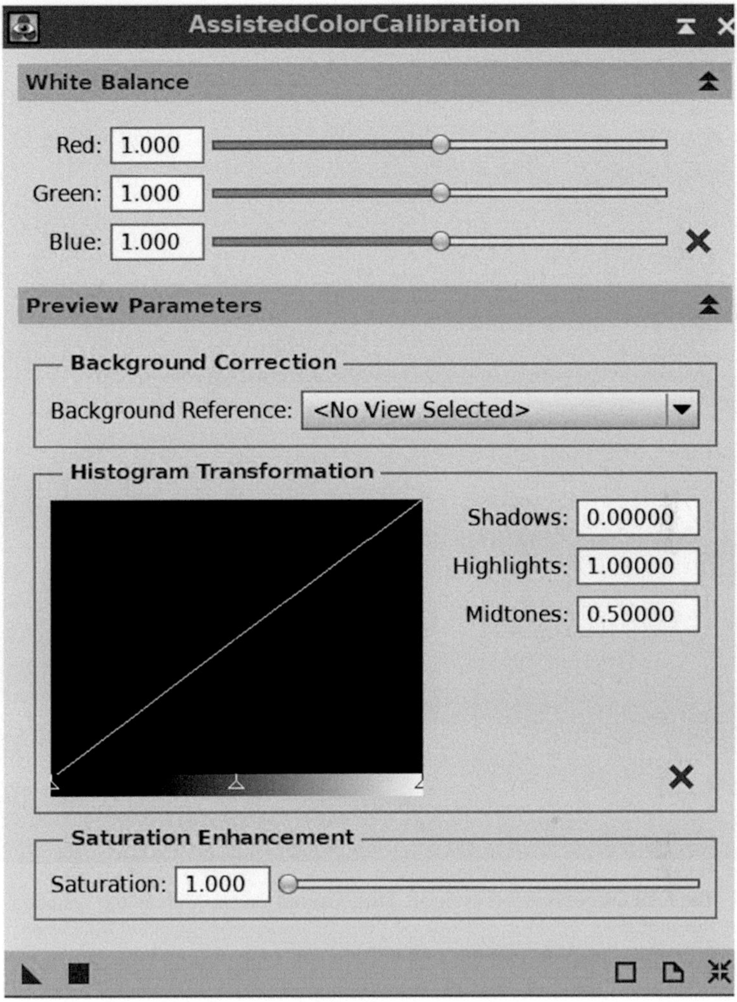

Fig. 4.46 The Assisted Color Calibration tool. Process → ColorCalibration → AssistedColorCalibration. (Image by the author)

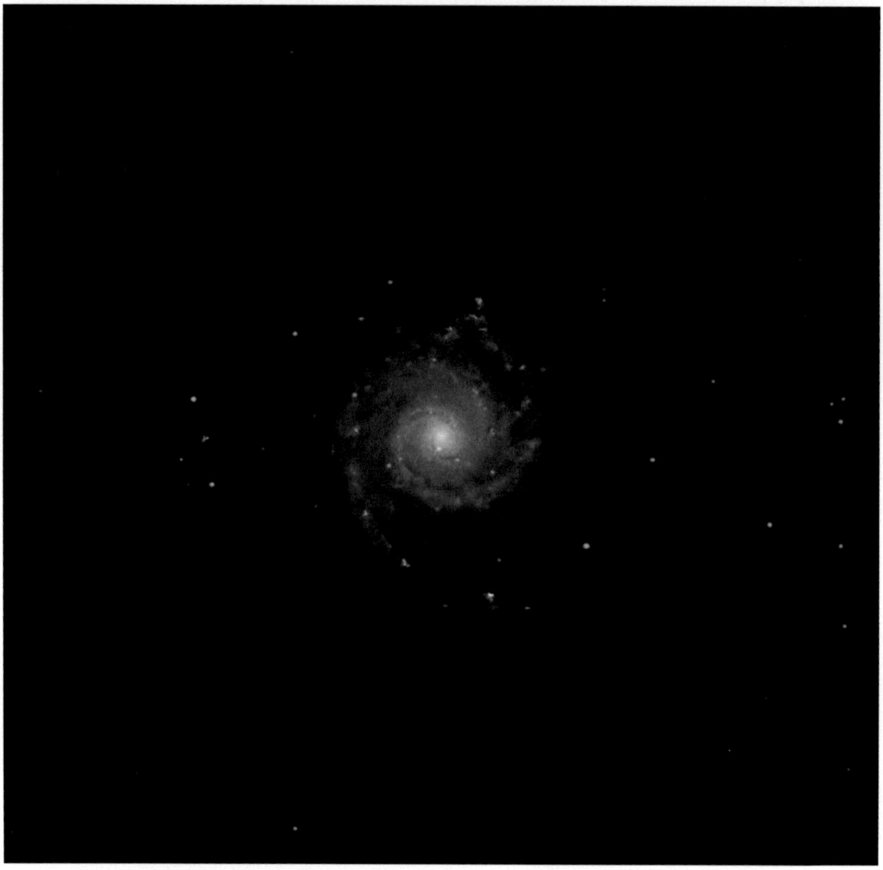

Fig. 4.47 The final image from this exercise: The Phantom Galaxy without the "phantom wheel". (Image by the author)

Tracking the Cause of the Phantom Wheel

The fact that I saw this phenomenon on frames shot in September, October and November tells me that the "wheel" was not caused by light pollution outside the telescope.

My suspicion went back to my calibration frames: the BFD or bias, flat and dark frames. Perhaps they were a big deal after all.

I had plenty of cloudy nights with nothing better to do than repeat them, so I did. In particular, I looked at the flat frames at various orientations of my camera (made no difference) and at various orientations of my light source. I wiggled and jiggled it, and slightly repositioned it. That made some difference. My "wheel of death" now wasn't so bad (see Fig. 4.48).

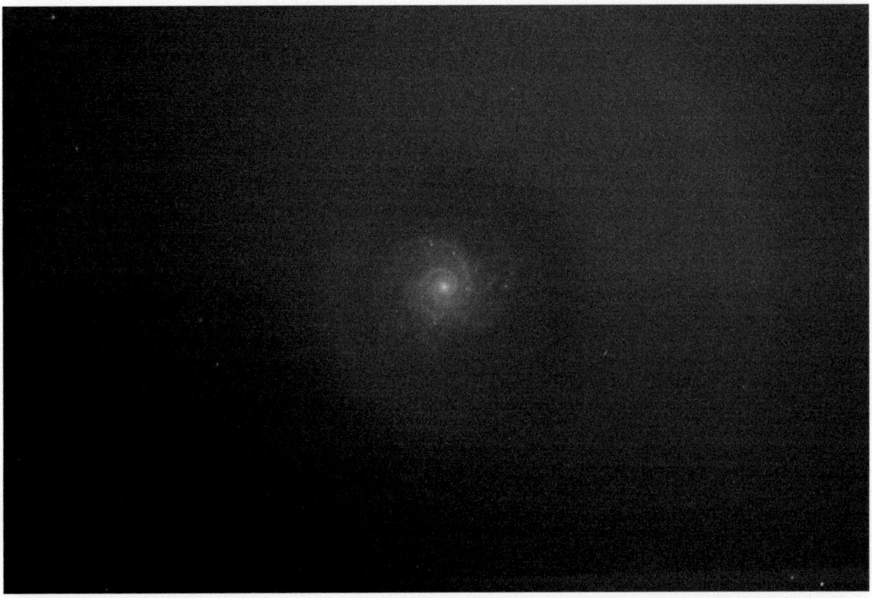

Fig. 4.48 The "phantom wheel" after calibrating with fresh BFD (bias, dark & flat) frames. (Image by the author)

I was emboldened to see if I could just mask the "phantom wheel" away without having to shoot a blank frame and subtract the "wheel" away.

As always, I played around with different masks. It's quick & easy to do this once you have had a bit of practice, so I'm pretty shameless about trial and error. I admit that I was not very quick at doing this six months before the date of writing these words, but that was six months ago.

The winning combination was a binary mask from the GAME script, whose edges were softened by making a range mask, rather like Fig. 4.35. The resultant mask is shown in Fig. 4.49.

With this in place, I used the Histogram Transformation tool in the manner of Fig. 4.39 to darken the area away from the galaxy. I did this delicately, and in several stages, because I didn't want a dirty great tidemark where the edge of the mask had been (Fig. 4.50). I also inverted the mask and stretched the galaxy a little (Fig. 4.53).

There's a little trick to enhance the spiral structure. This is more of an aesthetic thing than a "scientific" step. It's the Local Histogram Equalization Tool, shown in Fig. 4.51. You should use it live, with the round button at the bottom left. The Kernel Radius should be less than 100 to enhance small features and greater than 100 to enhance large features. Since this galaxy is small relative to the image on my camera chip, I used a value of 64. I didn't touch the Contrast Limit, but the "Amount" is a slider to add the percentage of new image to old image. If it is set to 1, you get only the enhanced image. At zero, you don't get any of the enhancement. In this case, I found that a value of 0.120 gave me a result I liked.

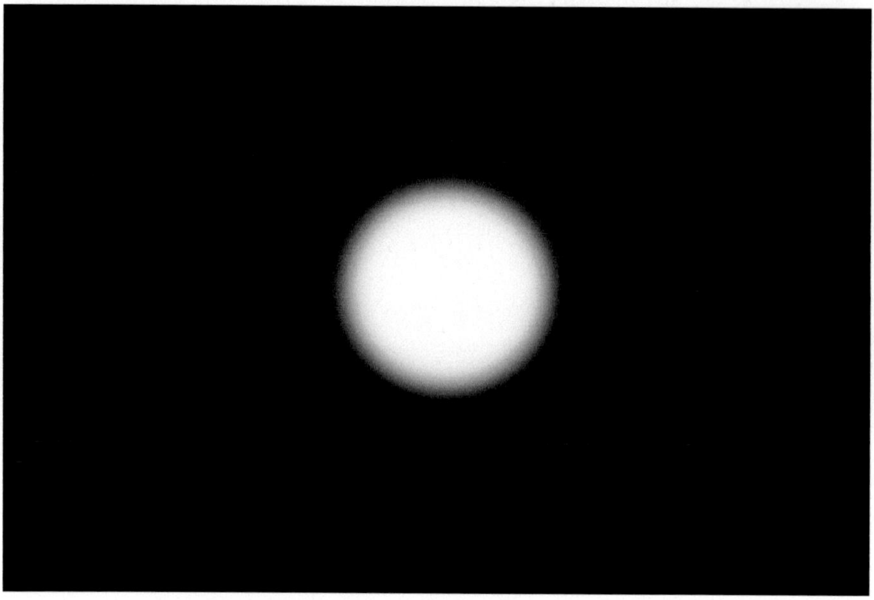

Fig. 4.49 The final image from this exercise after removal of the "wheel of death" and adjusting the color balance. (Image by the author)

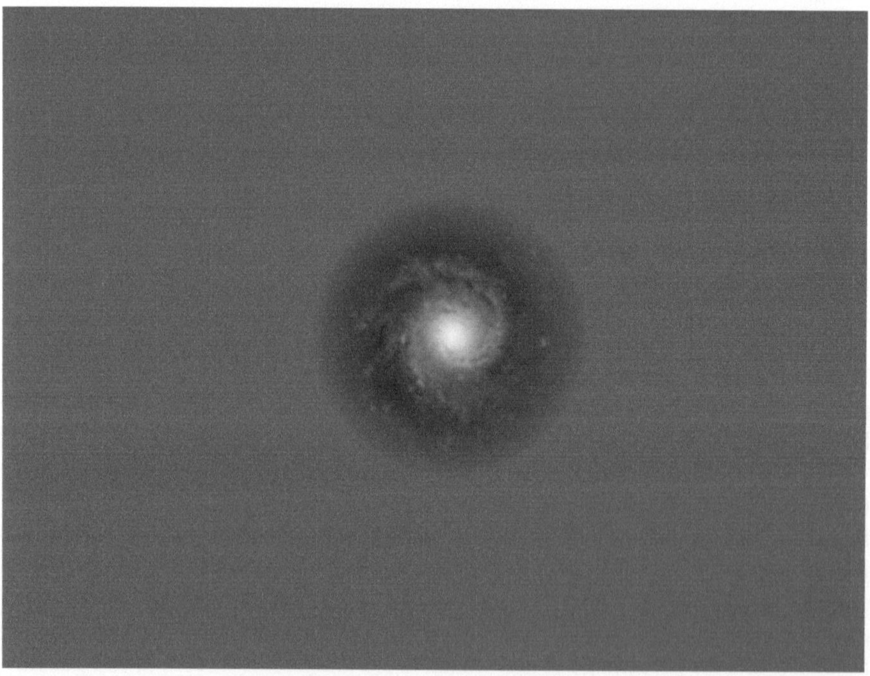

Fig. 4.50 The nucleus of the galaxy masked so that the non-galaxy image is not acted upon by the editing tools. (Image by the author)

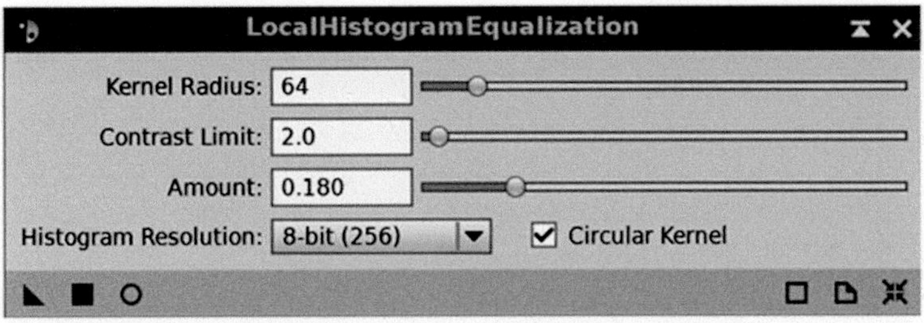

Fig. 4.51 The Local Histogram Equalization Tool. (Image by the author)

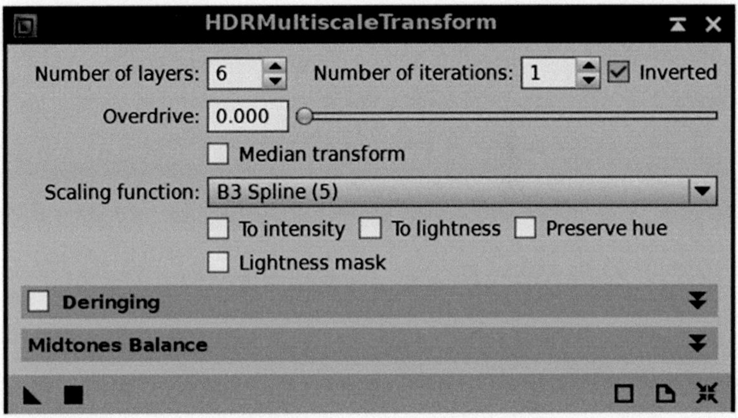

Fig. 4.52 The HDR Multiscale Transform tool. In this case, the default values worked fine. You drag the triangle at the bottom left onto the image to be edited. Process → MultiscaleProcessing → HDRMultiscaleTransform. (Image by the author)

One of the challenges when stretching galaxies is to stop the bright nucleus from overstretching. Fortunately, PixInsight has a tool for this: HDR Multiscale Transform (Fig. 4.52). It works by grading the brightness to make it more even throughout the galaxy. In this case I tried the default values and they gave a satisfactory result. If anything, local histogram equalization makes the nucleus even brighter, so this operation seems to be best carried out after local histogram equalization.

Now I have the image shown in Fig. 4.53.

I can still see a bit of a tidemark around the edge of the mask. But I had a good enough image now that I could try making a range mask based on the galaxy itself. This was a bit tricky, and required delicate adjustment of the slides in Fig. 4.34. The resulting mask is shown in Fig. 4.54.

I applied this mask to the image in Fig. 4.53. It worked, and the tidemark went away (Fig. 4.55).

Next, I sent the star-Free image of the galaxy to Topaz Denoise AI. Figure 4.56 shows the result.

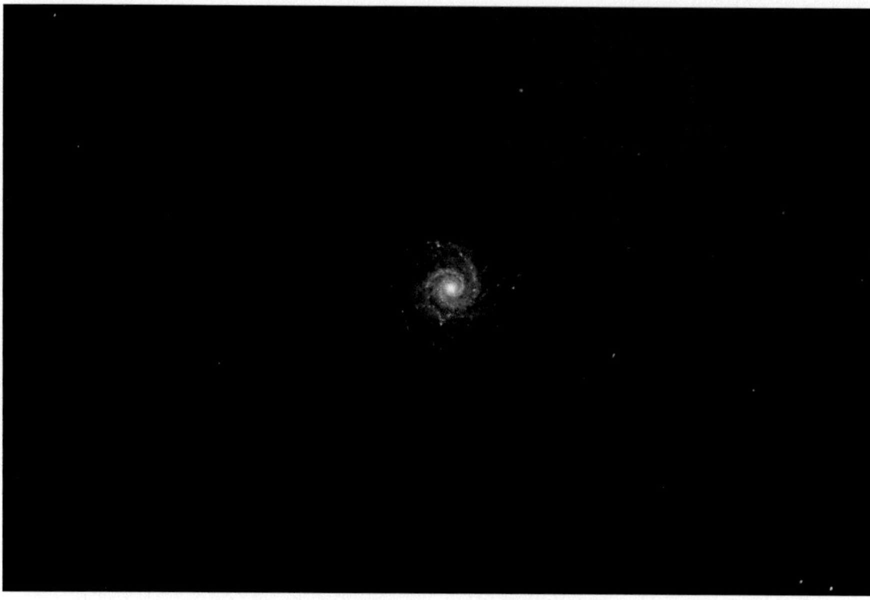

Fig. 4.53 A partly stretched image, after using my range mask of Fig. 4.49. The color saturation was increased per the method shown in Fig. 4.42. (Image by the author)

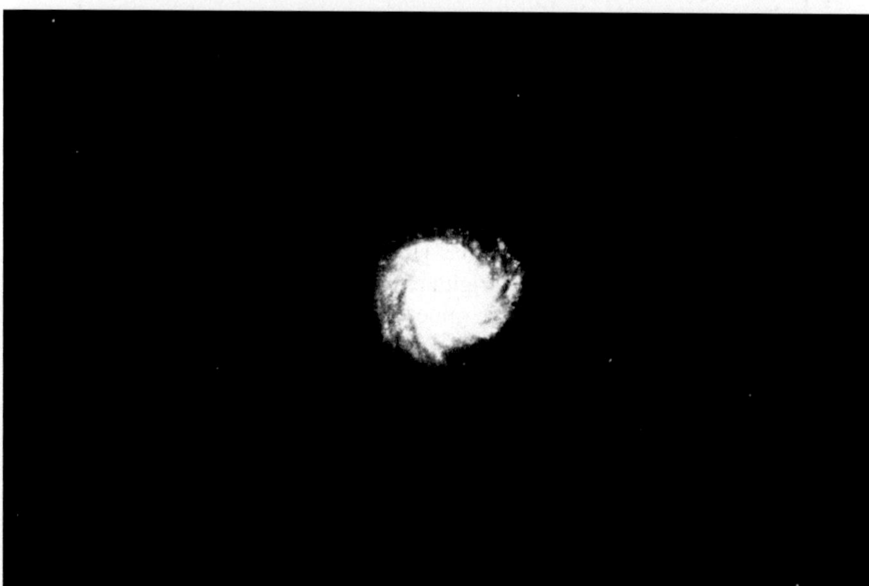

Fig. 4.54 I could now make a range mask based on the galaxy itself, and use it to help me eliminate the tide mark from the previous mask. Process → MaskGeneration → RangeSelection. (Image by the author)

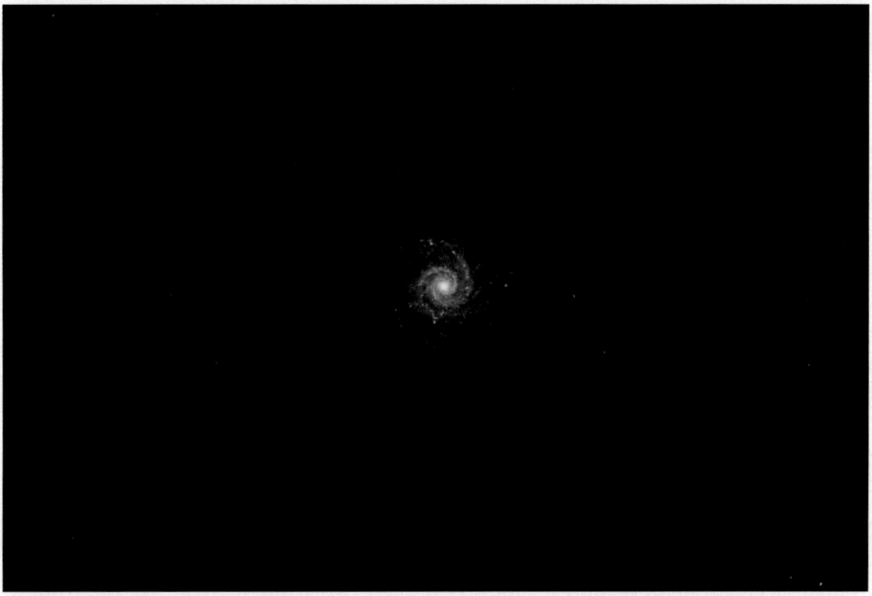

Fig. 4.55 The final image from this exercise after removal of the "phantom wheel" and using the Histogram tool (Fig. 4.32) to obtain a little more stretching. (Image by the author)

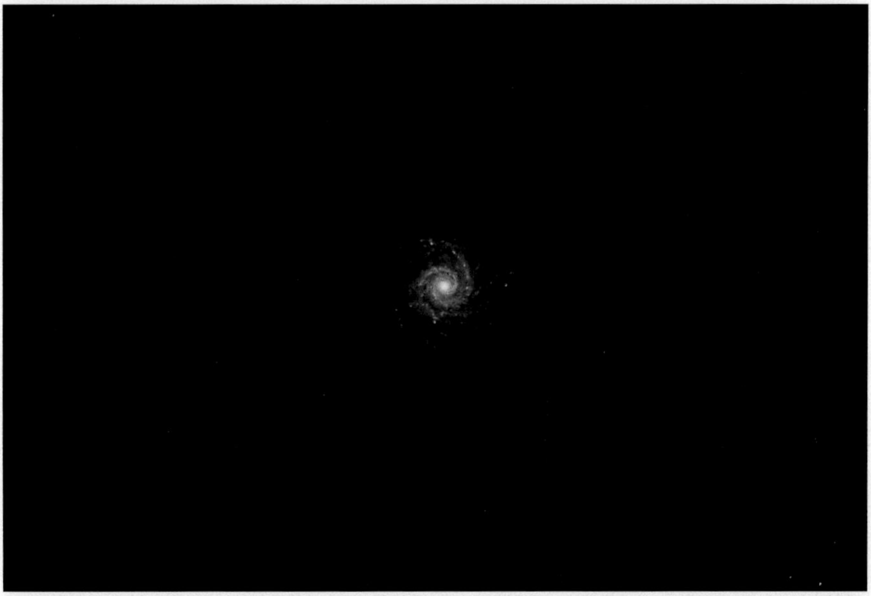

Fig. 4.56 The image from Fig. 4.55 after processing it with Topaz Denoise AI v. 3.7.2. (Image by the author)

I'm now happy enough to add the stars in by the procedure described above, where I increased color saturation and reduced the brightness with the Histogram Transformation tool, and flip the image through 180°.

The image now looks as shown in Fig. 4.57. Though I say so myself, I'm rather pleased with it. The colors are now those the spectrophotometric color calibration recommended, only saturated a little. I'm sure that the spectrophotometric color calibration process is only approximate, but I very much doubt if 2023 technology, or at least that which is accessible by amateurs, will give us a better idea of the 'true' colors.

Why do 'true' colors matter? A scientist lurks in me, never far below the surface, and I feel a bit of a need to depict nature as accurately as I can. This scientist, however, has to compete with my inner artist, who does not want her images to look boring. So I'm not averse to a little color intensification.

This is just my view of the world. Some people are more concerned with their art, and are less bothered about changing colors a little for artistic effect. That's fine by me. Sometimes, as in Figs. 4.5 and 4.47, my lack of skill doesn't give me the luxury of knowing the 'true' colors.

Next, I tried making my flat frames in a completely different way to see if the Wheel of Death died. It didn't. I loosely draped two white towels over the front of my telescope, in the hope of photographing the frames against a uniformly grey sky.

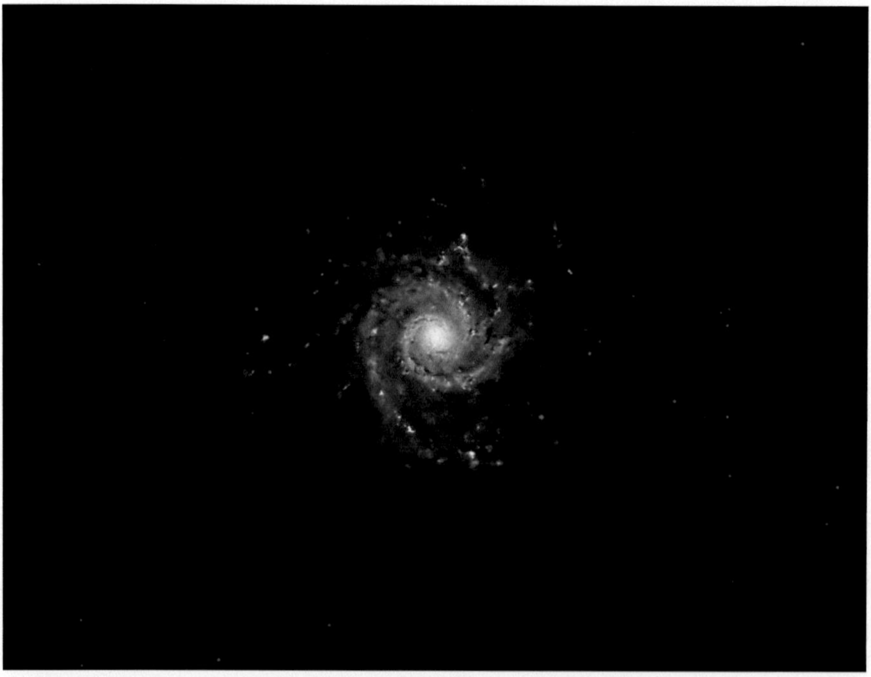

Fig. 4.57 My imaged processed with the new BFD files. (Image by the author)

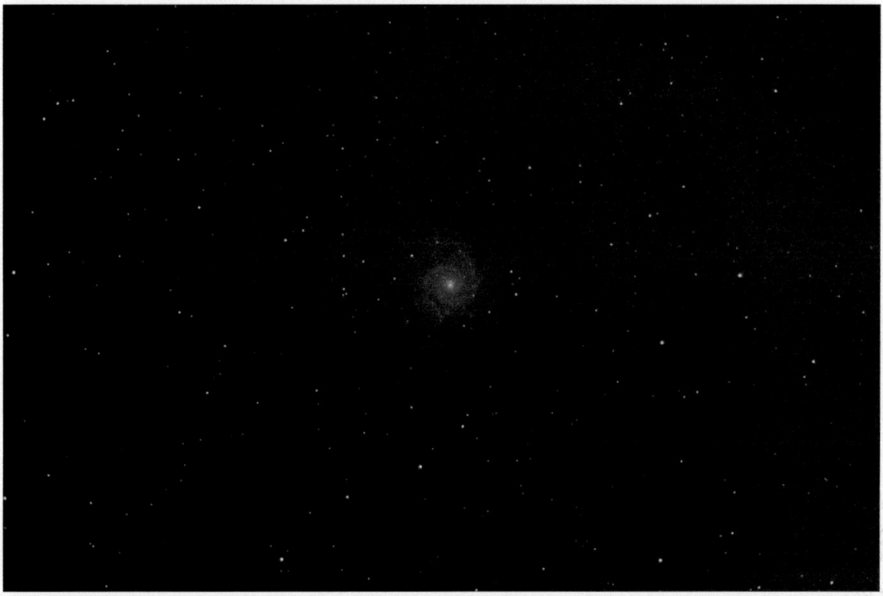

Fig. 4.58 A different method of flat frame generation, using diffuse skylight, did not remove the "phantom wheel". (Image by the author)

Alas for me, by the time I had set up, the wind blew the clouds away. So I had diffuse blue sky for my light source.

The color of the "phantom wheel" changed color (Fig. 4.58) but it is still there. Therefore, the method of generating flat frames has issues, but it is not my main problem.

I felt a need to track down the root cause of the "phantom wheel", for the obvious reason that I had no desire to repeat this misfortune with my next galaxy.

If in doubt, post a question on the forums. I did; and someone suggested looking to see where in my image train the reflections causing the phantom wheel were occurring.

At this point, my setup behind the scope consisted of a Celestron f/6.3 reducer/corrector, some connecting tubes, mostly of 42 mm diameter, a Pegasus Falcon camera rotator, a ZWO 2-inch filter drawer and the camera (Fig. 4.59).

It occurred to me that I had never asked the question, what should the distance from the back of the rotator/corrector to the camera be? The answer was 105 mm, but I had been using a distance nearer 175 mm. Not my best ever choice.

Anyway, I removed the ZWO camera, pointed my scope at some grey daylight sky, and photographed what I saw looking through the telescope, using my phone camera. Figure 4.60 shows what I saw.

I could see many reflections. The brightest were those coming from the filter drawer. They appear as the outermost ring in Fig. 4.60. I was not using that filter drawer for anything useful. I had a UV/IR blocker in it most of the time, but the

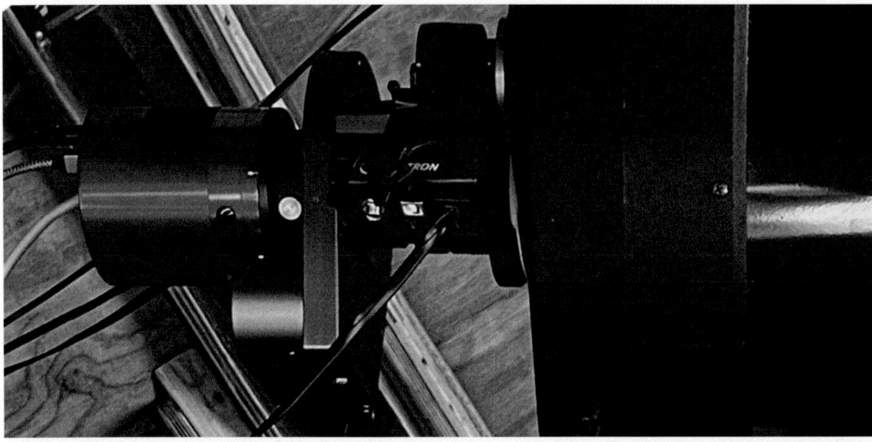

Fig. 4.59 The optical train I used to photograph M74. Behind the focus motor, you can just see a Celestron f/6.3 reducer connector, then a hodge-podge of tubes of various diameters tubes connecting the blue Pegasus Falcon camera rotator. Behind it is a ZWO filter drawer, and behind that is my camera, a ZWO ASI2600MC. (Image by the author)

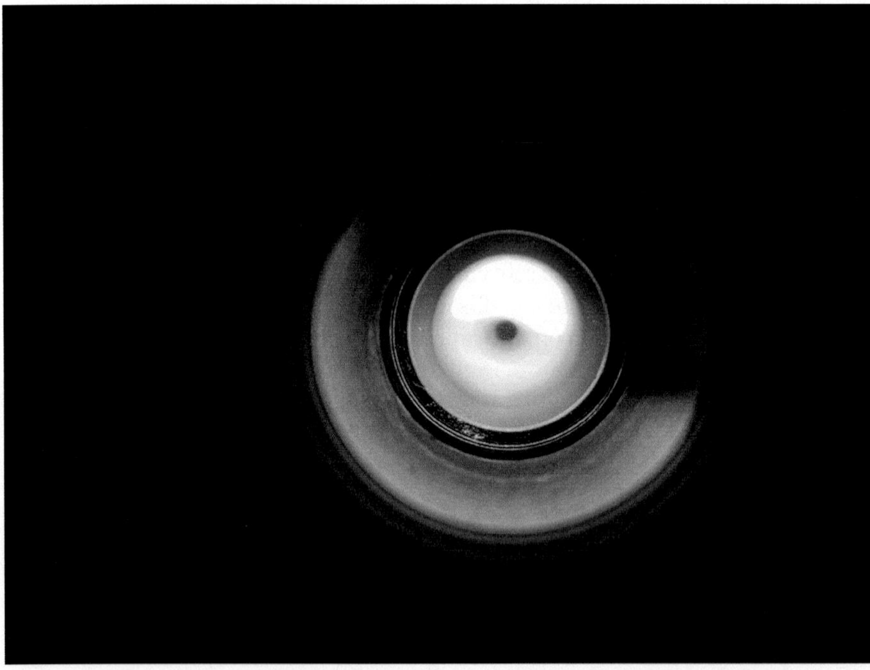

Fig. 4.60 The view through my optical train with the deep sky camera removed. The brightest reflections are the nearest. (Image by the author)

ZWO ASI2600MC has one of those anyway. So why did I even need to have the filter drawer there? I decided to remove it.

Without the filter drawer, there are still many reflections (Fig. 4.61).

At this point, I began to contemplate some drastic ideas to paint the tubes even blacker. Those largely proved to be non-starters by the time I had examined them. In the end, I decided to make the optical tubes a little wider, at 48 mm, to see if the reflections would then land away from my camera chip. The diameter of the aperture at the front of the camera is 42 mm.

In the UK, the firm with the best supply of tubes and thread adaptors is normally Rother Valley Optics. I turned to them & they sent me the relevant tubes and adaptors. All I had to do now was wait for another clear night when the Moon was away from M74, whose position in our skies is close to the Ecliptic. The new arrangement is shown in Fig. 4.62.

The resultant image was shot for about 80 minutes, in order to prove a point rather than to produce a nice image. This image still shows a phantom wheel around the Phantom Galaxy, but it is now much further out, and therefore easier to deal with (Fig. 4.63).

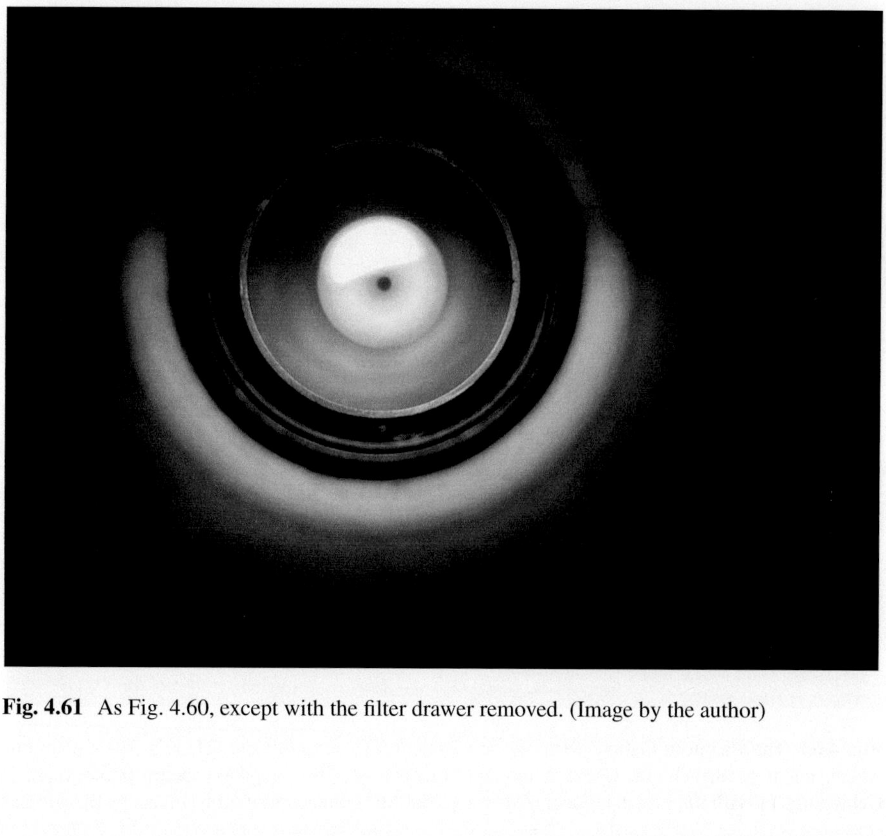

Fig. 4.61 As Fig. 4.60, except with the filter drawer removed. (Image by the author)

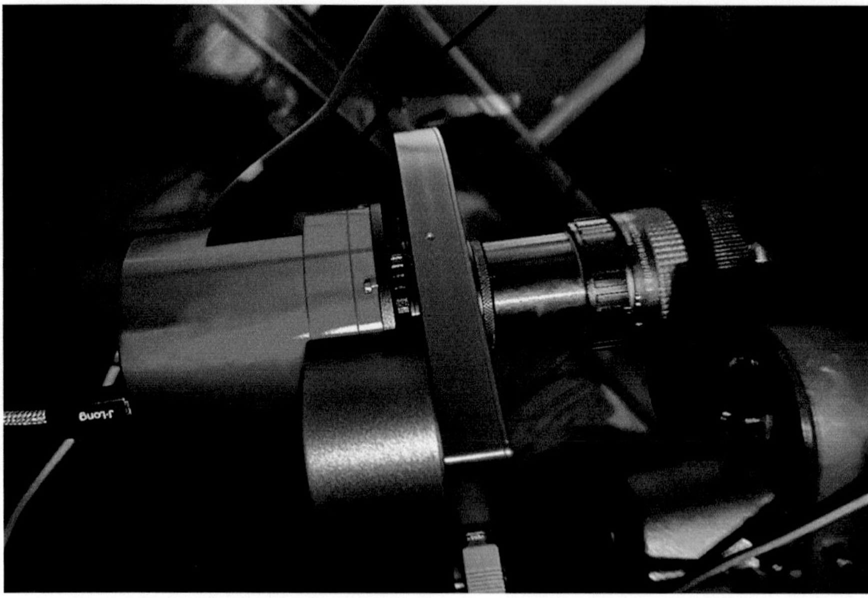

Fig. 4.62 My new optical arrangement, now at the correct corrector-to-camera distance, and without the surplus filter drawer. (Image by the author)

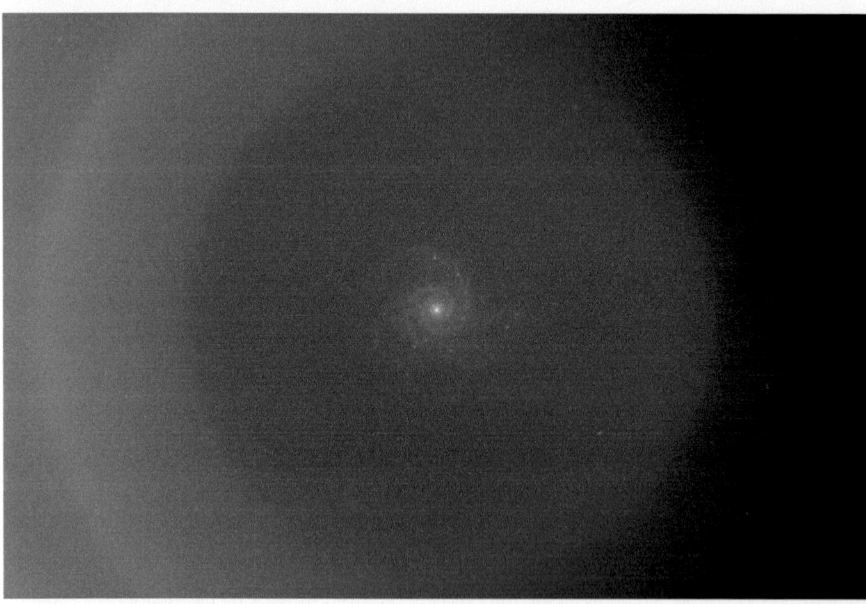

Fig. 4.63 The Phantom Galaxy with my new optical tube arrangement. There's still a phantom wheel, but it is further out, where it interacts much less with the galaxy being photographed. Celestron C11 with f/6.3 focal reducer, ZWO ASI2600MC camera, 82 × 60 s. (Image by the author)

Even though I have not fully dealt with the problem of phantom wheels, I now know its cause for certain, and can work further on minimizing it.

The moral of this sorry tale is: be aware of ghost images from local light pollution, and minimize them as best you can. As always, an ounce of prevention is worth several pounds of cure.

Chapter 5
Andromeda: The Green ValleyGirl

Why Green Valley?

An interesting idea that has emerged during the twenty-first century is that of a main sequence of galaxies obtained by plotting the measured star formation rate (color is often a good proxy) against the mass of the galaxy, a graph like that in Fig. 5.1 is obtained. The huge Sloan Digital Sky survey provided the data that made this kind of discovery possible. As so often, data drives scientific development, not theory. The idea of color separation came from Iskra Strateva and others [40]; and the idea of the plot seems to have originated with Brinchmann and others [41]. The basic concept has since been added to. The Green Valley idea is reviewed by Salim [42]. Green valley galaxies are ones which have low star [43] formation rates. They are not very common, but it is thought that we live in one [43]. The idea of starburst galaxies is discussed in a paper by Gomez-Guijarro and others [44], and reviewed by Heckmann [45]. Green valley galaxies are hypothesized to be on a journey from the main sequence to 'red & dead', which is where elliptical galaxies are found.

It couldn't be easier, right? M31 is the nearest large galaxy visible form the northern hemisphere, so it will be the brightest. Indeed, every time I express skepticism about it being naked-eye visible, someone indignantly contradicts me and says that they can see it. I think this is good evidence that it is the brightest big galaxy outside our own.

One problem lies in the word 'big'. It doesn't fit in my 11″ Schmidt-Cassegrain telescope. It would not have fitted in my previous telescopes, a 9.25″ f/10 SCT or the 8″ f/6 Newtonian. You could no doubt deal with this by using panoramic stitching if you are patient enough. Plenty of people do. But I decided to invest some of my pension lump sum in a 3″ f/5.9 apochromatic refractor, having worked out that it would be able to project an image of M31 onto my camera chip.

J. Clark, *Photographing Galaxies from Light Polluted Skies*, The Patrick Moore Practical Astronomy Series, https://doi.org/10.1007/978-1-0716-4506-2_5

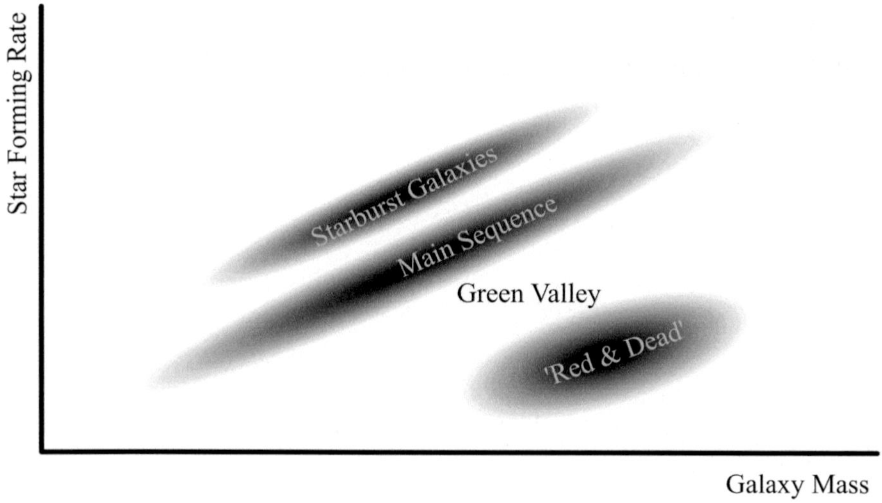

Fig. 5.1 A plot of Star formation rate measured for galaxies against their mass. Most galaxies are in the area called the main sequence. (Image by the Author)

Another problem lies in the idea that it's bright and therefore easy. Such things as photographing individual stars within the galaxy are no longer completely infeasible.

I fell into the trap of being tempted to have a go with my digital single-lens reflex camera and a telephoto lens, piggybacked onto my telescope. The result (Fig. 5.2) was dire.

The colors were impressively implausible. The stars look like the kind of rash about which you'd call your doctor, who would need to put on his or her best "don't-alarm-the-patient" poker face. The brighter ones are white with red fringes. They are not very round. Who ever saw a star like that?

The original frames are jpeg files. These are 8-bit images. First mistake. You run out of levels if you stretch an 8-bit image much (Figs. 5.10 and 5.11 show an example of this phenomenon.) In those days I didn't know about debayering, which is what you have to do to get a color RAW image out of your DSLR. RAW images are 14-bit according to Canon's website. That would have given me a lot more to play with when stretching the image. But it's too late now.

Also, I used non-astronomical software for processing. With hindsight, that's not a great idea.

The only thing I will say in favor of the image in Fig. 5.1 is that it reminds me how far I have traveled in two-and-a-half years. It is my earnest wish, dear reader, that I can help you climb the learning curve a lot faster than I did.

The next problem I encountered was the need to get round stars with my shiny new William Optics Zenithstar 73 III telescope. I had enough wit to buy their Flat73A adjustable field flattener with the telescope, knowing that I would not get a good picture without it. Figure 5.3 shows my first attempt. It still suffers from some

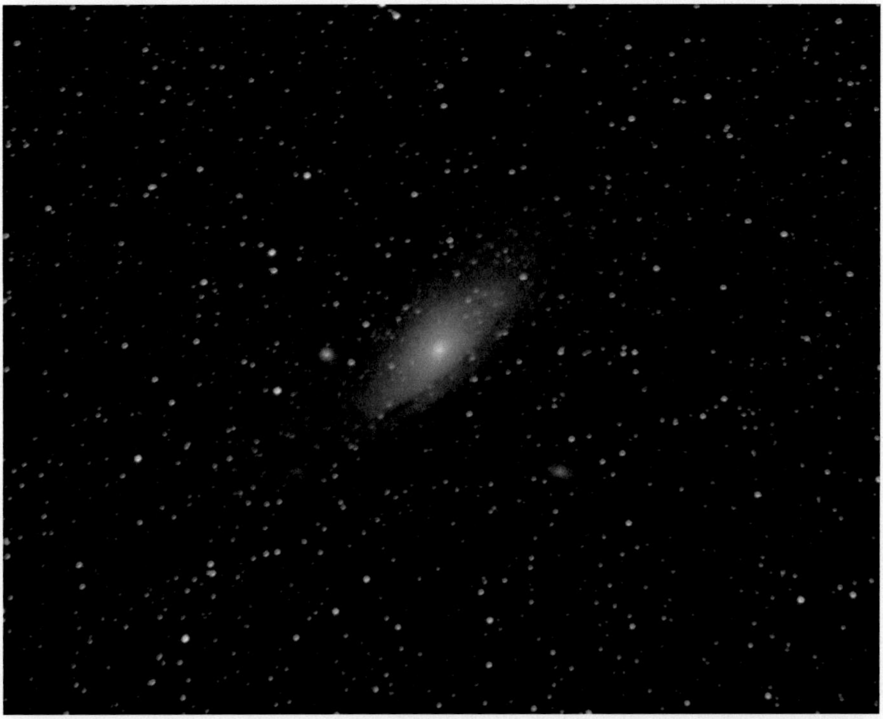

Fig. 5.2 Ten frames at 4 minutes each of M31, taken with my Canon EOS 1000D DSLR and a 75–300 mm zoom telephoto lens, all piggybacked onto my telescope. Zoom set to 200 mm. 10 × 240 s @ f/5.6. Canon EOS capture software, stacked in deep Sky Stacker and postprocessed in Gimp. September 25th 2020. (Image by the Author)

of the same problems, such as red-fringed white stars; and it has another problem. The stars in the corners are stretched into oval, if not downright linear, shapes.

This is because my flattener-to-camera distance was wrong. This proved to be tricky to adjust. You'd think you've got it with 2-second exposures, then when you spend several hours on one target, you find that you haven't. Take a deep breath. Go back & do it again. Did I say anything profane? Your Honor, I plead the fifth. Actually, I realized some time ago that it is not a good idea to express irritation vocally at 2 a.m. Not only would it wake the neighbors, but they'd all know perfectly well who woke them.

It did not help that I wear varifocal glasses, one of the prices I pay for accumulating all those years of wisdom. Varifocals are wonderful, but the image you see is only sharp if you look straight ahead. In desperation I ordered a pair of bifocals, designed to give me two clear distances: to the keyboard and to the computer screen. At any other distance they are useless, which is just fine, because now I can scan my eyes around an image and be confident that the corners are sharp.

At this point I began to take PixInsight a little more seriously. Since Andromeda is highest in the sky in the late summer and early autumn, I didn't really expect to

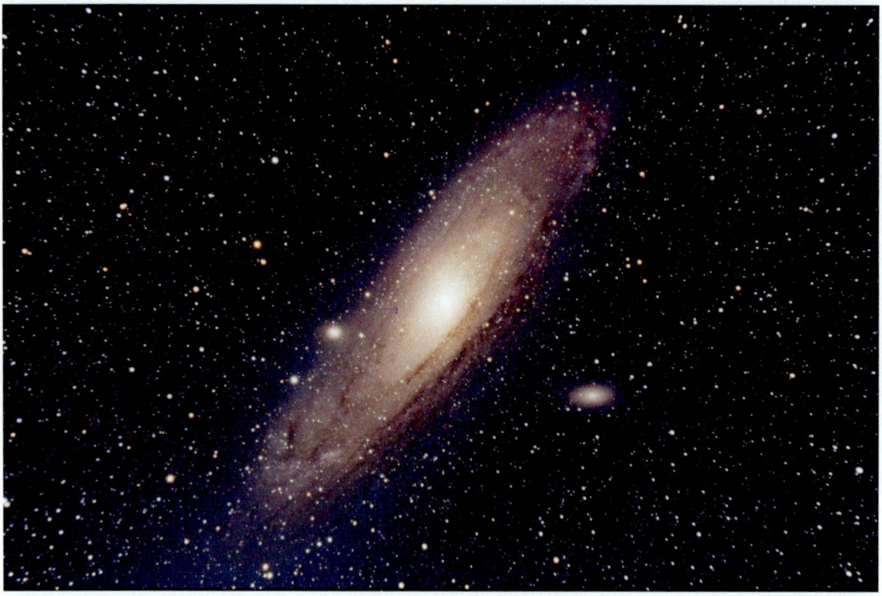

Fig. 5.3 My next attempt at M31, using the live stacking feature in SharpCap 4.0. Colors manually balanced. Notice how stretched the stars in the corners are. William Optics Zenithstar III telescope & Flat73A field flattener, ZWO ASI2600MC camera, 360 × 30 s frames, Gain = 100, T = −15 °C. January 17th 2023. (Image by the Author)

get another chance in January. But I was able to gather 4.3 hours of frames over a couple of evenings in February. At all latitudes above 48°N, M31 is circumpolar. At my latitude, 51.5°N, it is rather higher in the sky than it would be in, say, Chicago, let alone Dallas. So I got away with it. I stopped collecting frames when the galaxy had fallen to 25° above the horizon. I know. I also learned as a rule of thumb that you should not photograph objects that are fewer than 30° above the horizon. I think that might be a little conservative. I didn't see any atmospheric dispersion into red, blue and green. I think I probably would have seen this if I'd been using my SCT at much higher magnifications, but for low magnifications it seems to be OK to go down to 25° before you stop recording frames.

I also shot fresh dark, bias and flat frames. I suspect that cameras accumulate hot pixels over time, so it never hurts to re-take your dark frames every few weeks. The flats I repeated because I am now transferring my camera between two telescopes. New 'dust bunnies' are bound to keep appearing. That also meant new bias frames. The flats and bias frames don't take too long, but the darks take a while. I usually either do these on a cloudy night or after I have finished imaging, while I take a well-earned rest in the land of nod.

I stacked all these frames in PixInsight. By this time, I had figured out that it is much better to save the files as FITS files, because PixInsight and Siril both extract the focal length, date and RA and Dec from your FITS files and use them when

Fig. 5.4 The xisf file that PixInsight produced as a result of stacking four hours' imaging of M31. (Image by the Author)

processing the images. SharpCap gets this information by talking to CPWI, which is the software Celestron has written to control the mounts it produces, and writes it to the headers of the FITS files.

At first, you get rather a shock. The fruit of your hard labor is an image like the one in Fig. 5.4, in which you can see nothing. Nada. Zip zilch.

You have no idea at this point if your image came out. But, please, resist the temptation permanently to stretch it immediately and see what you have. You can do a temporary stretch in Siril. In PixInsight there is a button to do this along the top row of icons. It looks like one of those radioactivity warning symbols.

After performing the color calibration, de-blur and de-noising operations detailed in Chap. 3, you can run a script to stretch the image and see what you have. I imaged this this galaxy early in my project to write this book. I did some things that I now know better ways of doing. You will discover how I used these better tools in later chapters. There is no shame in climbing a learning curve to gain proficiency at anything. Nobody learns instantly. This is how I stretched the image.

Script → EZ Processing Suite → EZ Soft Stretch

EZ Processing Suite is a free add-on to PixInsight. To add it, do the following.

Resources → Updates → Manage Repositories

In the menu that appears, add the URL shown in Fig. 5.5.

You click the add button, type it in, then click OK. Now go through the menus like this.

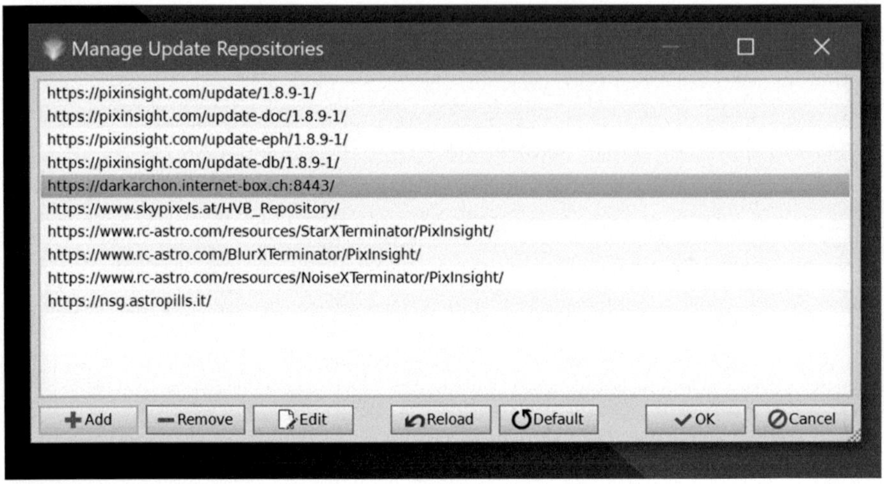

Fig. 5.5 To add the EZ processing suite of Scripts, manually add the URL highlighted. (Image by the Author)

Resources → Updates → Check for Updates

It will then show that here is one update available. Click OK and PixInsight will install it.

But you're not done yet. Make sure you save your work, then you have to stop, then restart PixInsight. The EZ Processing suite should now appear under SCRIPTS[1].

The EZily stretched image I produced looked like the one in Fig. 5.6.

At least I now know I have an image to work with. This stage reminds me of nothing quite so much as the good ol' days of the 1970s, when we all waited in the red "safe light" while watching a photographic print image emerge in the bath of developer chemical.

The first thing I can see is that I have solved the problem of elongated stars in the corners. I'll come back to this point shortly. The second thing I see is a gradient of light pollution, which is brightest in the bottom left and darkest in the top right. I usually find stray light pollution in my images.

First, I separate out the Stars with

Process → ObjectRecognition → StarXTerminator

I always check the "Generate Star Image" box before clicking the square button to make this happen. The first time you see this happen, it is a wonder to behold. The results are the images in Fig. 5.7.

[1] The process for installing the EZ suite has changed twice since I wrote this. It is likely to change again. You can keep abreast of all this by looking in the PixInsight forums, which are searchable.

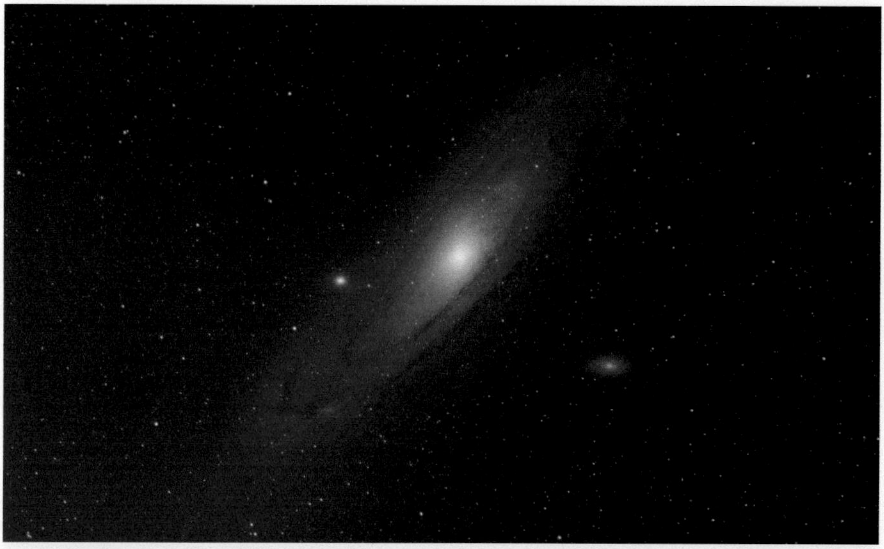

Fig. 5.6 My output from the EZ Soft Stretch script. (Image by the Author)

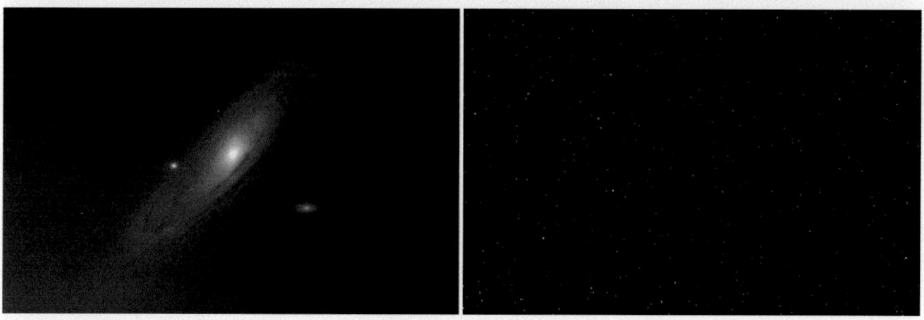

Fig. 5.7 The StarXTerminator process removes the stars and puts them into a separate image. (Images by the Author)

In Fig. 5.8, I show the top right and bottom left corners of the star image, zoomed right in.

I really have solved the problem of getting my field-flattener-to-camera distance right, or at least right enough that BlurXTerminator can make them look round.

Next, I run the process

Process → BackgroundModelization → AutomaticBackgroundExtractor.

The result is shown in Fig. 5.9. The background removed is also produced as a separate image that you could add back in if you really wanted to. I suppose it's really a check that you have only removed stuff you don't want from the image.

This is what I now have.

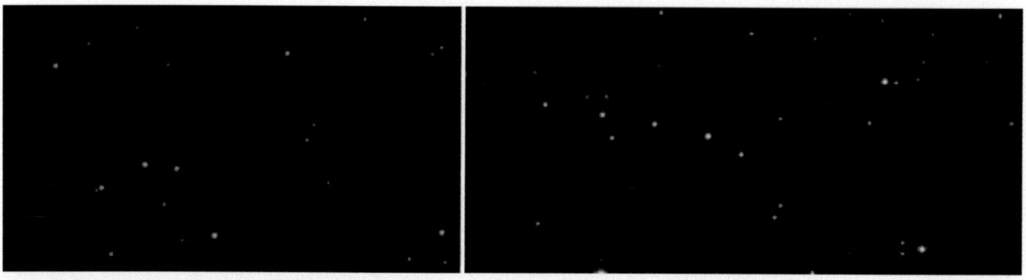

Fig. 5.8 Zooming right in to the stars in the bottom left (left) and the top right (right) shows that I really have solved the problem of getting my field-flattener-to-camera distance right. (Images by the Author)

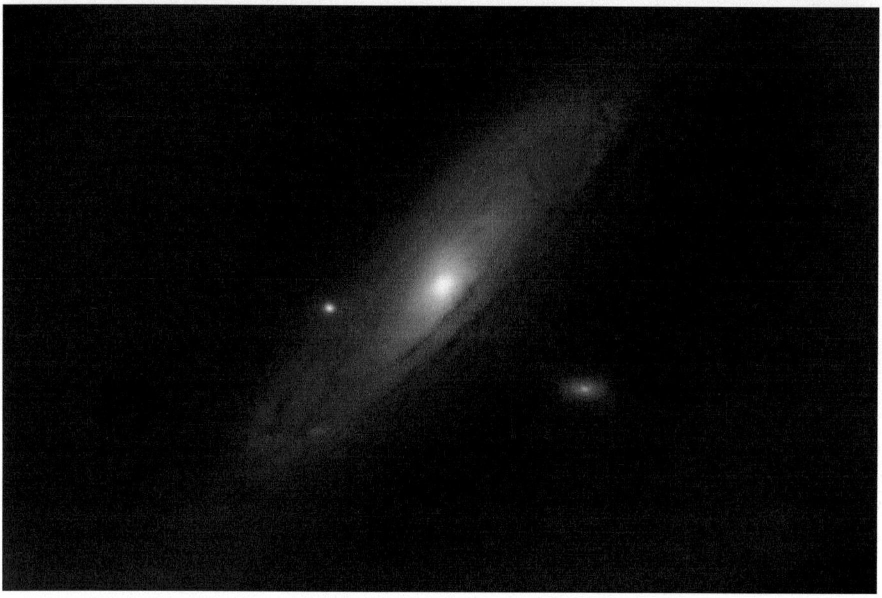

Fig. 5.9 The starless image with background automatically extracted. (Image by the Author)

It's not perfect, but it's a lot better. As an aside, Fig. 5.10 shows a JPEG file of what I managed to extract; and Fig. 5.11 shows a 16-bit image of the extracted background, which is now much smoother. This shows why it is important to have more than 8 bits in your images.

By the way, if, in Fig. 5.9, you look at the M31 nucleus, and at the nucleus of the elliptical galaxy to the right, which is M110, and then move your eye the same distance again along the same line, you can just see a galaxy. It is known in the Uppsala General Catalog of Galaxies as UGC 394. It is about 250 million light-years away, or 100 times as far as the main galaxies in the image. Therefore, with a 3-inch telescope, I have unexpectedly found something a quarter of a billion light-years away. You can read some rather cryptic information about UGC 394 here: http://simbad.u--strasbg.fr/simbad/sim-id?Ident=UGC+++394.

Fig. 5.10 A JPEG file showing the extracted background. There are no nasty surprises here. Saving as an 8-bit JPEG shows why 8 bits aren't enough. The colors no longer appear continuous. Had I saved a 16-bit TIFF file, they would have appeared smoother (see Fig. 5.11). (Image by the Author)

Fig. 5.11 A 16-bit TIFF file showing the extracted background, which is now much smoother. There are still no nasty surprises here. (Image by the Author)

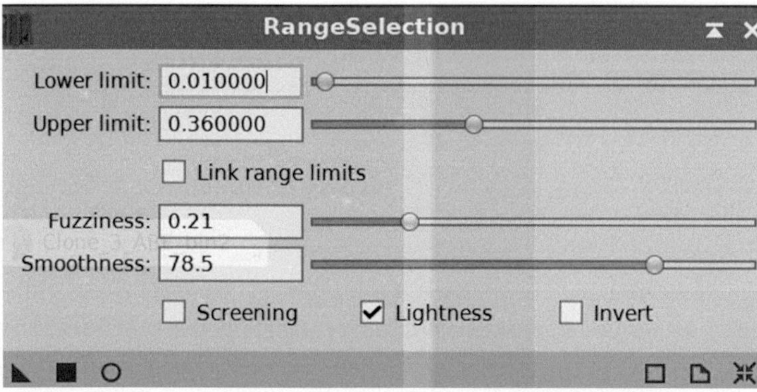

Fig. 5.12 The first range mask. Everything in Fig. 5.9 above a certain brightness is shown white. (Image by the Author)

There's still work to do, to get rid of the stray light at the edges of the image in Fig. 5.9.

There are many ways to skin this cat. I tried more than one, and did things inefficiently, but found that more efficient use of the tools didn't necessarily lead to a better result. I could have spent forever on this task. Here is what I did.

I masked off particular areas of the image. The idea is that when a mask is applied to an image, only certain areas of the image are processed by such processes as stretching.

First, I made what PixInsight calls a range mask. This is a two-tone image in which everything above a certain brightness is white, and everything else is black (see Fig. 5.12). The command sequence is

Process → MaskGeneration → RangeSelection.

The dialog box in Fig. 5.13 appears. By pressing the blue circle in the bottom left, you can experiment with the various sliders to see their effects.

So far, so good – I think. I have to make some more masks because I want to shield the target galaxies from my efforts at background removal. I can make masks of a chosen shape with.

Script → Utilities → GAME.

In this case, I want ellipses rather than arbitrary shapes. The GAME script menu is shown in Fig. 5.14.

All told, I made eight further masks, partly because I had to correct errors. Figure 5.15 shows the masks I made.

Sometimes I used the PixelMath feature, found at

Process → PixelMath → PixelMath.

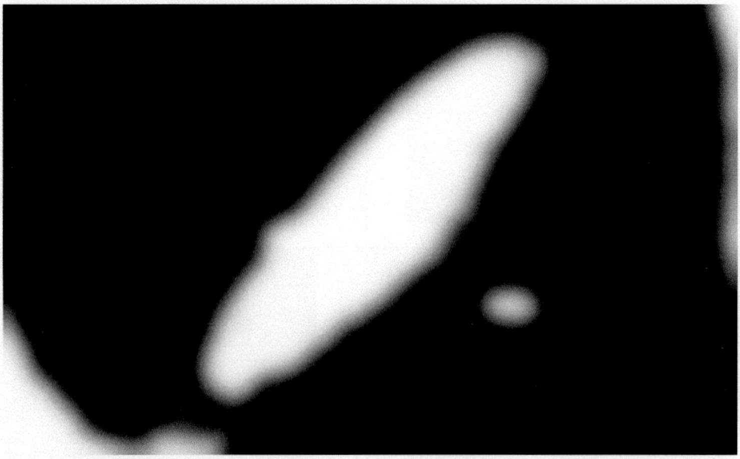

Fig. 5.13 The RangeSelection dialog. (Image by the Author)

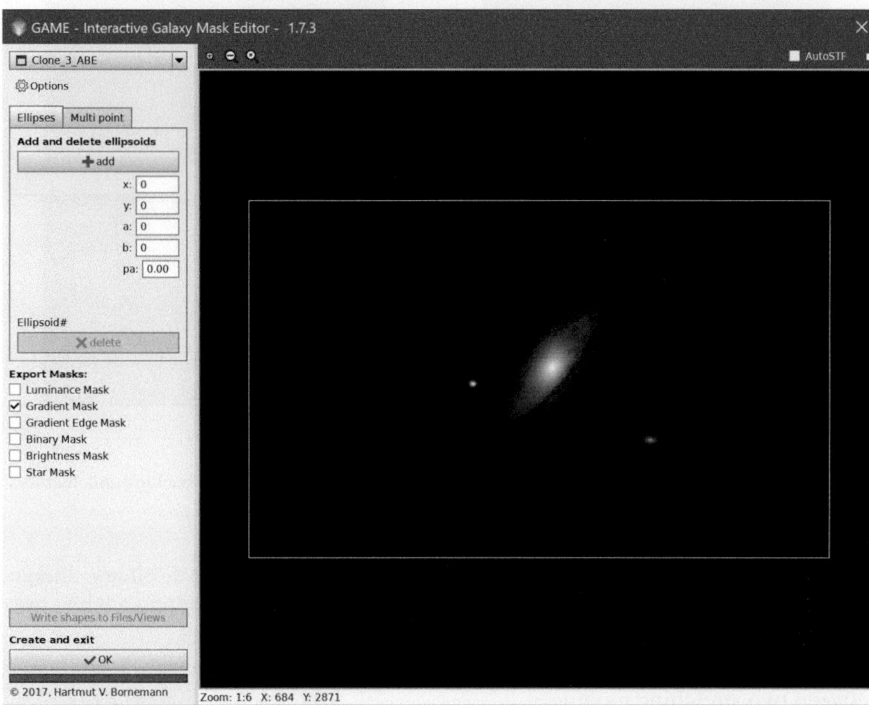

Fig. 5.14 The GAME script menu. If you click +add, it puts a shape template on the screen and you drag it to where you want it. It's a bit fiddly. The default size of the pop-up box is quite small, but you can make it bigger. This is worth doing to make the mouse clicks less hit-and-miss. (Image by the Author)

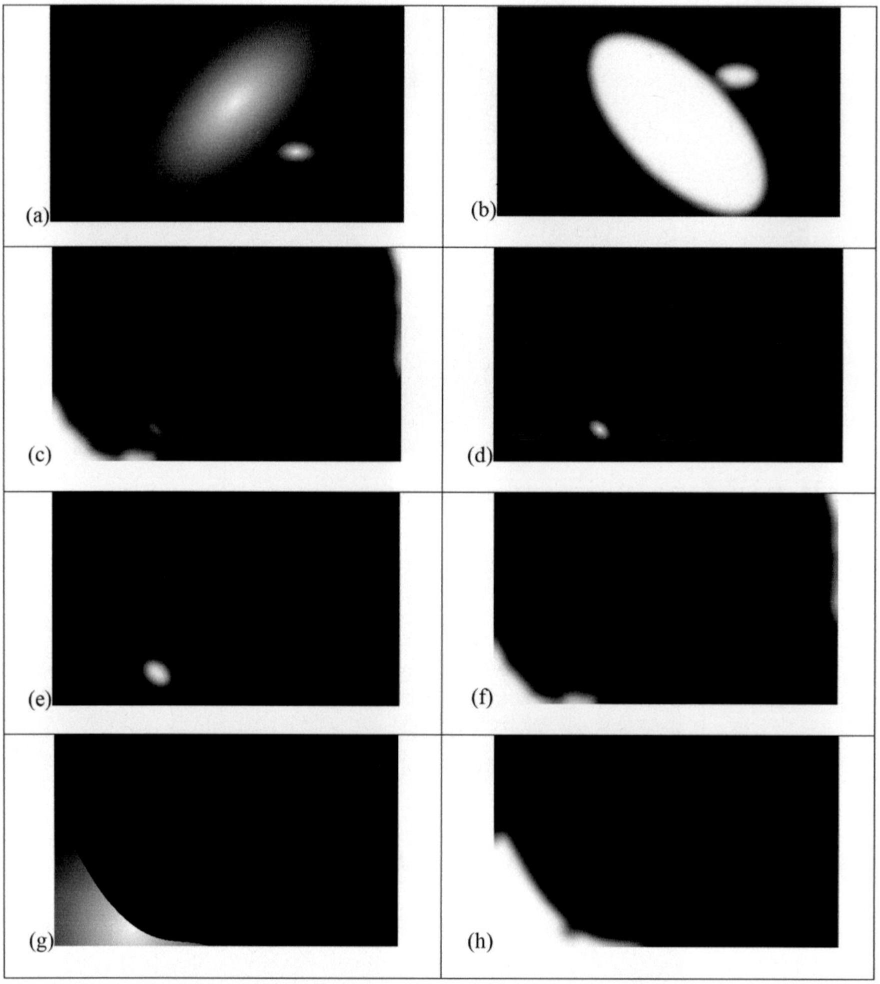

Fig. 5.15 The eight further masks I had to make to get rid of unwanted background features. (Images by the Author)

This process enables you to add and subtract images. If you have a 'binary' image, i.e. just a black & white image with no shades of grey, then subtracting a white area from an image leaves a black region in the resultant image.

Then, having subtracted of the unwanted fogging of the image I alternated between two stretching modules:

$$Process \rightarrow IntensityTransformations \rightarrow CurvesTransformation.$$

$$Process \rightarrow IntensityTransformations \rightarrow HistogramTransformation.$$

These are shown if Fig. 5.16.

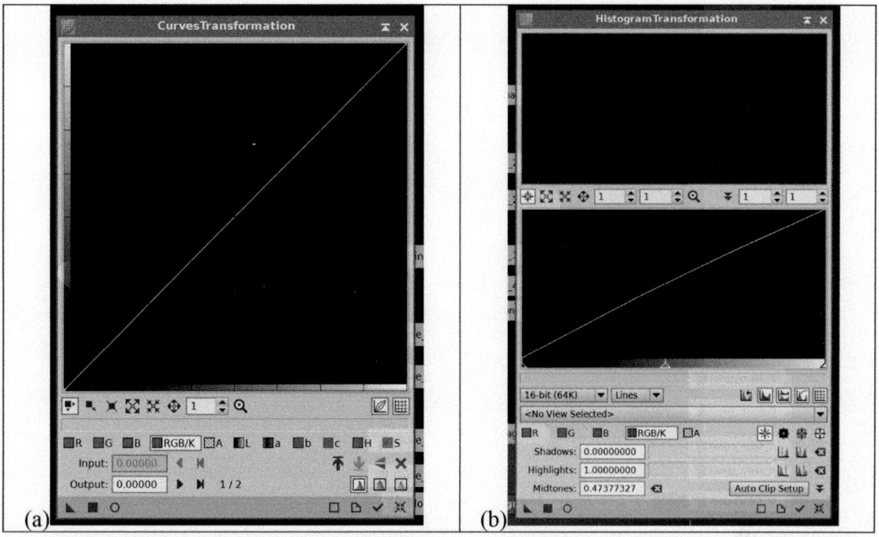

Fig. 5.16 Curves (**a**) and Histogram (**b**) transformations for stretching images. (Images by the Author)

In both cases, you get a live view of the effects of what you do by pressing the circle in the bottom left.

The curves tool is the more flexible. Both tools enable you to manipulate the R, G & B channels (red, green & blue channels) separately or together. In addition, the Curves tool allows you to manipulate other features. The one I used was the S curve to increase color saturation.

After all except the last two masks in Fig. 5.15, I ended up with the image in Fig. 5.17.

The image in Fig. 5.17 has the following defects: still some unwanted light, a green cast and a very green-looking M110. I used the final two masks shown in Fig. 5.15, dropped the green intensity using the Curves Transformation tool shown in Fig. 5.16, and made a mask to treat M110 separately, because it was so much greener than the rest of the image.

The resultant image is shown in Fig. 5.18.

From Fig. 5.18 I need to use the PixelMath tool to add the stars image. However, first I toned down the stars using the HistogramTransformation tool, to make them look less intrusive. Figure 5.19 shows the result.

After adding the stars back in, I got the image shown in Fig. 5.20.

I have but one remaining disappointment. Along the way I lost that cute faraway galaxy UCG 394. There's only one thing for it: I need to add it back in.

So I used CurvesTransformation to brighten up a clone of the image in Fig. 5.18, to make little UGC 394 visible. The result is shown in Fig. 5.21. Then I made a mask to cover UCG 394 while I re-darkened the rest of the image.

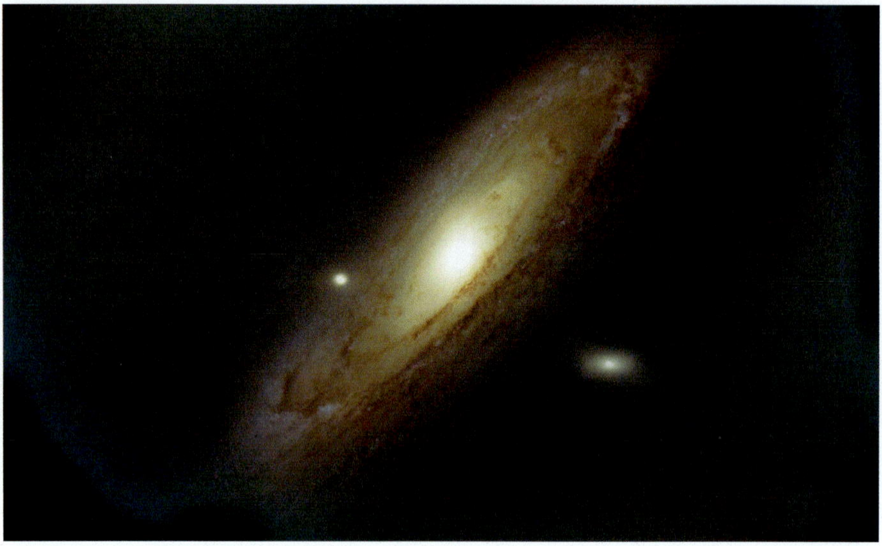

Fig. 5.17 An intermediate star-free image. It is still necessary to remove some unwanted background light. This was done with the seventh and eighth masks of Fig. 5.15. (Image by the Author)

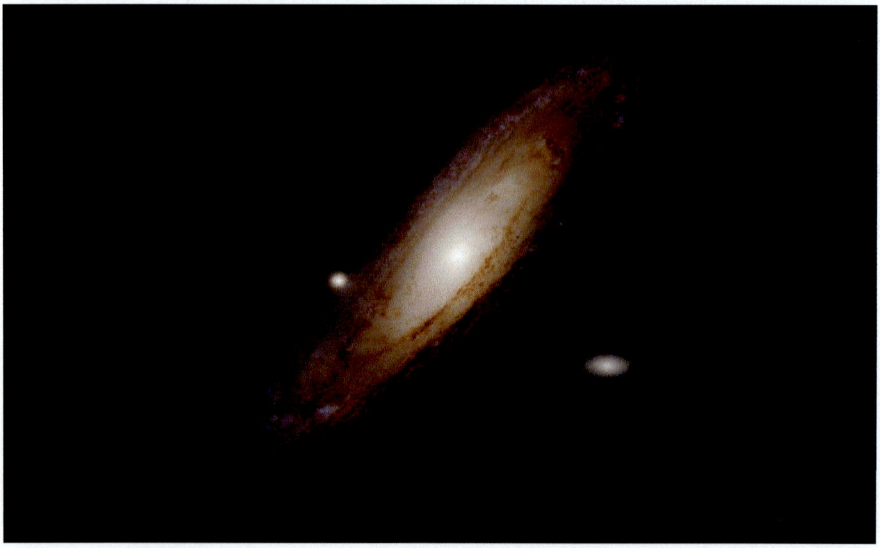

Fig. 5.18 I now have a nice-looking image of M31 and its attendant galaxies M32 and M110. (Image by the Author)

Unfortunately, I had overdone the brightening & ended up with the image in Fig. 5.22 after re-applying the CurvesTransformation with the mask of Fig. 5.23 present.

Fig. 5.19 Showing the stars as added. (Image by the Author)

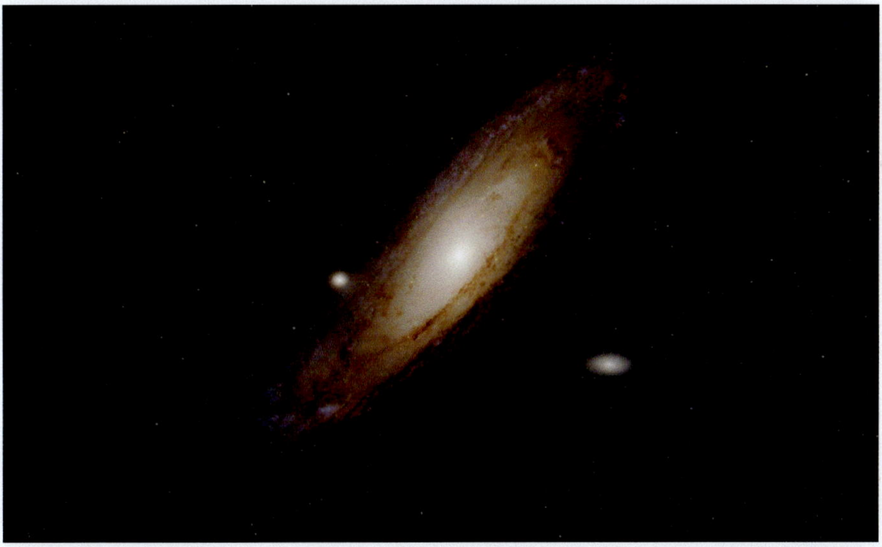

Fig. 5.20 The stars are now added back in. (Image by the Author)

Time to try another mask. This is shown in Fig. 5.24. I made a binary mask and decided to try my luck with PixelMath to see if I could subtract the bright spots away using PixelMath.

Success! I now have only UGC 394 in Fig. 5.25. Next, I used PixelMath to add it back in to the overall image. The result is shown in Fig. 5.26.

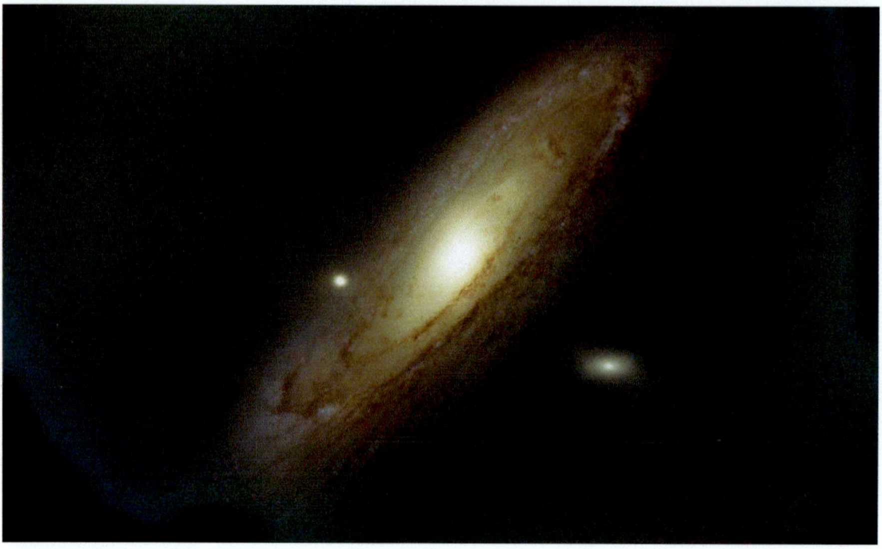

Fig. 5.21 I used a Curves Transformation to brighten this image enough to see UGC 394. (Image by the Author)

Fig. 5.22 I still have two unwanted bright regions in the middle of my image of Fig. 5.21 after I darkened the un-masked parts of the image. (Image by the Author)

Finally, I cropped this image to remove unwanted stuff at the sides. The result is shown in Fig. 5.27.

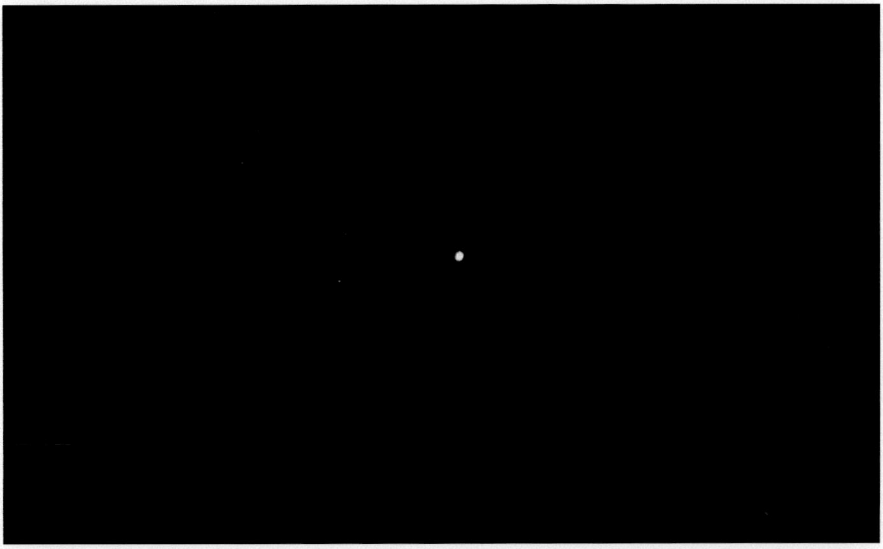

Fig. 5.23 The mask to protect UGC 394 while I manipulated the image. (Image by the Author)

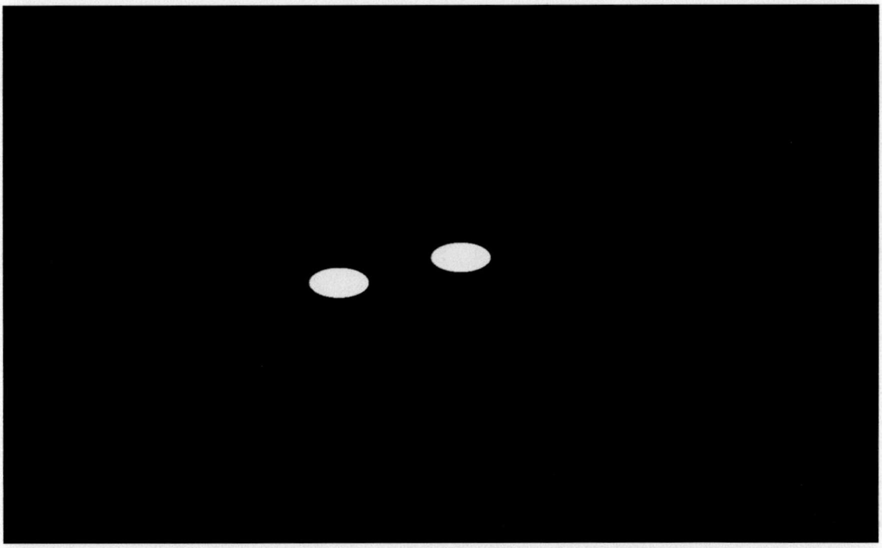

Fig. 5.24 The binary mask made to try to subtract away the bright spots in the center of Fig. 5.23 using PixelMath. (Image by the Author)

Please do not be discouraged by a feeling that all the stretching and masking is involved and tedious. It didn't take all that long. The first time you do it, it will take a while. The next time will be a lot faster. The viewer need be none the wiser about all the stray light you will have removed.

Fig. 5.25 UGC 394 on a black background ready to add to Fig. 5.20 using PixelMath. (Image by the Author)

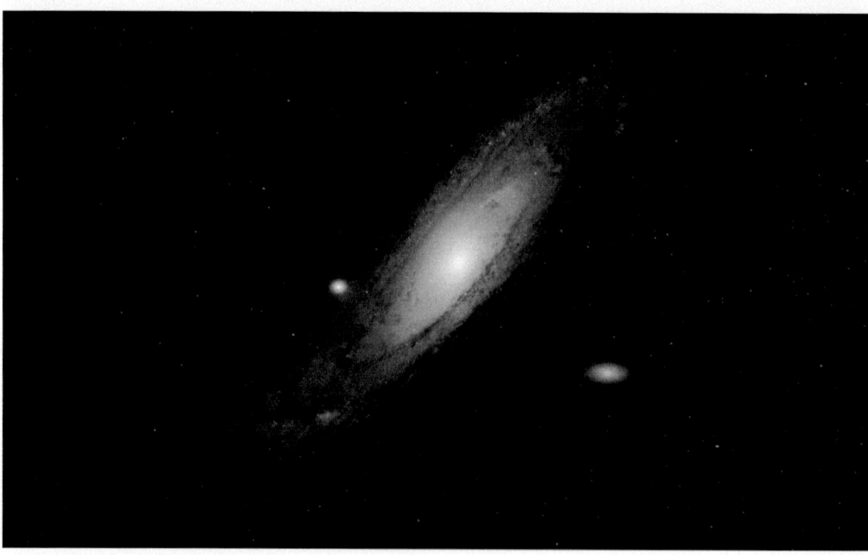

Fig. 5.26 UGC 394 is now added back in to my image. (Image by the Author)

The Stars in the Andromeda Galaxy

One of the most visible bluish features in M31, known as NGC 206, is shown in Fig. 5.28.

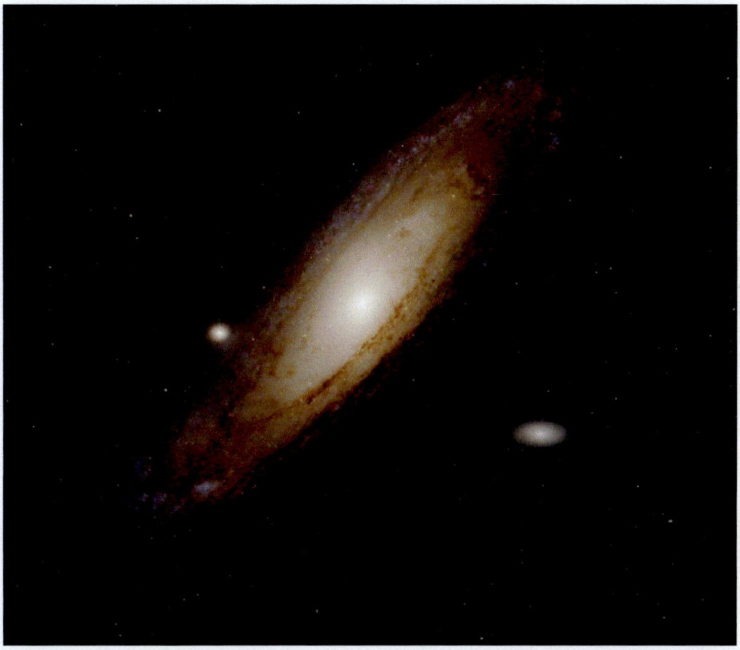

Fig. 5.27 My final image showing M31, M32, M110 and UGC 394. (Image by the Author)

In Hubble's day this was simply thought of as a star cluster. We've become a little more sophisticated nowadays, and refer to this cluster as an OB association. This is a star cluster with lots of O stars and B stars, i.e., the big, hot, blue ones, which live fast & die young, with lots of very sparse interstellar gas to form stars and to provide enough gravity to bind the group together. Once the gas has been used up, or perhaps blown away by supernova explosions or pulled away by absorption of small galaxies, the group disperses.

NGC 206 is the biggest OB association in the local group of galaxies.

It's tempting to think that we can make out individual stars in Fig. 5.28.

Edwin P. Hubble, of course, identified variable stars in M31 and used them to apply Leavitt's law to show that M31 had to be well outside our galaxy (Fig. 5.29). I have already told this story in Chap. 1.

Cepheid variables are unusually bright stars.

It occurred to me to wonder if the brightest of Hubble's Cepheid variables might be visible in my image of Fig. 5.27. Hubble in his 1929 Astrophysical Journal [47] article shows his images as negatives. It is straightforward enough to turn my image into negatives. In PixInsight, this process is almost, but not quite instantaneous. In Gimp, it takes long enough to drink several cups of coffee.

I found three of Hubble's variable stars, which he called V19, V42 and V9. Please bear in mind that my image was taken with a 3″ telescope in humid Welsh air 400 feet above sea level, whereas Hubble had all the time he wanted on a 100″

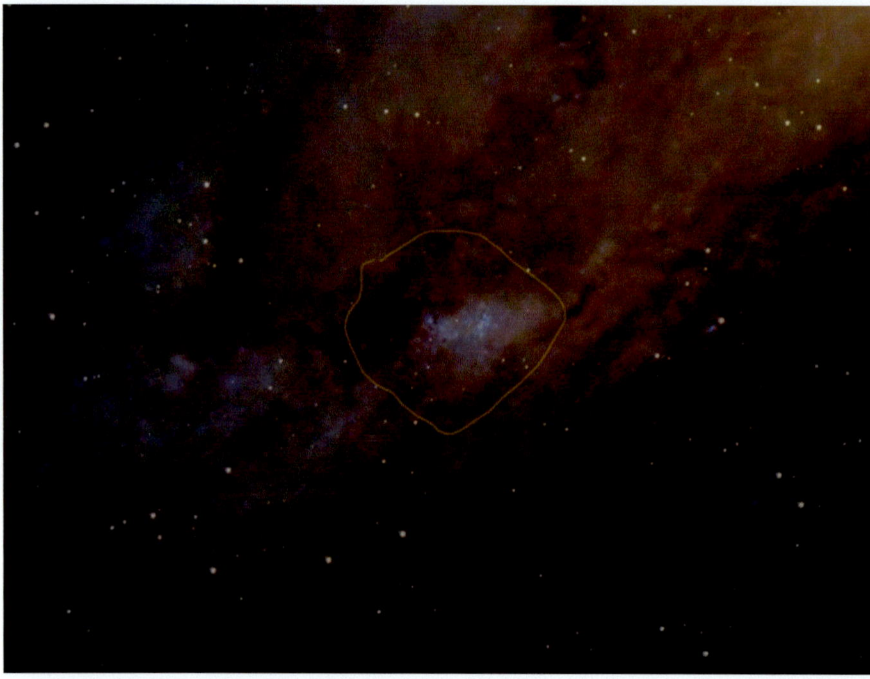

Fig. 5.28 NGC 206 is the bright, mainly blue, ringed feature. This picture is a blow-up of that of Fig. 5.27. (Image by the Author)

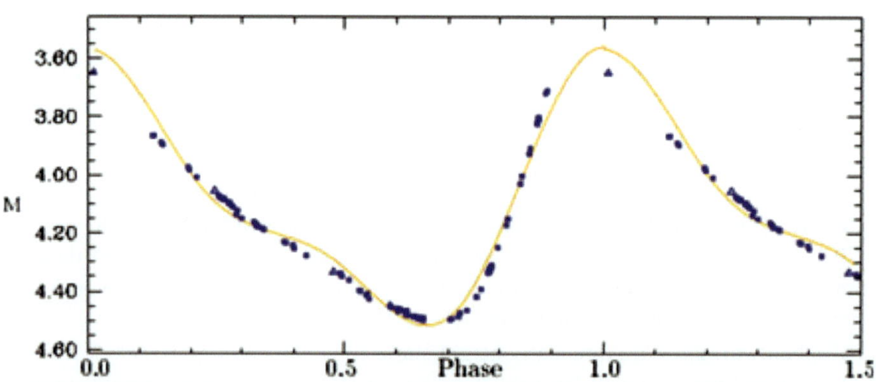

Fig. 5.29 The regular variation in brightness of δ Cephei, which has a period of 5.37 days and is just shy of 900 light-years away. Taken from HIPPARCOS data. (Image© 2024 G. Gangopadhyay and A. Kundu, under exclusive license to Springer Nature Singapore Pte Ltd. [46])

telescope atop a mountain in the California semi-desert. Therefore, I was very surprised to find these stars. My findings are shown in Figs. 5.30, 5.31, and 5.32.

The abbreviation ApJ means "Astrophysical Journal".

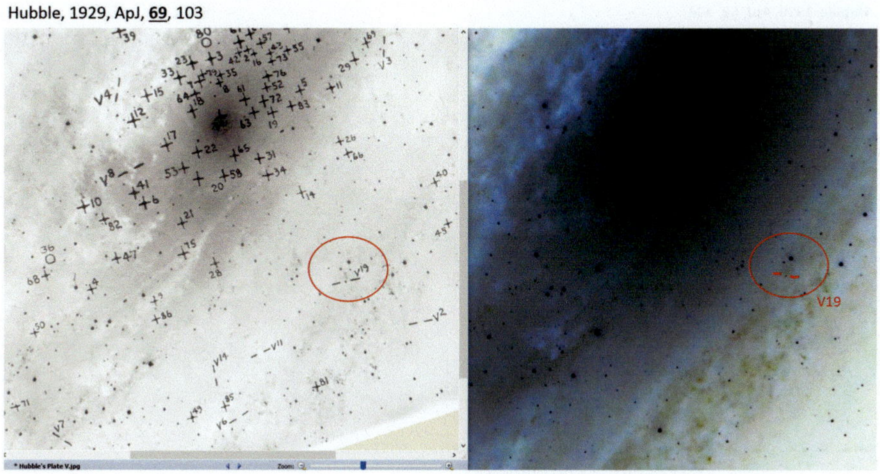

Fig. 5.30 Showing the non-Cepheid, very bright variable star V19 described by Hubble, which is in M31. Left: Hubble's image from ApJ, Vol 69, p.103 (1929). Right, Image by the Author. (Hubble's image is marked © American Astronomical Association)

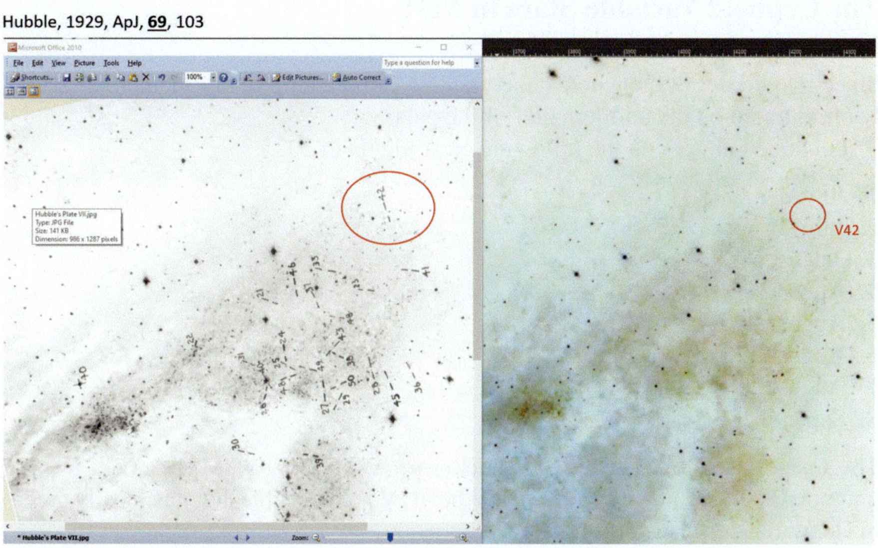

Fig. 5.31 Showing the Cepheid Variable star V42 described by Hubble, which is in M31. Left: Hubble's image from ApJ, Vol 69, p.103 (1929). Right, Image by the Author. (Hubble's image is marked © American Astronomical Association)

Now that I have discovered that I can photograph at least some stars in M31, the next stage is to see what I can find with my 11-inch telescope. I should be able to see 3-to-3½ magnitudes dimmer. At the time of writing, March 2023, M31 was very low in the sky. I had to wait until the following August to check this idea.

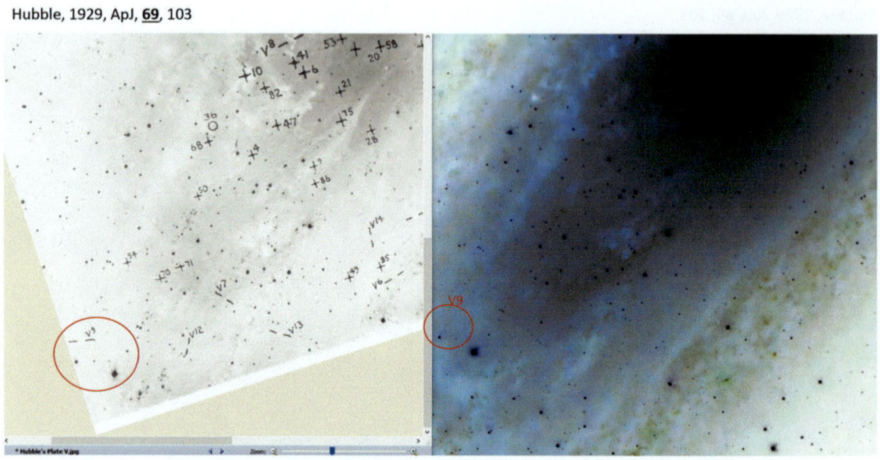

Fig. 5.32 Showing the Cepheid Variable star V9 described by Hubble, which is in M31. Left: Hubble's image from ApJ, Vol 69, p.103 (1929). Right, Image by the Author. (Hubble's image is marked © American Astronomical Association)

The Cepheid Variable Stars in M31

Since I could not fit the whole galaxy onto a camera chip in my 11-inch telescope, I concentrated on the northern half of the galaxy.

I obtained images on the following clear nights in 2023.

1. July21
2. July 25
3. August 5
4. August15
5. August 24
6. August 27
7. August 29
8. September 2

The images were taken with a monochrome ASI ZWO1600MM camera using a photometric filter known as a Bessell V-band (V is for Visible, but it is mostly green) made by Baader Planetarium.

A typical example is shown in Fig. 5.33. I made it negative, partly because I could see more; and partly because Hubble did the same.

It was going to be a lot easier to compare the stars with those of Hubble's 1929 paper [47] if I did the opposite to what galaxy photographers normally do, and brighten the stars but darken the galaxy. Then, in a negative image like Fig. 5.33, the stars would be dark & the galaxy faint. Sometimes, once I became really familiar with these stars, I didn't even bother to add the galaxy back in to the image. I just examined the stars.

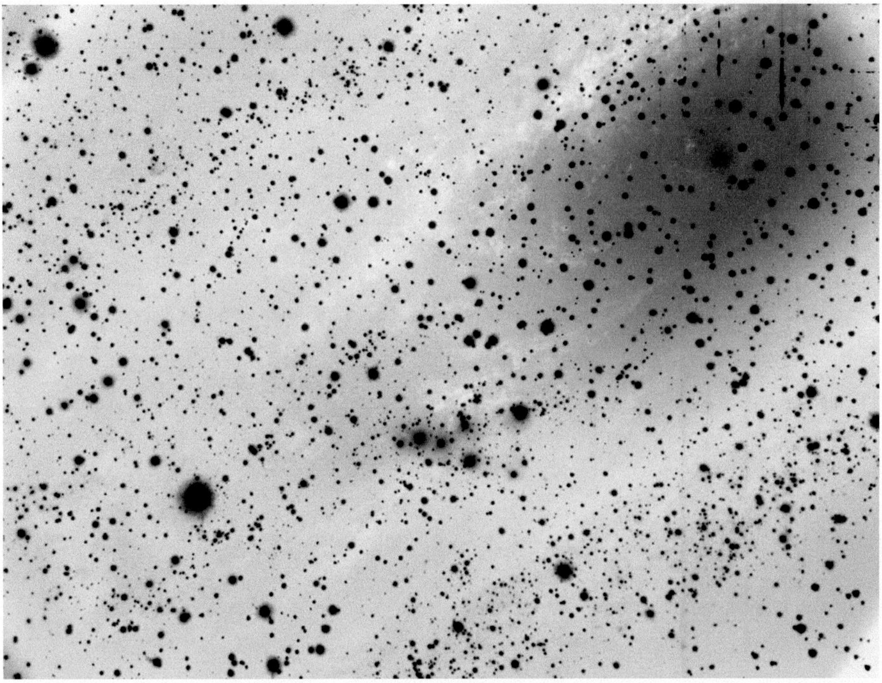

Fig. 5.33 A typical image of the northern end of the Andromeda Galaxy, shown in negative. Typically, I exposed for 3 hours in 60 s frames at a gain of 300 and a temperature of −15 °C. The weather did not always let me gather data for the full three hours. Celestron C11 telescope on a CGX mount. (Image by the Author)

Figure 5.34 from August fifth shows that I identified seven of Hubble's Cepheid variables by comparing my image against his 1929 Plate V in reference [47]. I eventually worked out that the star marked with a question mark was not a variable star.

The only star that I could see every time I looked was V9. Hubble [47] reported a period of 32.56 days for this variable.

On that basis, I would expect this star to have roughly the same the same brightness relative to its neighbors on August 24 as on July 21. That does indeed happen. On July 21 it is dimmer than the star immediately above it and to its right (Fig. 5.35).

By August 15 it is brighter (Fig. 5.36). By August 24, 34 days after July 21, in Fig. 5.37, V9 is about the same brightness relative to its nearest neighbor as on July 21.

In Fig. 5.38 I can see that July 25 and August 27, the 'phase' of V9 is approximately the same, as we would expect. This makes me confident that I have correctly identified V9.

The full set of images to compare is shown in Fig. 5.38.

To sum up, I have cracked the problem of how to photograph M31, and have gone on to discover that my 11-inch SCT telescope and ZWO ASI1600MM camera

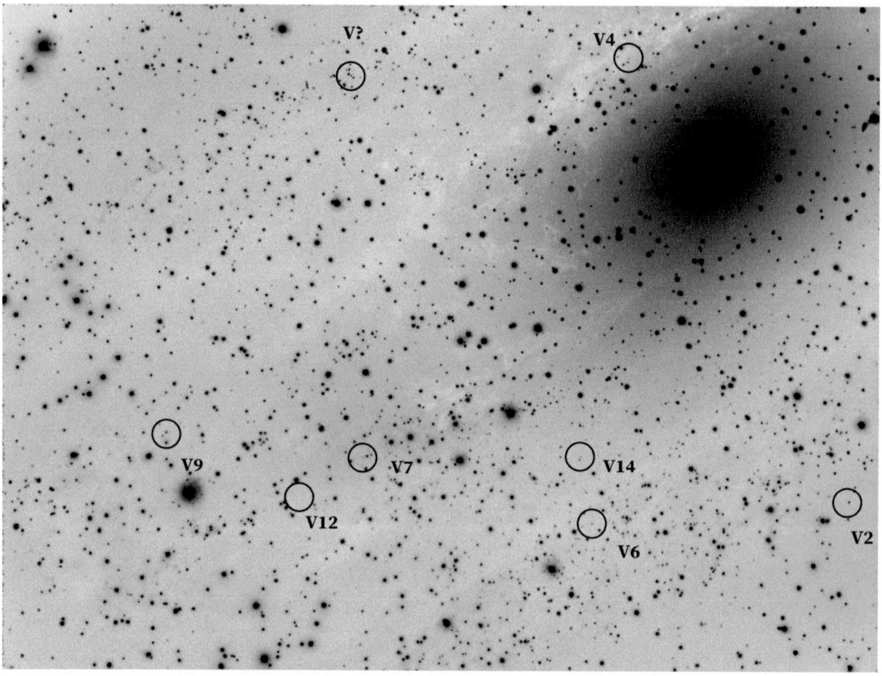

Fig. 5.34 Showing that on 2023 August 5, I identified seven of Hubble's Cepheid variable stars by comparing against Hubble's image. (Image by the Author)

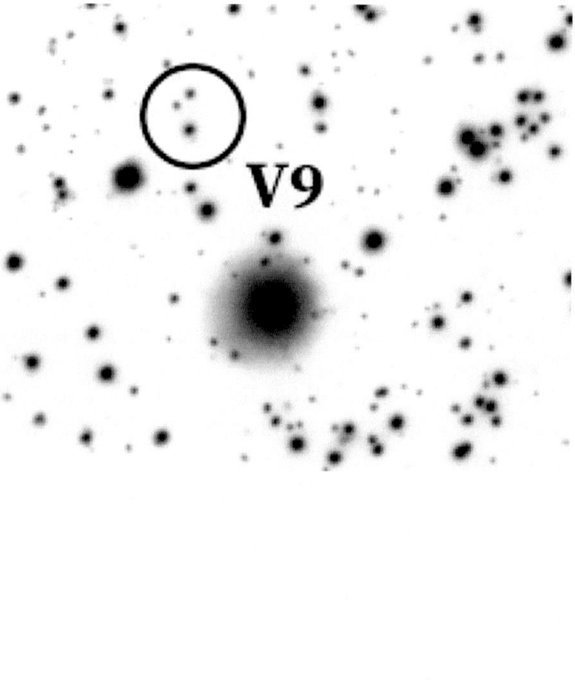

Fig. 5.35 Hubble's Cepheid variable star V9 on 2023 July 21. (Image by the Author)

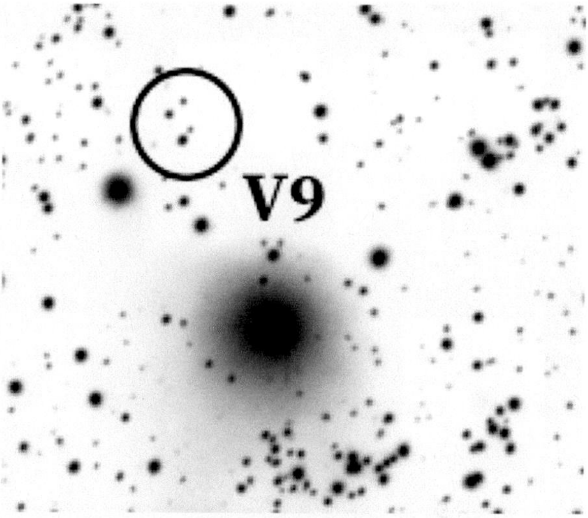

Fig. 5.36 Hubble's Cepheid variable star V9 on 2023 August 15. (Image by the Author)

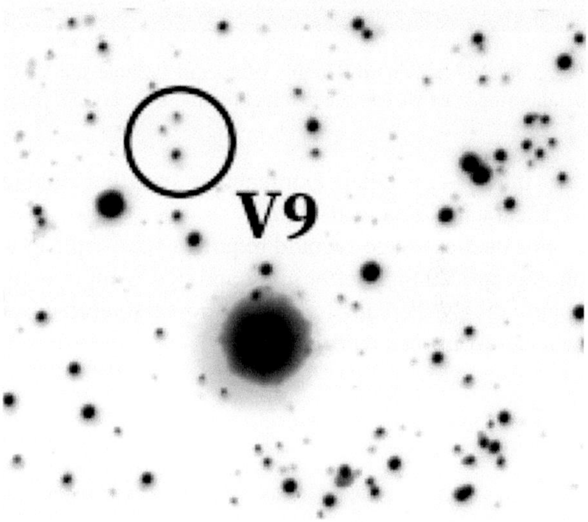

Fig. 5.37 Hubble's Cepheid variable star V9 on 2023 August 24. (Image by the Author)

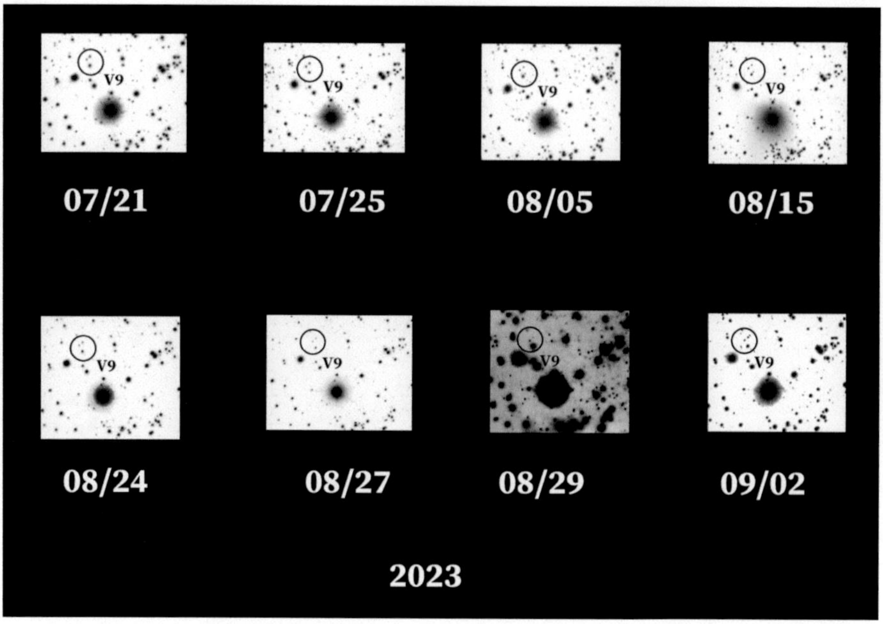

Fig. 5.38 My tracking of Hubble's Cepheid variable star V9 over six weeks. By comparing it to neighboring stars we can see that its brightness is varying. (Images by the Author)

have the power to detect extragalactic stars that Hubble needed a 100-inch telescope to detect a hundred years ago.

For the next stage I would require a year with six weeks of mostly clear skies to photograph V9 night after night and figure out how to measure its brightness curve quantitatively. Perhaps I could interest a remote observatory located in a desert or near-desert in such a project?

Chapter 6
The Fab Four

These four galaxies are all close to Usra Major, even if one of them is technically in Canes Venatici[1]. They are the lowest-hanging fruit in the northern Hemisphere. Southern Hemisphere dwellers will no doubt be able to obtain images from Telescope Live (see Chap. 8).

The Fab Four are M51, M101, M81 and M82. Their locations are shown in Fig. 6.1. They are great targets to cut your teeth on.

I tried them very early in my galactic career. Figure 6.2 shows my first ever photo of a galaxy through a telescope, or, more accurately of the pair of galaxies known as M51. I can no longer remember how I processed the image. I suspect that I did it in Registax, an old but good program for Solar System work. It was done with a monochrome camera. I certainly did not guide the telescope.

I was encouraged enough to persevere, and we all have to start somewhere. I found the galaxy with a little bit of difficulty using the finder scope and the GOTO on the mount's hand controller. I do remember that it stayed clear all night, in late December, so I had plenty of chance to experiment, because it was dark from about 4 p.m. until 8 a.m. here.

Getting More Experience with M51, the Whirlpool Galaxy

In subsequent years, I made more attempts at M51, and gradually improved. I'm showing you these examples so that you can see that I needed practice. I got no real help from anyone. I was on my own. At the time of the picture in Fig. 6.2, I was working, and my astronomical focus was on my previous book until I finished it in 2021, so the fact that I wasn't getting any help was really a side issue.

[1] Pronounce this "KANNays Vena-a-TEE-chee". The pronunciation of Latin varied somewhat over the centuries. I am recommending an approximation to medieval church Latin.

J. Clark, *Photographing Galaxies from Light Polluted Skies*, The Patrick Moore Practical Astronomy Series, https://doi.org/10.1007/978-1-0716-4506-2_6

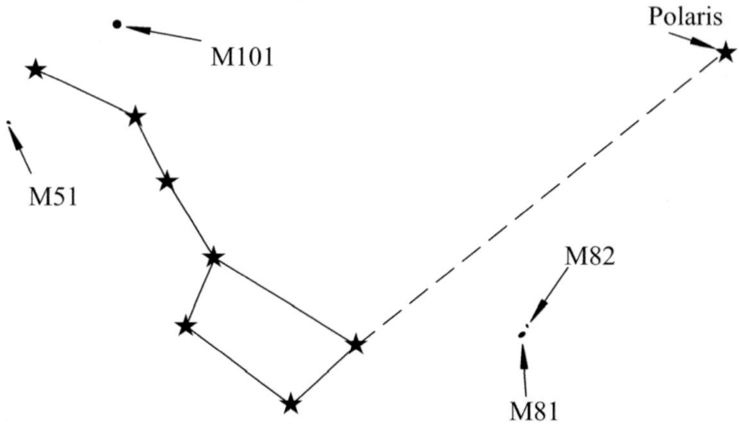

Fig. 6.1 The locations of the best-known galaxies around Ursa Major. (Image by the Author)

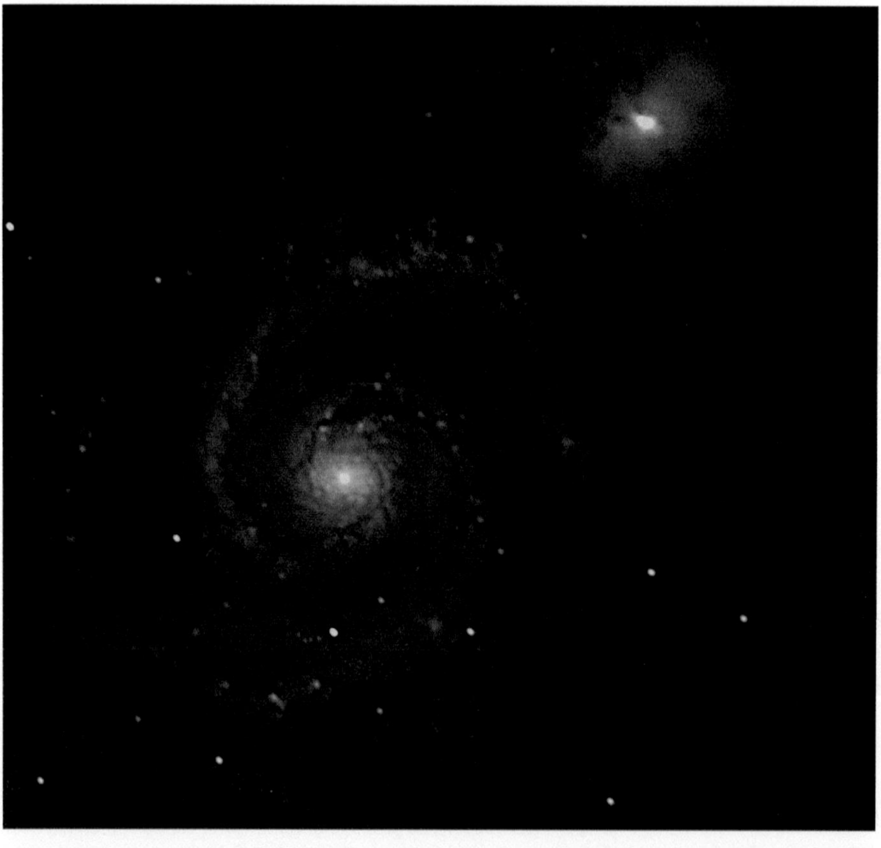

Fig. 6.2 Here is one of my first ever photographs of a galaxy, M51. Celestron C925 XLT telescope on a GCEM Mount, 120 × 15 s with an Atik E414EX monochrome CCD camera at −10 °C, captured as a SER file using SharpCap v. 3.2. December 25 2019. (Image by the Author)

I had an unsuccessful go at shooting M51 in color in 2020, which looked very implausible. My next attempt appears to have been in January 2022. By then I had a Celestron C11 telescope on a CGX mount, and a one-shot color camera, a ZWO ASI294MC. I processed the image in Siril. It's getting better, but I still don't believe the colors (Fig. 6.3).

In January 2023 I had another go. I was quite pleased with this one (Fig. 6.4). By this time, I had traded up to a ZWO ASI2600MC camera. I also exposed for quite a bit longer: 259 minutes or 4 h 19 m. The colors are much more believable, but the stars aren't sharp, which presumably means that the galaxy isn't sharp either.

So that was the learning curve.

My next attempt at M51 was a year later, over the nights of January 5, 6 & 7, 2024. I collected 435 minutes or 7 h 15 m of data.

The image with a small amount of stretching from the EZ Soft Stretch script, a free add-on to PixInsight, is shown in Fig. 6.5 after I had removed the stars, as detailed in Chaps. 3, 4, and 5.

All these images were captured with my Celestron C11 11-inch SCT with Celestron's own f/6.3 focal reducer/corrector, on a Celestron CGX mount. Image capture was with SharpCap 4.1, and processing was with PixInsight 1.8.9-2. The camera was a ZWO ASI2600MC one-shot color camera, with the chip held at −15 °C (5 °F) and the gain set to 100.

I will show images in monochrome where possible, even though they were actually color images. There is a lot of vignetting visible in this image. The flat frames have not removed it all. Therefore, part of my task is to figure out how to remove it the hard way.

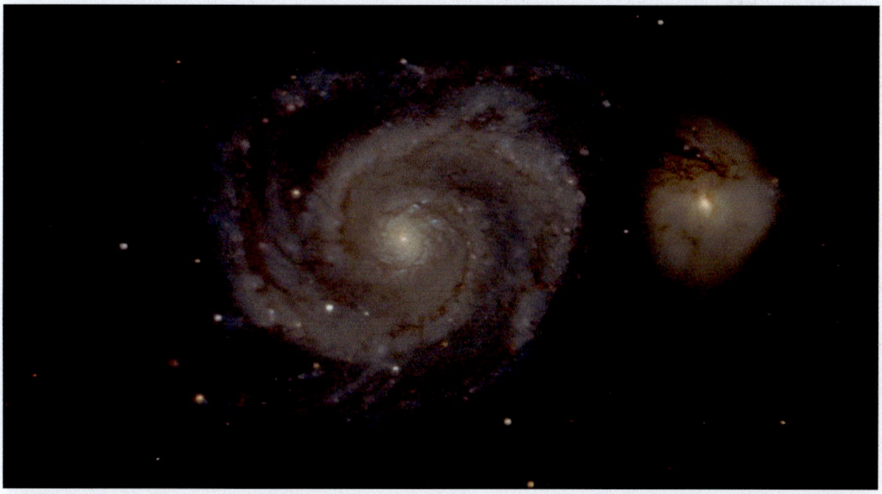

Fig. 6.3 A color attempt. Celestron C11 telescope on a CGX mount, and a one-shot color camera, a ZWO ASI294MC. Processed in SharpCap 4.0 and Siril, probably version 1.0.6. (Image by the Author)

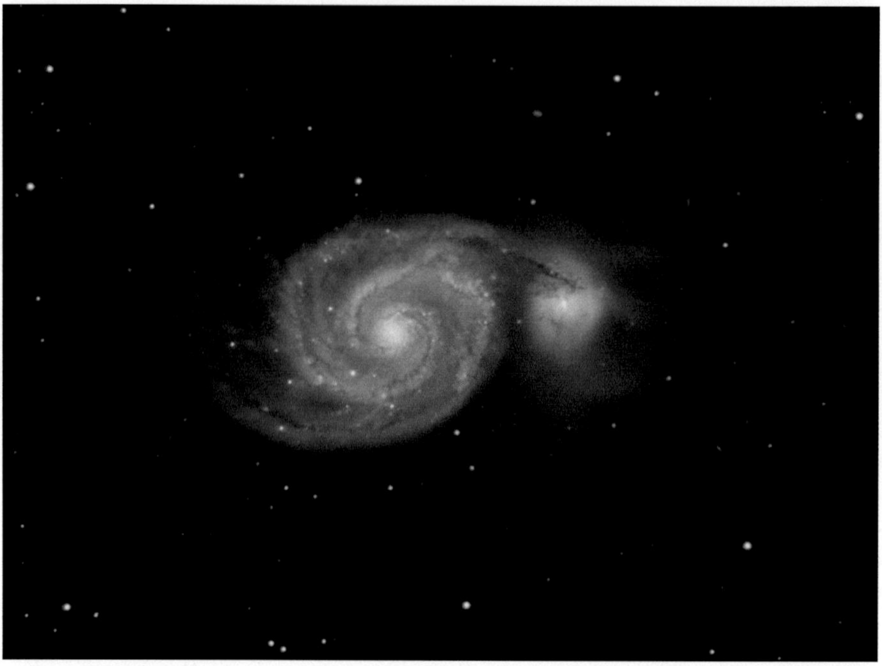

Fig. 6.4 By January 2023 my attempts are beginning to look a lot better. Processed in SharpCap 4.0 and Siril, version 1.0.6, and somewhat in PixInsight. (Image by the Author)

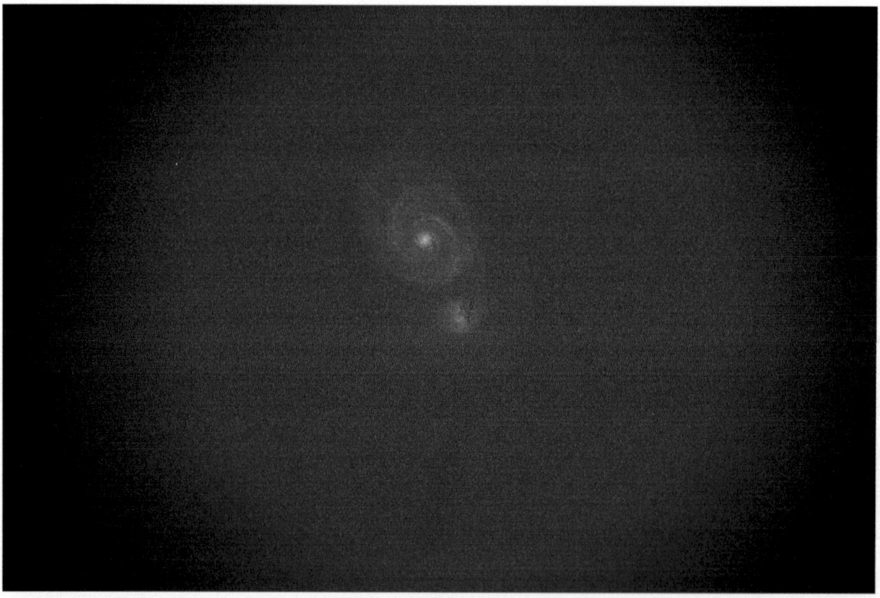

Fig. 6.5 M51 with gentle stretching, enough for StarXTerminator to work. (Image by the Author)

I will need to stretch the image more. I also need to enhance the color. I will have to ask you to take my word for that for now, as you can't see the insipid colors in Fig. 6.5.

I tried several approaches. The first was to try to create masks to get out of trouble. You can draw masks of any shape with

$$Scripts \rightarrow Utilities \rightarrow GAME$$

The Game Script menu is shown in Fig. 6.6. GAME is an acronym for Galaxy Mask Editor. You can create ellipses or arbitrary shapes, though the latter takes a bit of practice. I used the ellipse tool to draw the outline in Fig. 6.6. You can make a sets of different types of mask. I tried a binary mask, which I later gave fuzzy edges in

$$Process \rightarrow MaskGeneration \rightarrow RangeSelection$$

(see Fig. 6.7).

You drag the identifier tab on the left-hand side of mask you want, in this case the fuzzy-edged version, onto the left-hand bar of the image you want to mask. Its identifier than turns brown. You invert the mask, CTRL-SHIFT-I, and what you will see is shown in Fig. 6.9.

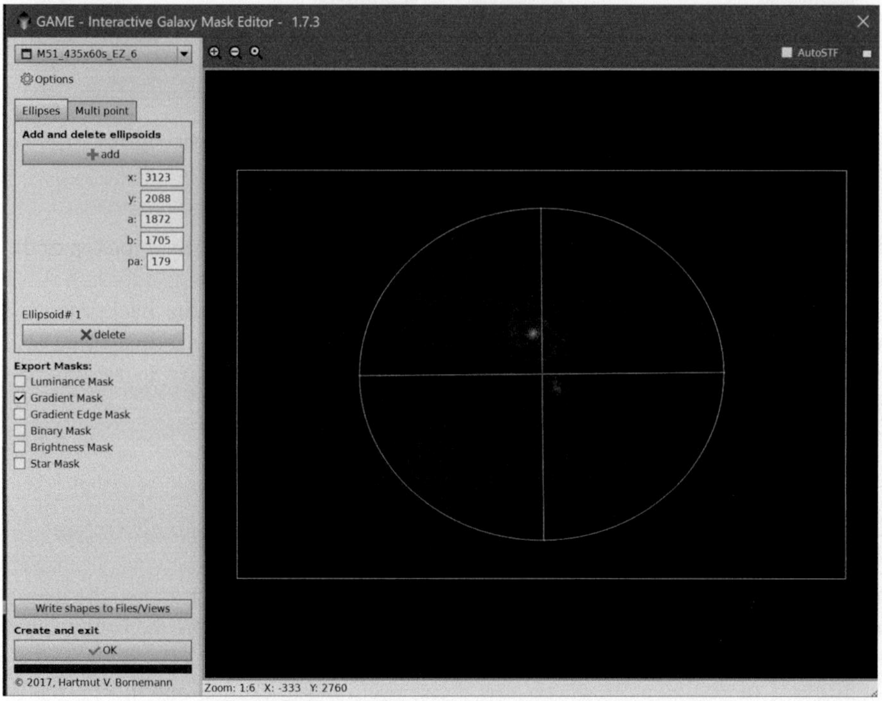

Fig. 6.6 The GAME script for drawing masks. (Image by the Author)

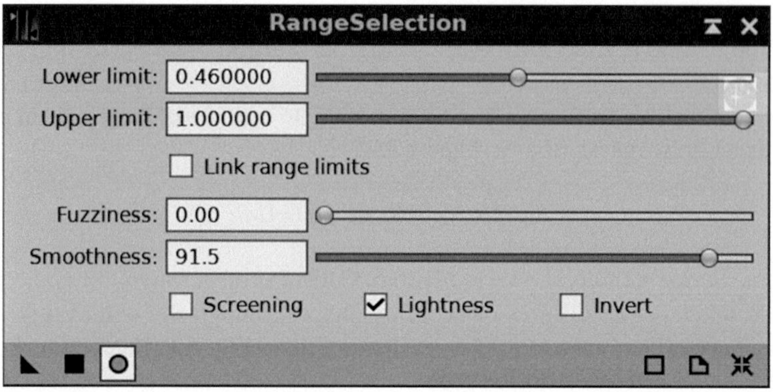

Fig. 6.7 The Range Selection tool as used to create the right-hand image in Fig. 6.8. (Image by the Author)

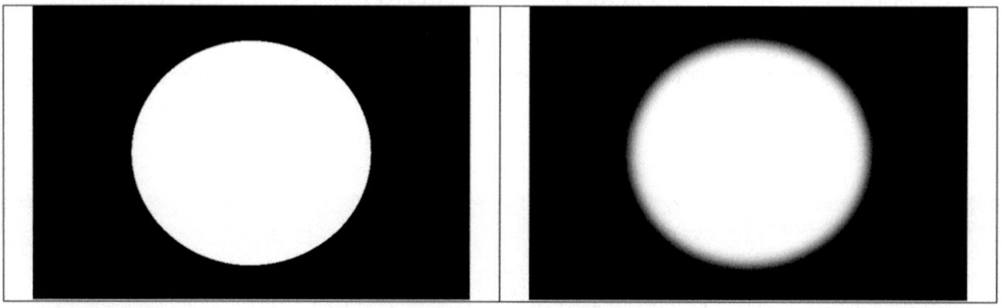

Fig. 6.8 Left: The binary mask made by the GAME script (Fig. 6.6). Right, the fuzzy-edged version created with the RangeSelection tool (Fig. 6.7). (Image by the Author)

You can manipulate, and remove, the mask using the Mask menu at the top of the PixInsight window.

Unfortunately, this approach didn't really work to remove the background. I mention it to save you from making the same mistake. You can see that I had also used the GraXpert script, now a tool in the Process menu, to try to remove the background.

Back to the drawing board.

I tried

Process → *BackgroundModelizationn* → *AutomaticBackgroundExtractor*

with equally little success. Now I tried GraXpert on its own, and stretched the image a little. Figure 6.10 shows that I have now separated the galaxy from the background somewhat.

Right. I'm getting somewhere now. I can use a mask like the one in Fig. 6.9 to reduce the grey ring to invisibility.

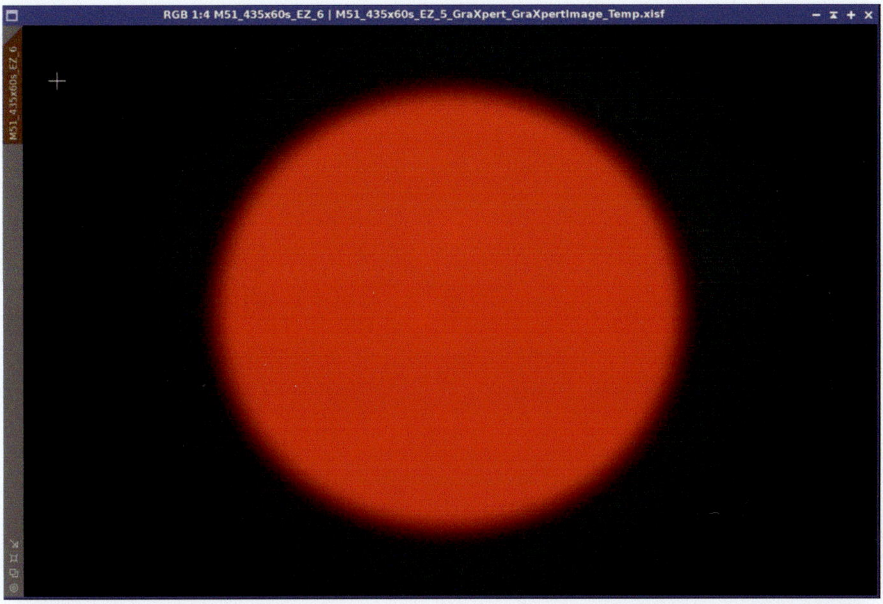

Fig. 6.9 An image with a masked region. Any processing will not affect the area shown red. Note that the identifier tag on the left is brown. This means that the mask is active. (Image by the Author)

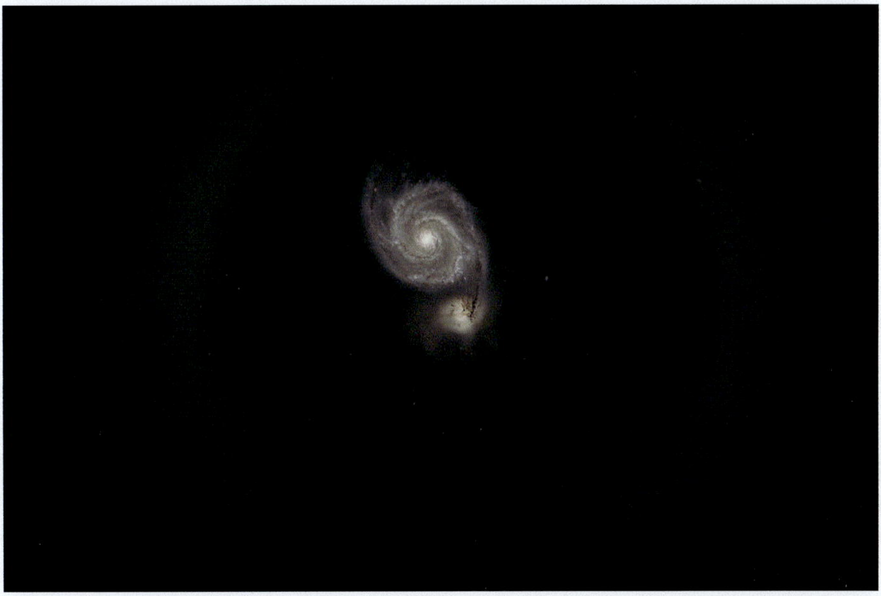

Fig. 6.10 I have isolated the background into a ring away from the galaxy, so I do not need to worry about damaging the galaxy while I operate on the ring. (Image by the Author)

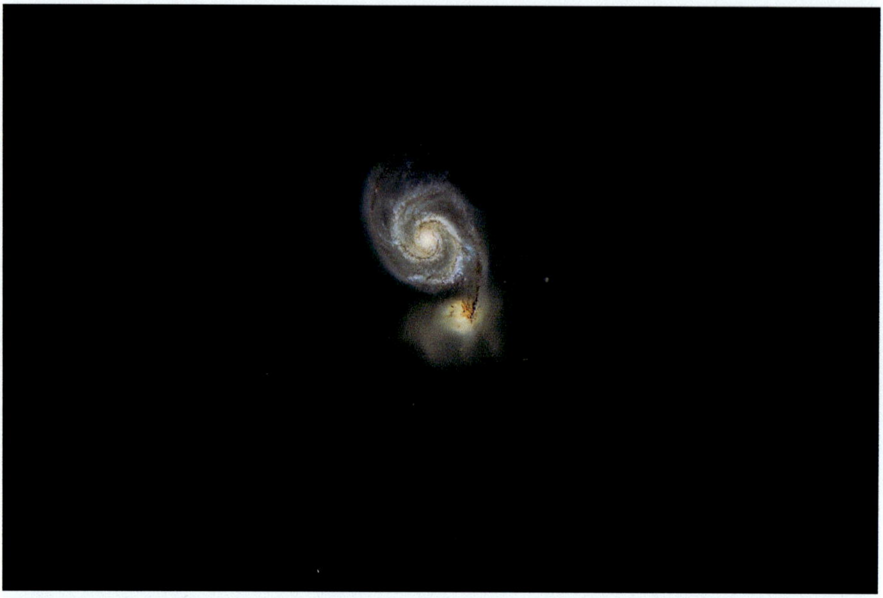

Fig. 6.11 The image is now stretched enough and the annoying background has been removed. (Image by the Author)

This I did, and stretched the image some more with

Process → IntensityTransformations → HistogramTransformation

And the result is shown in Fig. 6.11. I am now happy; and ready to out the stars back in. The stars have been dimmed a little with the same histogram Transformation tool.

I added the stars using the Pixel Math expression

M51_435×60s_EZ_6_stars + M51_435×60s_EZ_34.

PixelMath is found at

Process → PixelMath → PixelMath

I set up the PixelMath window as shown in Fig. 6.12. In particular, I create a new image with the same parameters as the target, in this case *M51_435x60s_EZ_34*, onto which I clicked to make it active. To run this operation, you have to click the solid square at the bottom left, as shown in Fig. 6.12.

I have an image I am happy with. I then use

Process → Geometry → Rotation

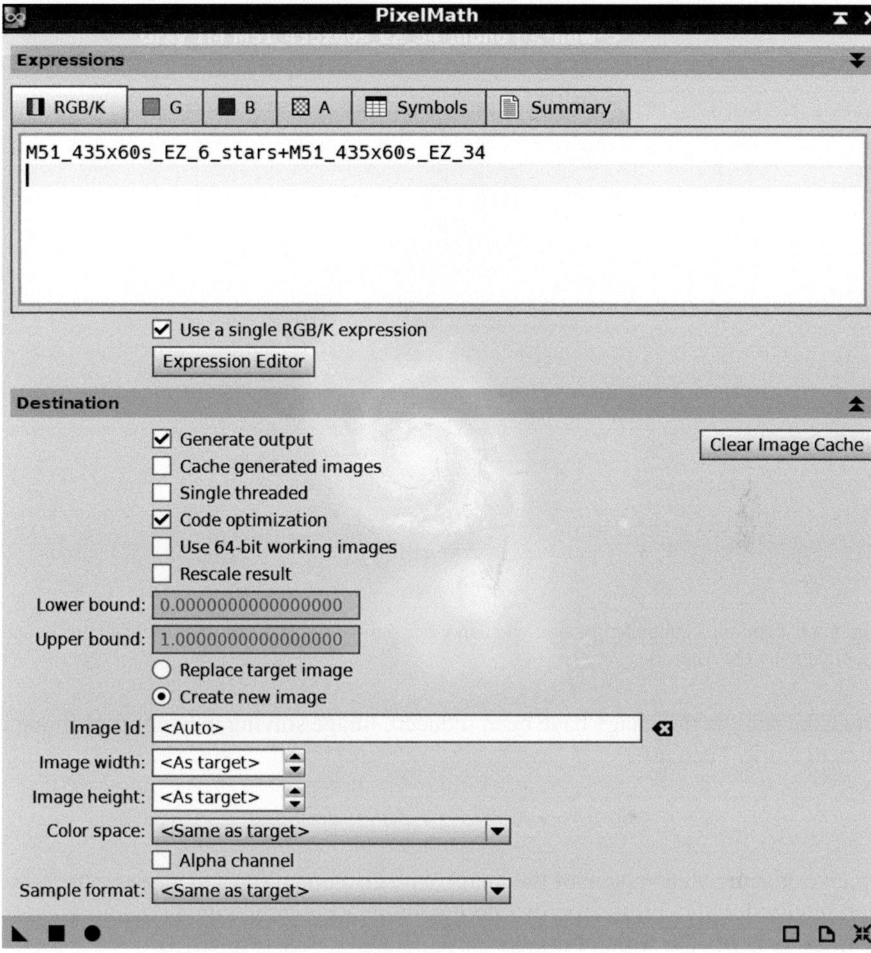

Fig. 6.12 The PixelMath window. (Image by the Author)

To rotate the image until north is up, and crop with

Process → Geometry → DynamicCrop.

You have to close, not park, the Dynamic crop window when you are done.

To do the rotation, you take an early stretched version of the image where the stars have not been reduced in intensity, and image solve with

Script → Image Analysis → ImageSolver

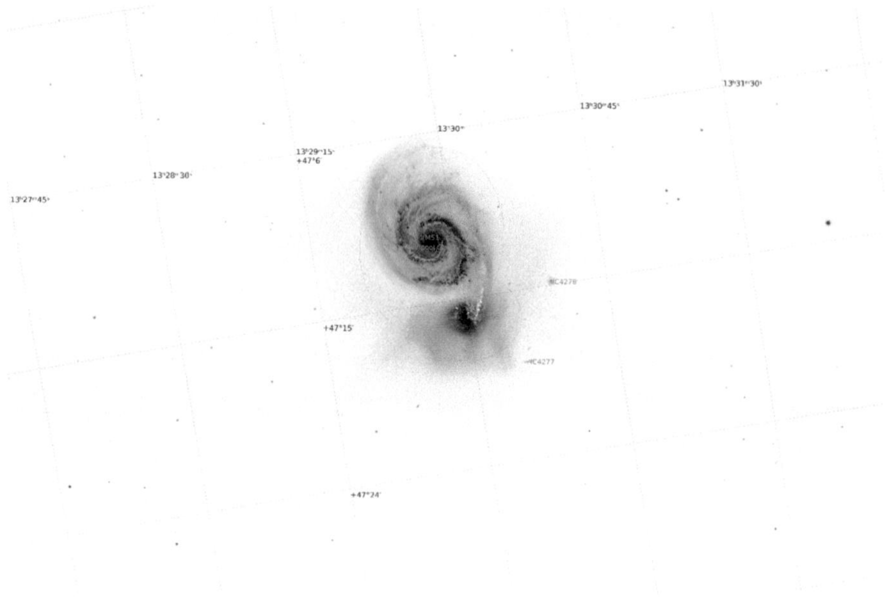

Fig. 6.13 From this annotated image, you can work out the rotation required to make north face up. (Image by the Author)

Once the stars in the image have been reduced, image solving may not find enough stars to work. Then use

Script → Render → AnnotateImage

To give an annotated version of the image with grid lines. You can measure the rotation required with a protractor, or I do this more accurately with a computer-aided drafting system, into which I import the annotated image (Fig. 6.13).

The final result is shown in Fig. 6.14.

I have increased the color saturation with

Process → IntensityTransformations → ColorSaturation.

I apply this gradually in several steps because it is easy to overdo the color saturation.

My stars are now pin-sharp. My colors are credible and I can see how much the smaller galaxy has been disrupted by its interaction with the larger one.

Interpreting What I Can See of M51

The two galaxies are also known as NGC5194, the bigger spiral, and NGC5195, the smaller disrupted galaxy. They are also sometimes known as M51A & M51B respectively. Since that's a more easily remembered nomenclature, I will use it here.

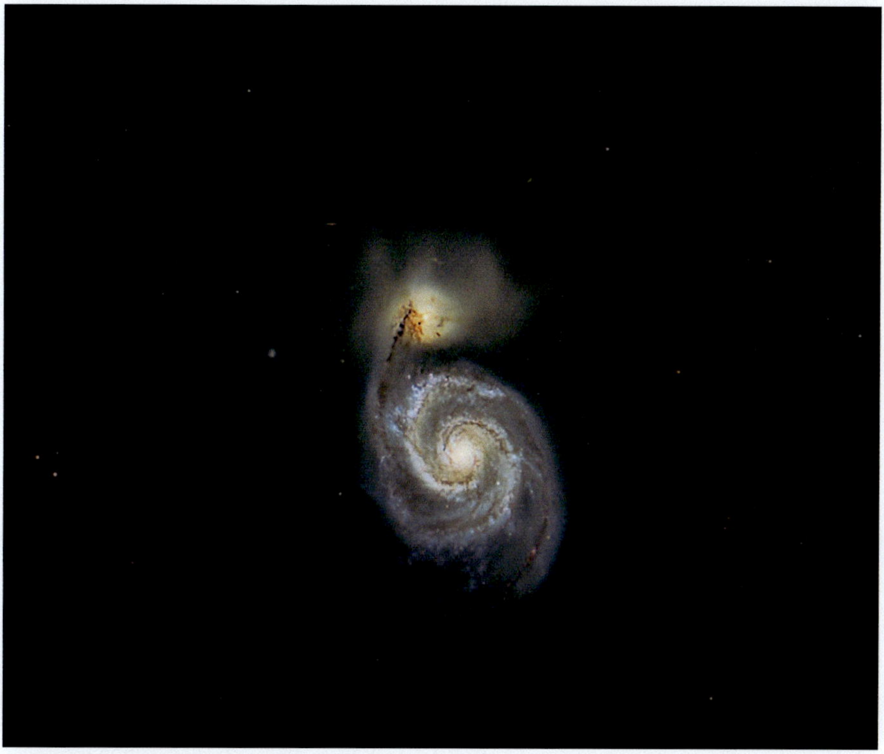

Fig. 6.14 My final image of M51. (Image by the Author)

First, let me explain why some regions are blue and others are yellow.

Figure 6.15 shows a ninety-minute exposure of the Pleiades. The brightness is dominated by a handful of very bright blue stars, but most stars are red, orange or yellow. The theory of stellar evolution tells us that blue stars are massive, burn very brightly and die young – after millions, not billions of years – often spectacularly. So, any region where the color is dominated by blue stars has to be a star-forming region because these stars don't last long.

The spiral arms of M51A are predominantly blue, and therefore are star-forming regions. The nucleus of M51A is yellow and therefore has not recently been a star-forming region. There is no more than a hint of blue in M51B in Fig. 6.14. It is basically yellow. So, to a first approximation, the interaction between M51A & M51B has killed off the star formation in M51B.

The reddish-brown material is dust, predominantly microscopic silicate particles, particles of water ice, and particles of polycyclic aromatic carbons (which is basically what soot is made of). It is notable that the dust "lanes" of M51A are on the insides of the spiral arms. One of them extends across M51B, showing that, from our viewpoint, M51B is behind M51A. Just imagine a whole galaxy-length stream of microscopic particles. That's an awful lot of particles!

It will come as no surprise that the M51 system is almost the archetype for the study of galaxy collisions. It was first studied in a pioneering simulation in the early

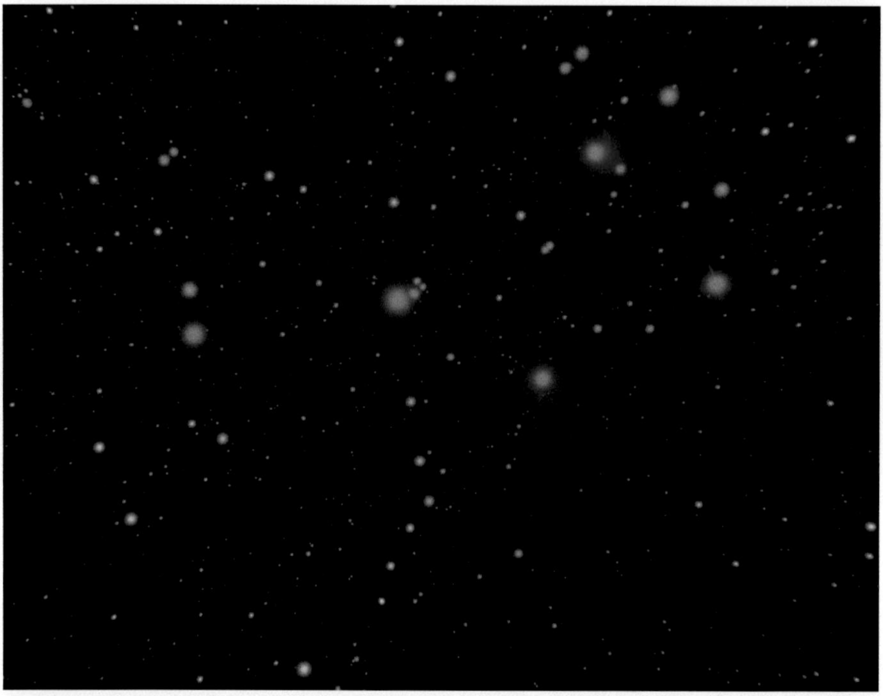

Fig. 6.15 Ninety minutes of the Pleiades, M45. 90×60s, ZWO ASI2600MC camera, Sharp Cap 4.1, From Live Stack, Stars desaturated in Siril 1.2.1, William Optics ZenithStar Z73 telescope, 2023 01 20. (Image by the Author)

days of computers by the Toomre brothers [48]. It was not only an excellent feat of astronomical research, but was also an extraordinary feat to get the computers of the early 1970s to produce the graphics. We take that capability for granted now, but it was barely available back then. The methods I had to use a few years later in the mid-1970s to produce computer-generated graphs as an undergraduate would now seem laughable.

Toomre & Toomre assumed that the overwhelming majority of the mass of the galaxy is treatable as a point mass at the center. The stars orbit this center, but do not noticeably attract each other gravitationally. As two galaxies approach one another, the stars begin to be gravitationally attracted by two galactic nuclei, and no longer orbit stably in peace and quiet. The Toomres found that each galaxy creates a "bridge" and a "tail" in the other. The bridge is of stars moving towards the other galaxy. The tail moves away. I suspect that the mechanism is similar to that by which the Moon causes one high tide on the ocean surface near the Moon and one on the opposite side. This happens because of a combination of gravitational and centrifugal forces.

After trying many possibilities their simulation of the M51 system has M51B (NGC5195) with one third the mass of M51A (NGC5194). It is taken to approach M51A on a parabolic path. Their final results are shown in Fig. 6.16.

Part (b) of Fig. 6.16 shows the view that we see of this pair of galaxies. There is an obvious tail-and-bridge pair on NGC5194 (M51A). The tail goes up, and the bridge goes down in the diagram. The actual tail and bridge are clearly visible in Fig. 6.14 in my photograph. The tail of NGC 5195 (M51B) is also identifiable. But what about the bridge of the smaller galaxy? It is much harder to see.

I reprocessed my image with extreme stretch using

Process → IntensityTransformations → HistogramTransformation.

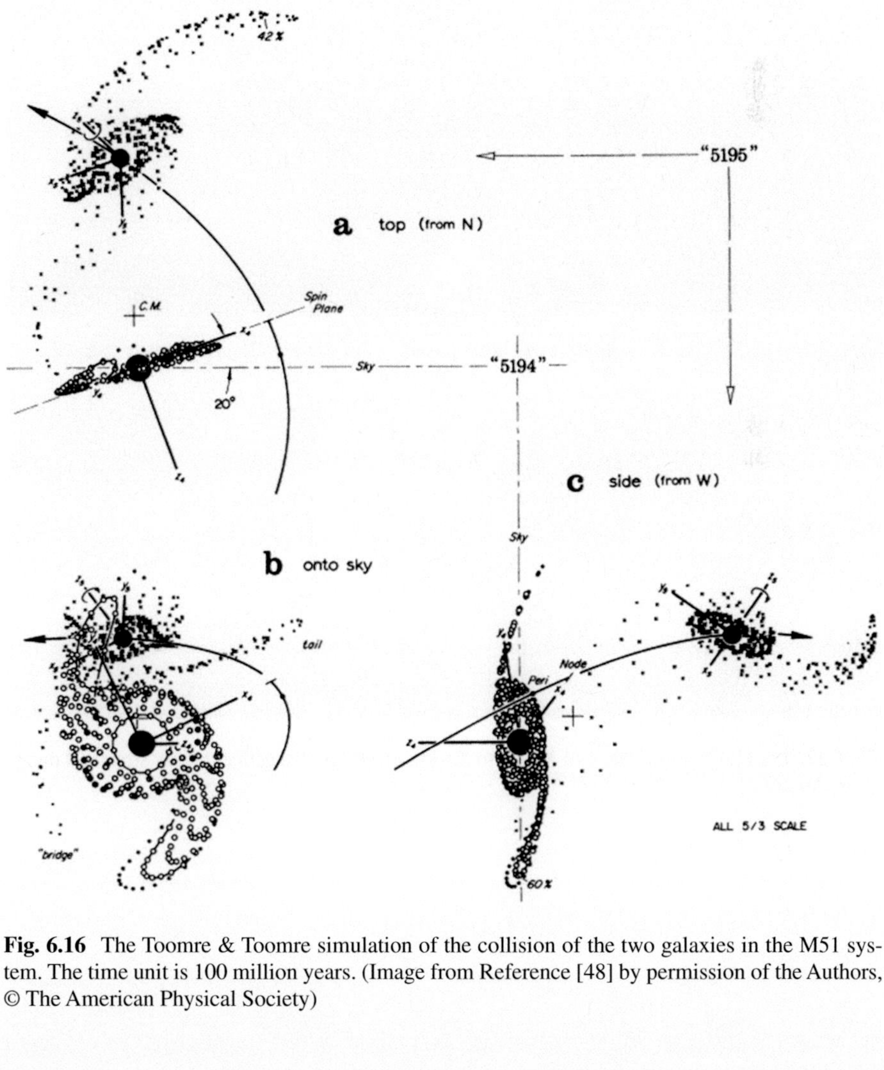

Fig. 6.16 The Toomre & Toomre simulation of the collision of the two galaxies in the M51 system. The time unit is 100 million years. (Image from Reference [48] by permission of the Authors, © The American Physical Society)

I also made it into a negative using

$$\textit{Process} \rightarrow \textit{IntensityTransformations} \rightarrow \textit{Invert}.$$

The result is shown in Fig. 6.17. This image looks horrible, but that is not the point. The point is to find the bridge of M51B (NGC5195). It is definitely visible as far down as level with the nucleus of M51A, and it extends further until it is level with the downmost spiral arm of M51A. I don't think that's just wishful thinking on my part: I can clearly see it on my screen in a 32-bit tiff version of the image.

It was much more satisfying to be able to observe for myself that the Toomre predictions are broadly correct than to read about them.

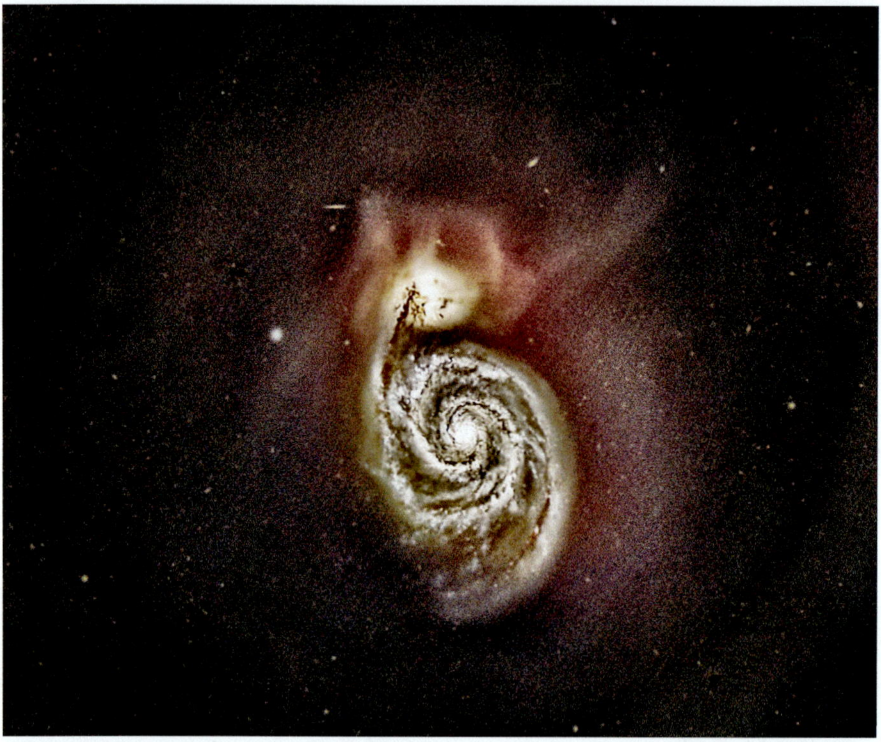

Fig. 6.17 Looking for the bridge of M51B to the left of M51A. (Negative, highly stretched image by the Author)

The Pinwheel Galaxy M101: An Exercise in Photojournalism

M101 and I were each minding our own business when word came through on Facebook that a supernova had been spotted in M101. At the time, late May 2023, our horrible climate gave us six weeks of clear skies. I could not believe my luck. So as soon as I could, I told the galaxy I was imaging that, "I don't still need you, I won't still feed you, when you're M64," and moved the scope over to M101, which was then nearly overhead from here. Onto Facebook I posted a "rush" shot, taking 6×60s frames from the live stack in SharpCap 4.0, just to get something out quickly, but then I made a decision to maintain social media silence for the rest of the night.

Back then, my default modus operandi was still to register & stack in Siril. So that's what I did. Version 1.0.6 of Siril put out a file called result.fit, which I moved into PixInsight. There I ran the photometric color calibration, because I had also not yet got spectrophotometric color calibration to work. I ran BlurXTerminator and NoiseXTerminator, and used EZ Soft Stretch to begin the stretching. After that I could run StarXterminator, which I did The result is shown in Fig. 6.18.

I ran

Process → BackgroundModelization → AutomaticBackgroundExtractor,

which actually did a good job of identifying and removing the green color with purple edges.

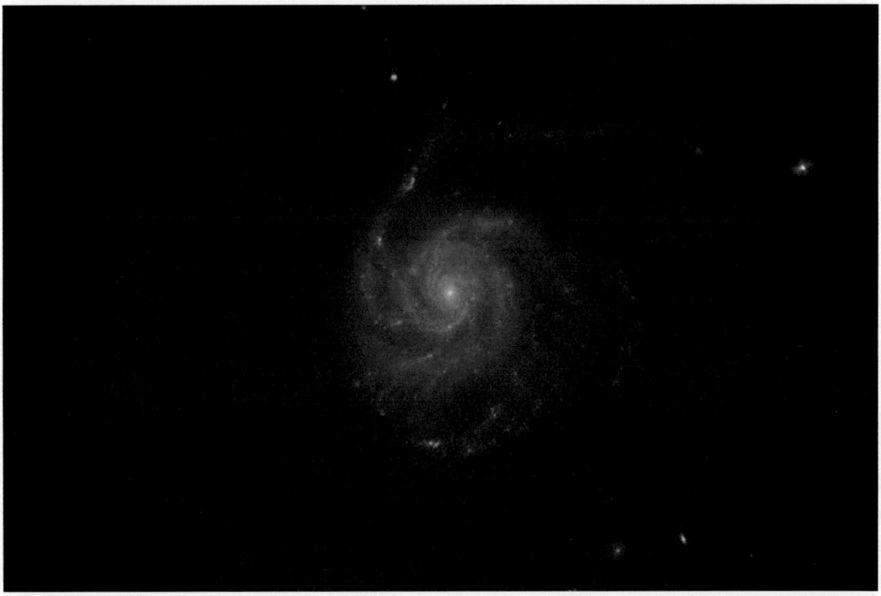

Fig. 6.18 M101 with the stars removed. The galaxy looks nice, but the background is intrusive. (Image by the Author)

I then ran

$$Process \rightarrow IntensityTransformations \rightarrow HistogramTransformation$$

To enhance the image and suppress what little background was left.

The only other thing I did was to try to remove what looked like a green color cast using

$$Process \rightarrow ColorCalibration \rightarrow AssistedColorCalibration.$$

I ran this tool several times with the very gentle green suppression shown in Fig. 6.19, dropping the triangle at the bottom onto the target image every time. Although this tool has other capabilities, I did not use them because the tool has no live preview.

I didn't do anything with the foreground stars, for a reason. StarXTerminator quite rightly removed the supernova with them, and I did not want to change that. So I tolerated unsuppressed stars. I moved the image into Gimp to draw the two white lines indicating the supernova. I also rotated & cropped the image in Gimp. That's actually not a great idea: Gimp is very slow at rotating images.

The processing was completed by a few minutes past six the following morning, resuting in the image shown in Fig. 6.20. I thought I would be cheeky and send the image to BBC Sky at Night magazine.

For that purpose, it was important to have an image that is timely and looked good rather than one that brought out every last detail in the galaxy. That's the art of photojournalism.

The BBC took my bait, and loaded the image onto their website, and eventually printed it in their July 2023 edition. They also sent an interviewer for their radio show Science in Action to talk to me as I captured the galaxy again. I felt that I had to warn the neighbors that my back yard might be a bit noisy that night. The interviewer was supposed to talk to me for half an hour, but he stayed for two hours.

The proof of the pudding was in the eating. Someone else thought my image looked good.

Fortunately, I had obtained a "before" picture of M101 in the small hours the previous February 15th. You can clearly see that it's not there in Fig. 6.21. That helped my case with the journalism, because I had a very good story that the new star really was new.

I photographed the galaxy & supernova several times over the next week. It brightened noticeably and then appeared to have roughly constant brightness.

Is the supernova really brighter than the whole galaxy as is frequently claimed? Figures 6.22 and 6.23 show how I answered this question. During a live stack, I turned the brightness down until only the nucleus of M101 and the supernova were visible. Therefore, the only thing bin M101 brighter than the supernova is the very nucleus of the galaxy. The supernova outshines everything else.

Eventually, I asked myself what would happen if I were to stack all my raw frames together. Would I see more of the galaxy? The answer was emphatically

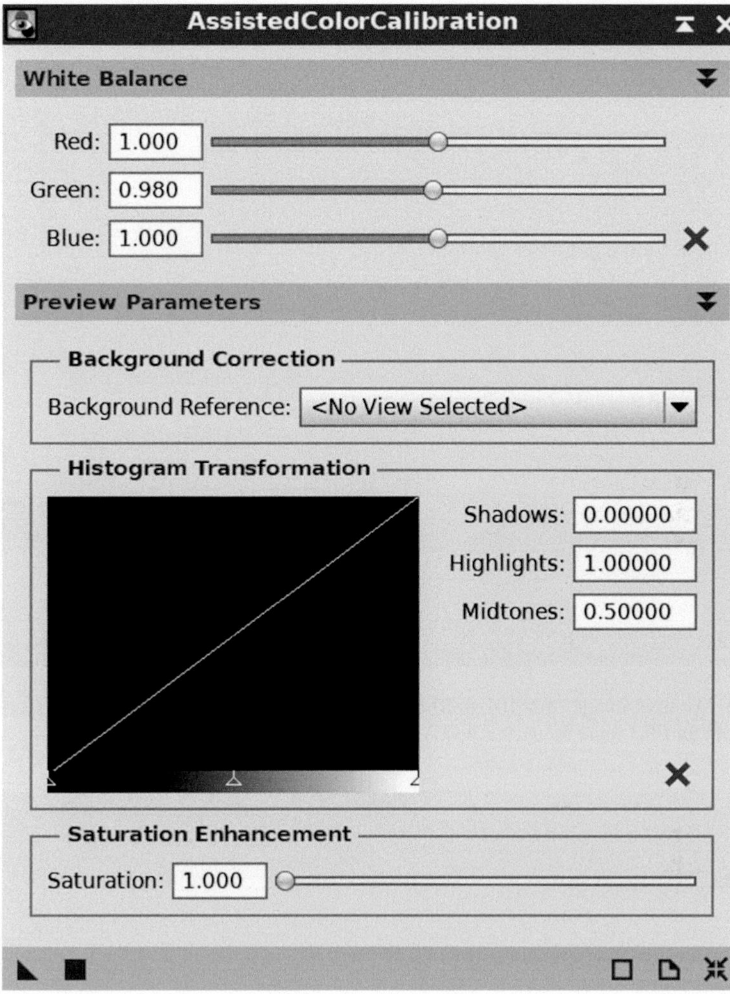

Fig. 6.19 The settings I used in the Assisted Color Calibration tool to remove a green color cast. (Image by the Author)

"yes" (Fig. 6.24). With 594 minutes of frames (9 h 54 m) some fainter spiral arms are visible, and more of the galaxy is filled in where there appeared to be gaps in my other images of M101.

From this I drew the conclusion that there is no "correct" exposure time for a galaxy. The more frames you collect, the more you will see. The background light in Fig. 6.24 is visible. The signal-to-noise ratio would slowly increase the more I collected data, but this is a slow process. To double your signal requires four times as many 60 s frames. Another factor will eventually limit you: there is always some nebulosity in the Milky Way galaxy. Eventually you will detect this.

There are two ways you can decide when you have enough data. One is artistic: what makes a satisfactory picture? The other is scientific: what can I see, and what

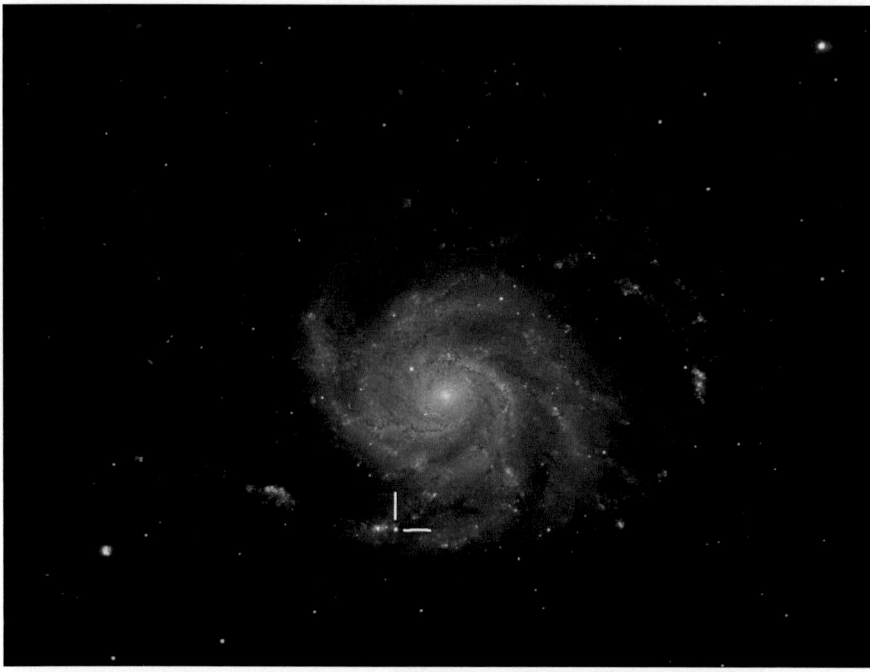

Fig. 6.20 My final image of M101 & SN2023ixf. 69 × 60s. 11-inch Celestron SCT on a CGX mount, with an f/6.3 focal reducer. ZWO ASI2600MC camera at −15 °C. (Image by the Author)

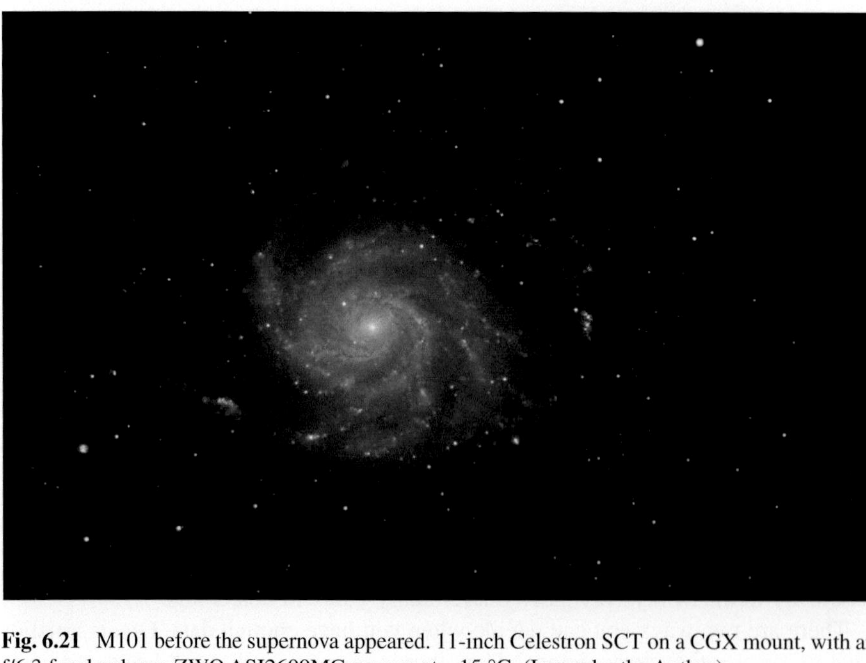

Fig. 6.21 M101 before the supernova appeared. 11-inch Celestron SCT on a CGX mount, with an f/6.3 focal reducer. ZWO ASI2600MC camera at −15 °C. (Image by the Author)

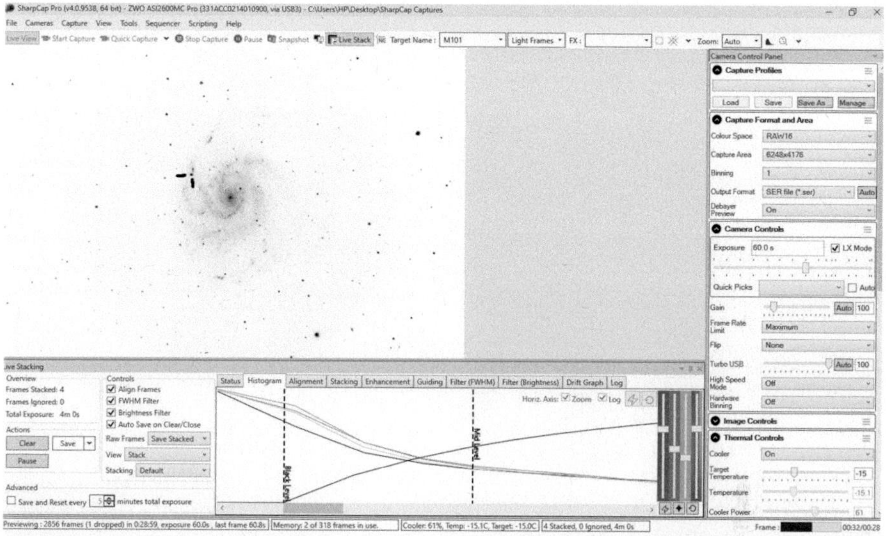

Fig. 6.22 The full galaxy being stacked live in SharpCap. The Supernova is marked by black lines. (Image by the Author)

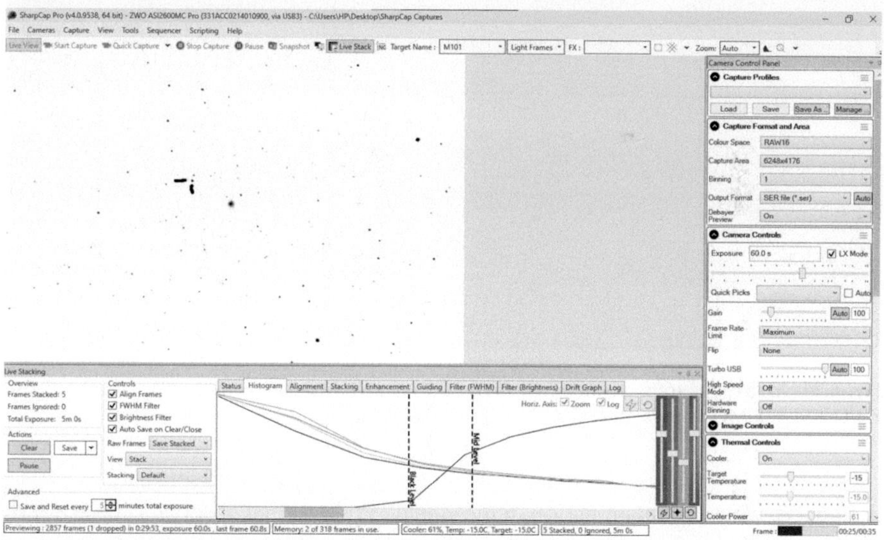

Fig. 6.23 Now the brightness has been turned down. Only the supernova and the galaxy nucleus are now visible. (Image by the Author)

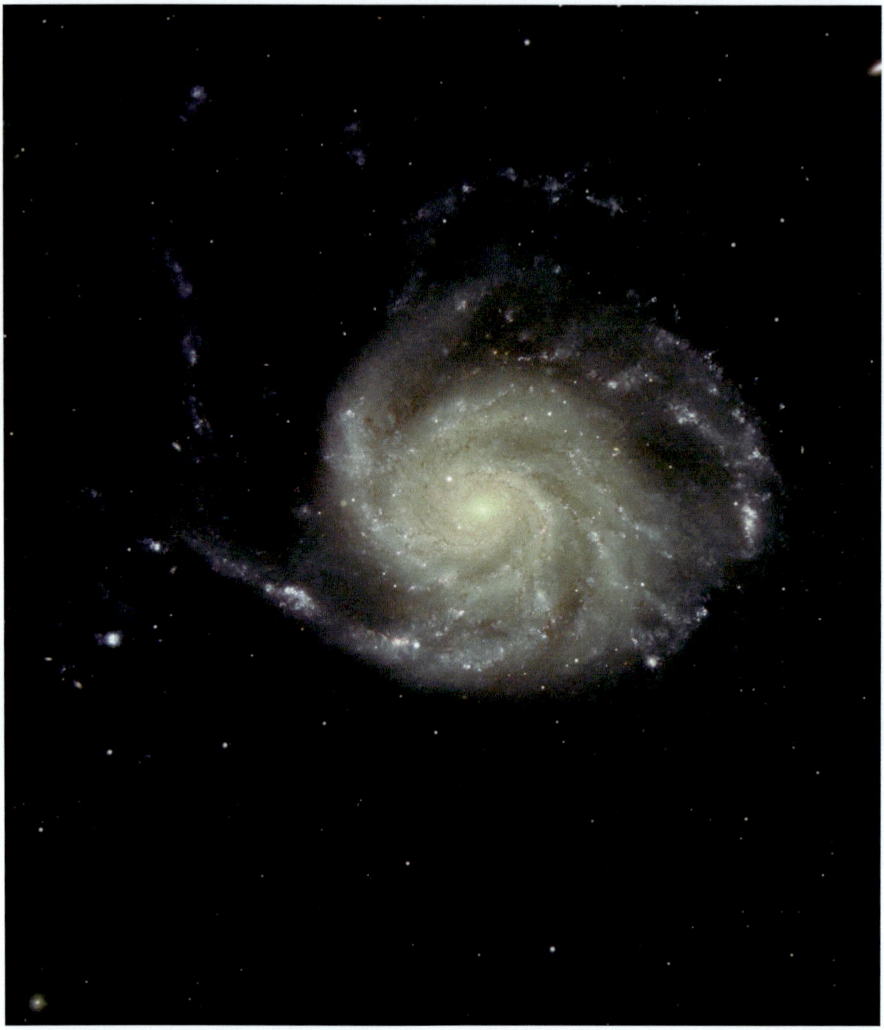

Fig. 6.24 M101, 594×60s. 11-inch Celestron SCT on a CGX mount, with an f/6.3 focal reducer. ZWO ASI2600MC camera at −15 °C. (Image by the Author)

is signal & what is contamination of the signal? Professional observational astronomers spend their whole lives worrying about, and working to uncover, contamination of their data. There are no easy answers to either the scientific or the artistic question.

To conclude this section, I want to discuss the question of whether my fellow astronomers & I witnessed an event that happened 21 million years ago, M101 being 21 million light years (Mly) away.

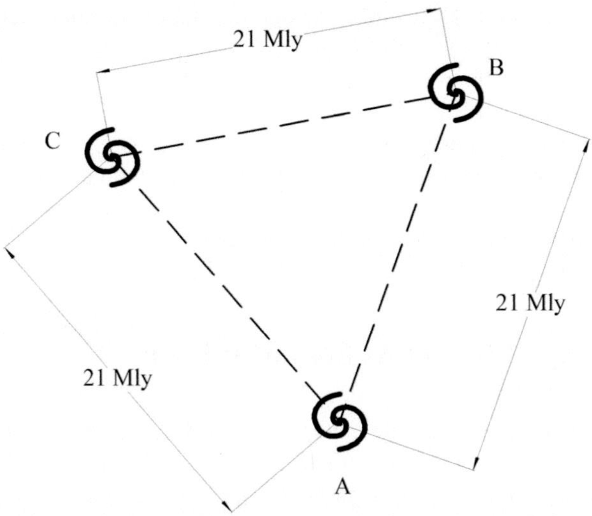

Fig. 6.25 Three galaxies each 21 million light years (Mly) apart from one another. (Image by the Author)

In Fig. 6.25, let's think about three galaxies, each 21 Mly from the other two. One night, an observer in galaxy C spots two supernovae, one in each of galaxies A & B. Observer C perceives these supernovae SN-A and SN-B as simultaneous.

In Galaxy B, Observer B spots SN-A, and knows from geological records that there was a bright supernova in his or her own galaxy 21 million years before.

In Galaxy B, Observer A spots SN-B, and knows from geological records that there was a bright supernova in her or his own galaxy 21 million years before.

In other words, what Observer C records as simultaneous, Observer B records as SN-B first, then SN-A. Observer A, on the other hand, records these supernovae as SN-A first, then SN-B. There is no such thing as absolute simultaneity.

The speed of light is constant at approximately 300,000 km/s, or 186,000 miles/s, for all observers, no matter what their speed. This was discovered experimentally by Michaelson & Morley in 1887, and theoretically understood by Einstein in 1905. You can never catch up or overtake light. No matter how fast you go, the light will always recede from you, and move towards you at 300,000 km/s. Therefore, you can't be some super-observer that sees when the supernovae "really" exploded. That's impossible. The relative times of these events is not absolute: it depends on where you observe them from.

Einstein also showed, remarkably, that for the light itself, time does not elapse, because otherwise there would be a paradox. The light would recede from itself at 300,000 km/s. It can no more do that than you & I can be in two places at once.

The only logical conclusion from this is that we see supernovae exploding live. The light that reaches us is no "older" than when the supernova emitted it.

Wow! What the heck?

I entirely agree that that's bizarre, and that it violates common sense. That's one of the things that makes science such fun.

Science is highly counterintuitive. Who ever said that the idea of double helices of DNA transmitting genetic information from parent to child was common sense? Who ever said that Newton's law of universal gravitation is common sense? Both these discoveries were great surprises when they were first made. Common sense is a guide, of sorts, to the everyday world that we can see, touch, taste, smell & hear. But it fails us once we move beyond the everyday.

Bode's Galaxy M81: Not As Easy as It Looks

In some ways, M81 is very easy. It's at a very high declination, close to the pole star so it doesn't move around the sky much. From the UK, and indeed from most of Europe and the North American continent, it is circumpolar, and high enough to observe for much of the year – at least as long as you don't live in the southern USA, Hawaii or Mexico. And, since it is a big, nearby galaxy, it is bright.

The difficulty lies in the colors, as I will show.

One clear but moonlit night, I managed to get 440 minutes' worth of 60 s frames. That's 7 h 20 m. It was late November, so the night was long. Equipment was the Celestron C11 with f/6.3 reducer/corrector, on a Celestron CGX mount, with a ZWO ASI2600MC camera. Images were captured with SharpCap 4.1.

After my usual preprocessing routine, followed by gentle stretching with EZ_stretch and star removal, Fig. 6.26 shows my starless image.

The galaxy is there all right, but so is a lot of other stuff. There's a dreadful brown cast to the image, even after spectrophotometric color calibration, and a hint of some blue reflected noise to the left of the image.

I tried range masks like Fig. 6.27, obtained from

Process → MaskGeneration → RangeSelection.

This mask protects Holmberg IX, an irregular galaxy to the right of M81, so it seems to offer some advantages in getting rid of noise.

Unfortunately, it leaves M81 looking implausibly oval. I show this in monochrome in Fig. 6.28. There was not much color in this image, so you are not missing much.

Therefore, I went back to the drawing board and just tried stretching the background away. One thing that happens is that the nucleus then gets overstretched. So you make a mask like the one shown in Fig. 6.29 using the GAME script shown in Fig. 6.6. It only covers the nucleus. Then I use the tool

Process → IntensityTransformations → HDRMultiscaleTransform.

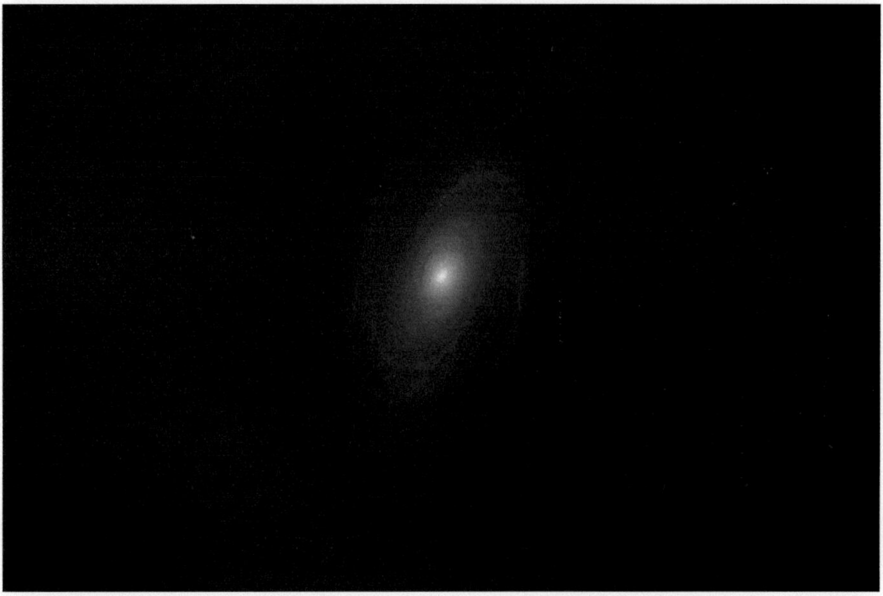

Fig. 6.26 My lightly-stretched starless image of M81. (Image by the Author)

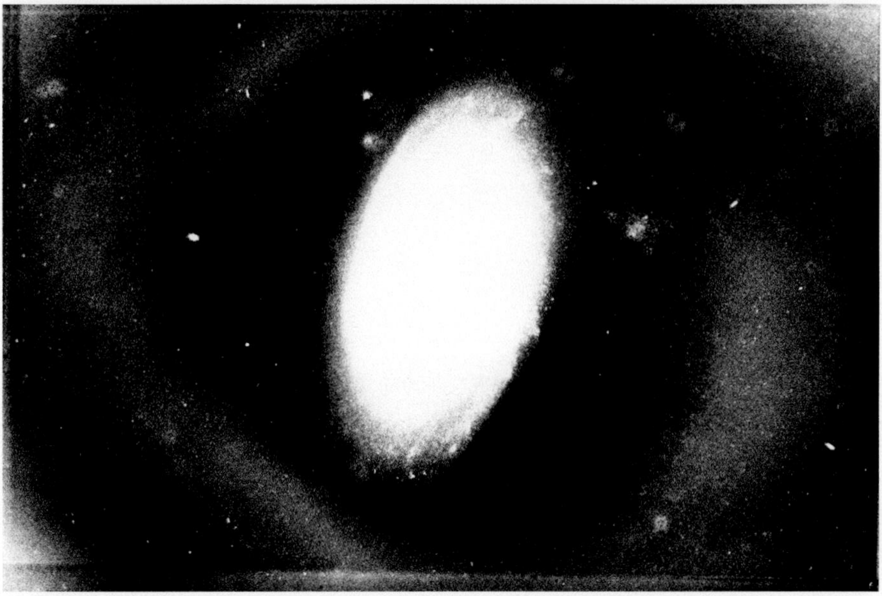

Fig. 6.27 A range mask used to operate on the image in Fig. 6.26. (Image by the Author)

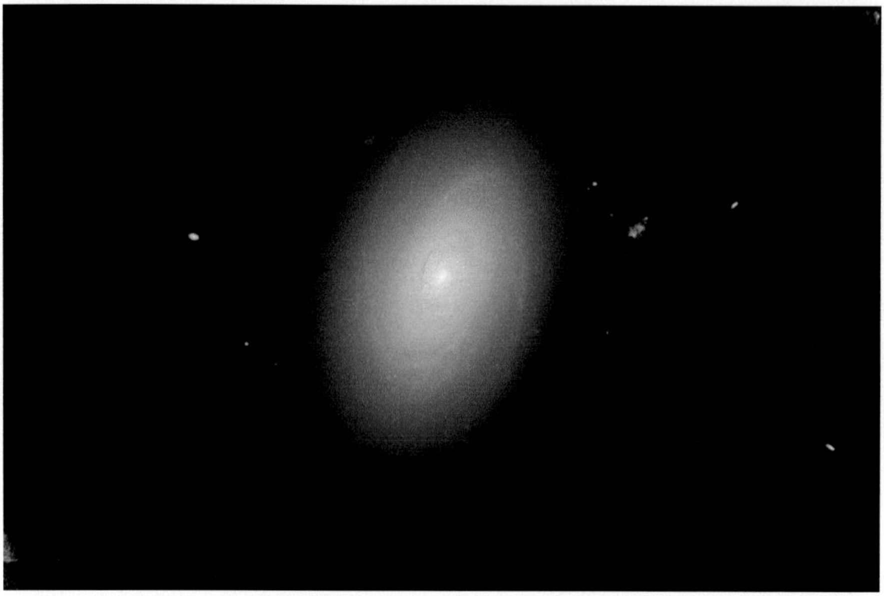

Fig. 6.28 M81 after applying the mask shown in Fig. 6.27 and stretching. Image by the Author

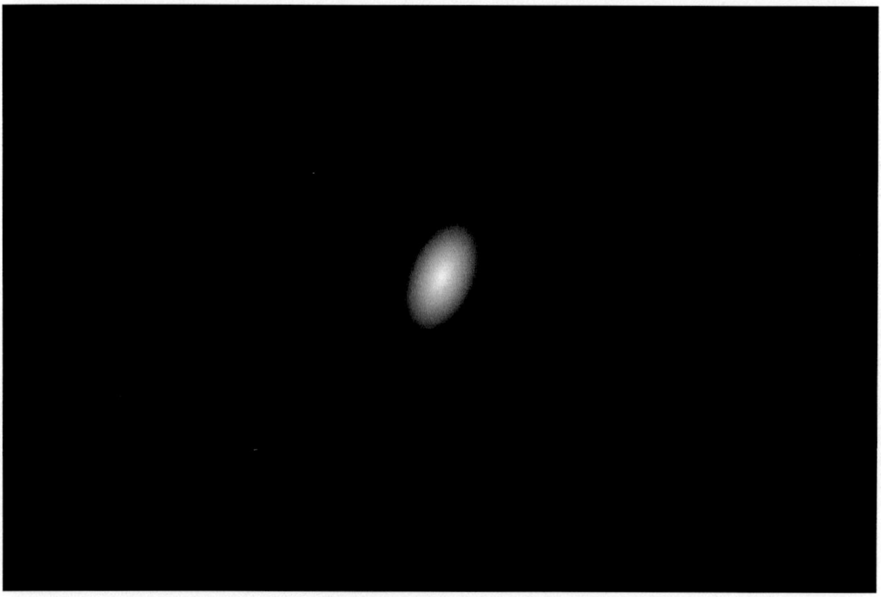

Fig. 6.29 A mask made with the GAME script. (Image by the Author)

I normally use default values unless they obviously overdo the suppression of nucleus brightness.

The resultant image looked OK but the colors were anemic. So I used

Process → IntensityTransformations → ColorSaturation.

As always, I applied this gently in several steps.

Then a little stretching with

Process → IntensityTransformations → HistogramTransformation

And I had an image to which I was happy to add stars, using the PixelMath tool shown in Fig. 6.12. Figure 6.30 shows what I now had.

This just needed to be rotated to north-up and cropped for showing to the world.

But it bugged me. I kept thinking that it's too brown. I stewed on it for a few weeks while I went about my life, and came back to it a little over a month later. I tried two things.

First, I hypothesized that the brown cast was caused by light pollution from somewhere. It might have been moonlight that scattered off particles in the air, or it might have been something else. I don't know. It's too late to find out. And does it really matter what the cause was? The issue was how to deal with it. One useful little trick in PixInsight is that you can run your mouse over the image highlighted with blue along the top bar, which PixInsight calls the target image, and the RGB

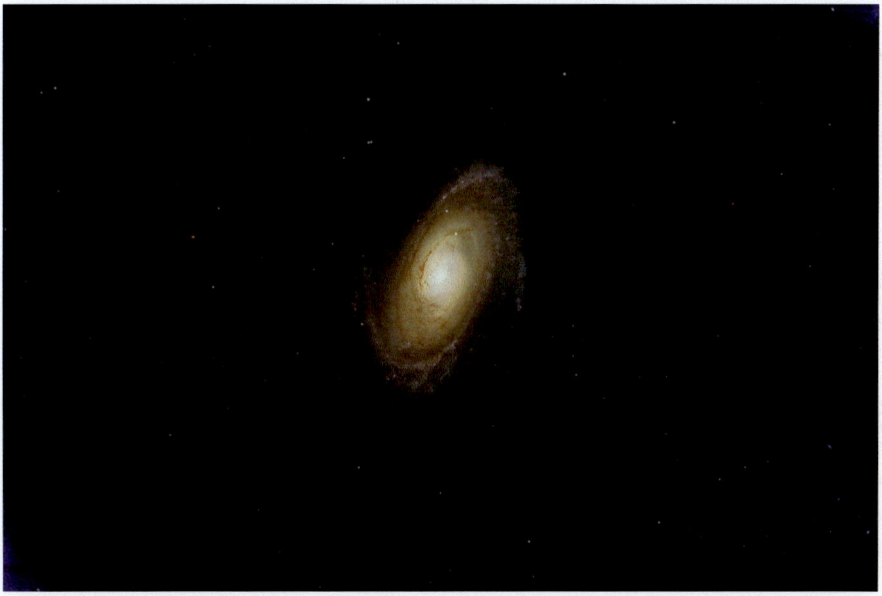

Fig. 6.30 M81 after following the steps detailed in the text above. (Image by the Author)

brightness values at the pixel to which you are pointing are given at the bottom of the main PixInsight window.

I zoomed in on the brown background in the first starless image I had, shown in Fig. 6.26. The brown background was of course noisy, but by eyeball I guessed that a typical RGB value was

Red: 0.1500,
Green: 0.0800,
Blue: 0.0600.

I made an image of exactly that color combination in every pixel. I did this with the Pixel Math Tool, which we have already seen (Fig. 6.12). It is found at

Process → PixelMath → PixelMath.

You make some changes to the menu shown in Fig. 6.12.

1. Uncheck the box "Use a single RGB/K expression"
2. Change the color space from "< as target>" to "RGB color".
3. Manually type in the image width & height as your other images, in my case 6248 × 4176 pixels
4. In the RGB/K, G and B tabs at the top, type in the red, green & blue values above.

Press the square box at the bottom left and your image will be made. I renamed mine to Manual_Background.

Now go back into Pixel Math, and set it back to the values in Fig. 6.12, except that the sum you now make it do on the starless image with which you are happiest is

$$\$T - Manual\_Background,$$

where $T is your target image. Pixel Math understands $T to be the target image on which you have clicked.

Incidentally, Pixel Math is a hugely powerful tool with all sorts of capabilities. By exploring it, and trying a few things, you can learn a lot about its capabilities.

This, of course, was a completely hit-and-miss process. I had no idea whether it would work, but it did work quite well. It had a green cast, which I had to remove manually. This I did using

Process → ColorCalibration → AssistedColorCalibration.

Now I added the stars back in and rotated the image to make north face up. The resultant picture was a little bit too noisy for my taste, so I moved a tiff version of the image into an app I have called Topaz Denoise AI. I probably could have figured out how to do this in PixInsight, but the Topaz app is really easy to use, so that is what I did.

The result is shown in Fig. 6.31.

I also tried another trick. By trawling around looking for inspiration, I came across a script called AutoColor. This script is a no-cost add-on, one of many from

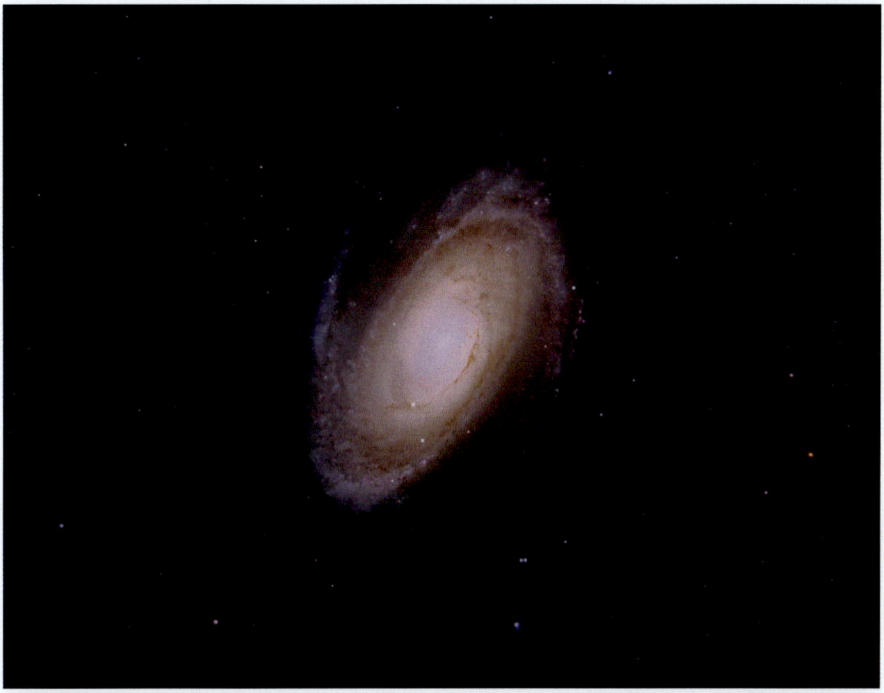

Fig. 6.31 My M81 image after subtracting off the brown background. (Image by the Author)

Fig. 6.32 The masks I used. I made them in the GAME script, then ran Range Mask to make the edges fuzzy. (Image by the Author)

Hartmut V. Bornemann. You can find a list at https://www.skypixels.at/pixinsight_scripts.html. The website appears to be in German, but much of it is also given in English. Anyway, AutoColor combines "BackgroundNeutralization and ColorCalibration" according to the comments at the top of the script. (You can find the source code of most PixInsight scripts un the Program Files section of a Windows PC, and no doubt in similar directories on MacOS and Linux computers.) To find this script, go to

Scripts → *Utilities* → *AutoColor*.

I ran it on my masked image, Fig. 6.28, and continued to make and use masks to suppress background noise. Figure 6.32 shows the masks I made.

Eventually I obtained an image I was reasonably happy with, except that it was much too blue. So I ran AutoColor a second time, but learned that you get silly results if you try to run it on a cropped image. I therefore had to do the re-run on an uncropped image, and re-crop it. The final result is shown in Fig. 6.33. I was now convinced that masks could work for this image of M81, about which I had earlier had doubts.

I am also in no doubt that the Spectrophotometric Color Calibration bombed out on this job. Sometimes it works very well, but my faith in it is shaken. There's no harm in using your brain as well as these tools in image processing software. We should all critically evaluate what we're doing with software.

Which do I like best, Fig. 6.31 or Fig. 6.33? The former is more brilliant. The "brilliance" is clearly an aesthetic issue. At first, it seduced me. But, on a second

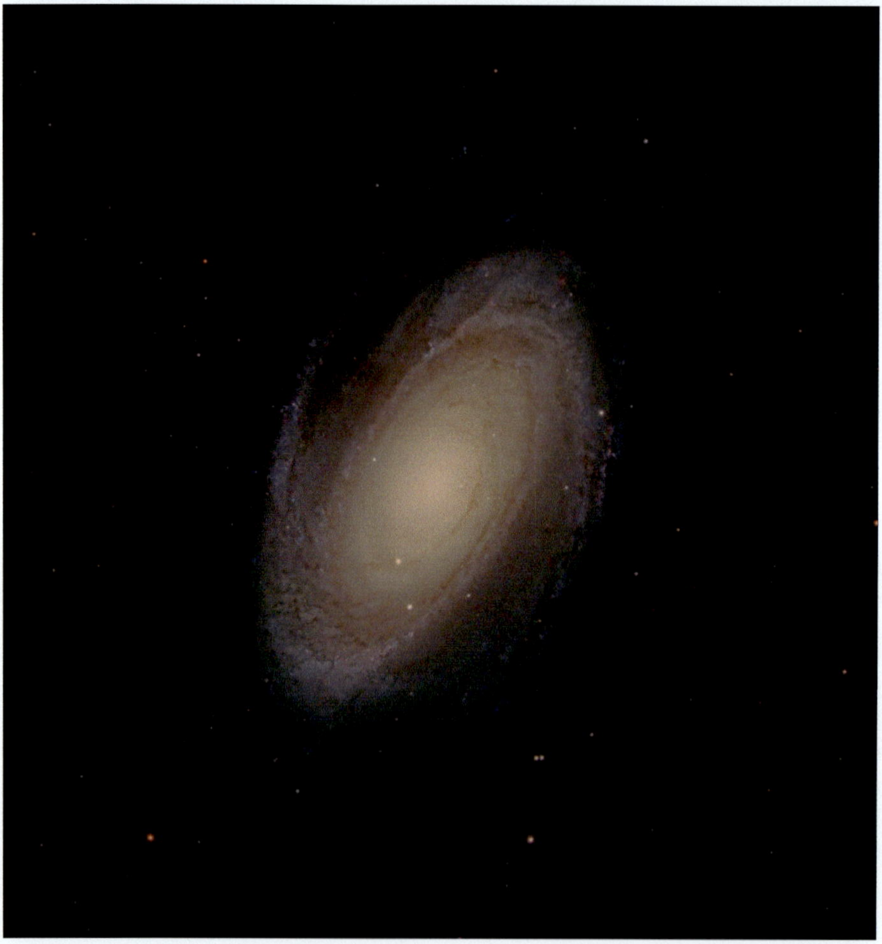

Fig. 6.33 Using masks and the AutoColor script to make an image of M81. (Image by the Author)

look, more detail is visible in around the edges in Fig. 6.33, but more is visible in the center in Fig. 6.31. I can also see the small galaxy Holmberg IX in Fig. 6.33. It is a faint galaxy to the center left. It was discovered in 1959, but in 2024 we can see it with amateur equipment. That's progress for you.

So my preference eventually shifted to Fig. 6.33. You, dear reader, are fully entitled to disagree. There's no arguing about taste. Sometimes I like one better, sometimes the other. I can also learn more about what's in the galaxy by displaying them simultaneously.

For example, it looks from Fig. 6.33 as if the spiral arms may extend further than Fig. 6.31 would have us believe.

Which, if either, has the "right" colors? I'm afraid I don't know. Does it even matter? Obviously, it would matter if the galaxy had a green nucleus and maroon stars. I don't mean that. I mean does it matter if the colors aren't "exact" in a way that assumes that "right" colors even exist?

I don't claim to be doing science here. The image in Fig. 6.33 is a work of art. There are ways of doing science with color. There are agreed standards for color filters for photometry, i.e. measurement from photographs. But that's not what this image is about.

Adding Hydrogen Alpha Colors

A spectral color at 652.3 nm, the hydrogen alpha, or H-α, often written Hα or Ha, is emitted by atomic hydrogen if it has been ionized and is recapturing an electron. Ionized means that an electron is knocked free of the proton in the hydrogen atoms' nucleus that normally binds it. The recaptured electron loses its escape energy in stages, as electrons always do. One of those stages is, the jump from energy level 3 to energy level 2, emits a photon to lose the energy. This photon always has wavelength 652.3 nm.

It is often said that Hydrogen alpha emission is a good proxy for areas of rapid star formation. It's a fairly good proxy, but not a desperately accurate one.

The idea is that hot young stars ionize hydrogen, i.e. strip some or all of the electrons from it. There is some truth to this, but a more realistic picture is that only molecular gas clouds can condense under their own gravity to make stars. Molecular hydrogen, the H_2 molecule most of us learned about in high school, unfortunately dos not emit photons unless it is very hot, so a molecular gas than can actually be detected is what people use. The easiest one to detect in galaxies is the carbon monoxide molecule CO, which can be detected with radio telescopes [49]. Figure 6.34 shows how H-α-rich regions and CO-rich regions only somewhat coincide.

Therefore, the claim that H-α shows regions of star formation isn't complete horse feathers, but it should be taken with a pinch of salt.

That said, the only option for amateur astronomers is to photograph the H-α light. Alas most of us can't afford very big radio telescopes.

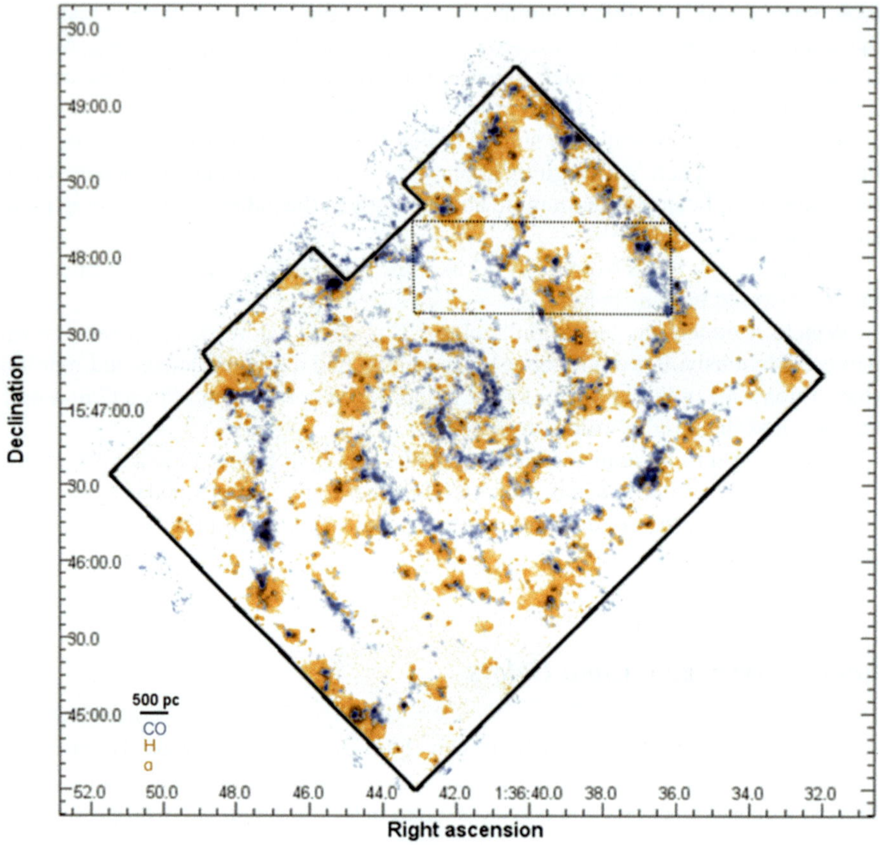

Fig. 6.34 Showing how Ha (yellow) and CO (blue) regions don't quite coincide in the center of M74 (NGC628). (Source: Kreckel et al., Reference [49]. Used with permission. © American Physical Society)

Commercial H-α filters come in three flavors: 652.3 ± 7 nm, 652.3 ± 5 nm and 652.3 ± 3 nm. The narrower the band transmitted, the costlier the filter is. None of these filter types reduce the amount of light transmitted enough to look at the Sun. They all leak at least some background light.

I have tried a 7 nm filter and a 3 nm filter. The comparison is shown in Fig. 6.35. There's quite a bit more light-leakage with the 7 nm filter. But something else interesting also happens.

Around the H-α nebulae in the spiral arms, the 7 nm filter picks up more apparently H-α material than the 3 nm filter. Is this extra material actually from star-forming clouds? I don't have a good answer to that question. This is another instance where you should think about what you have actually photographed. Are the quasi-continuous chains of H-α you see in some galaxy photographs on Pinterest, Astrobin,

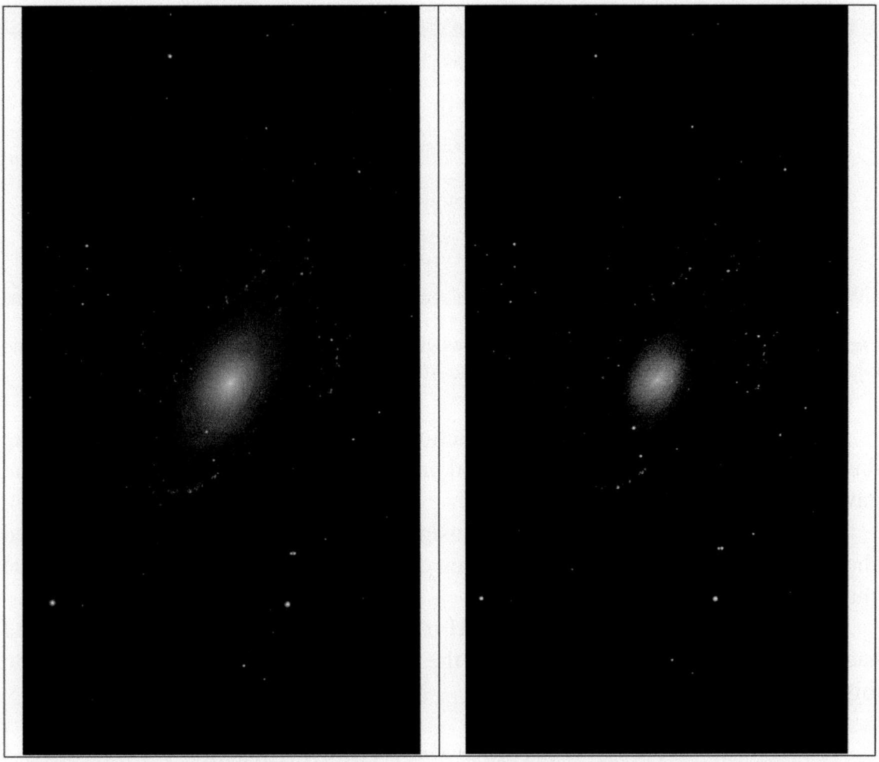

Fig. 6.35 Comparison of Ha for M81 photographed through a 7 nm filter (left) and a 3 nm filter (right). Taken with the red channel of images from a ZWO ASI2600MC color camera, with the respective filters in place. (Images by the Author)

Facebook and similar web sites real? Are we seeing more of the gas clouds in the left-hand image of Fig. 6.35, or are we just seeing the spiral arms?

I will now show you how I added the H-α from the 3 nm filter to the image in Fig. 6.33. First, obviously you need to take the image with the filter in place. I usually place it immediately in front of my camera. Figure 2.24 shows this. You have to place modern coated filters with the coated side toward the telescope for the filters to work. Then you capture frames in the usual way.

With a one-shot color camera, you only really need to process the red channel in WBPP, but you can always split the channels afterward with

Process → ColorSpaces → ChannelExtraction.

This tool is shown in Fig. 6.36. You need to select the RGB option and drag the right-angle triangle at the bottom left onto the target image. It then produces three monochrome images with names suffixed with _R, _G and _B respectively added to the original mane of the frame.

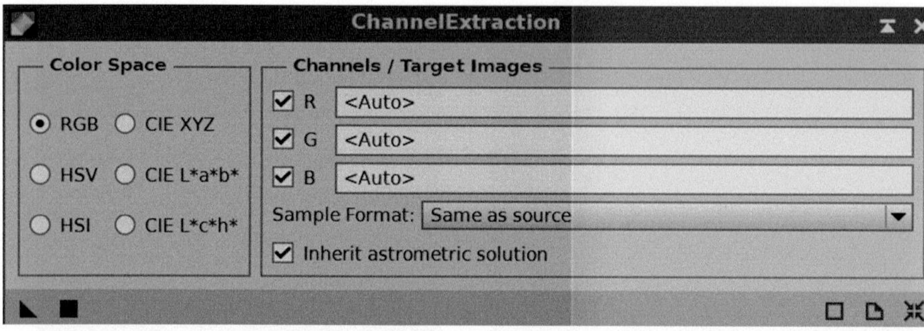

Fig. 6.36 The Channel Extraction tool in PixInsight, showing how I selected the RGB option and told it to make all three channels. (Image by the Author)

I also chose to do this channel extraction to my target image Fig. 6.33. Before this, I duplicated the image (always good practice to avoid irreversible damage) and gave my target image the name "main".

As an alternative, Pixel Math in PixInsight understands that "main[0]" is the red channel of "main", "main[1]" is the green channel of "main", "main[2]" is the blue channel of "main".

Anyway, I also made a copy of my H-α image and called it H. But not before I had made sure it is aligned with, and the same size as, the main image. You do this with

$$\textit{Process} \rightarrow \textit{preprocessing} \rightarrow \textit{StarAlignment.}$$

You fill out the form in this tool, using the View options, selecting your reference image, and selecting your target image(s) to align. I did find out the hard way that (a) you have to check as well as click on target images, and (b) that if you don't click on the target image in the PixInsight window, it aligns the wrong image. There may be a bug there that will have been fixed in a future release of PixInsight by the time you read this. I'm using Release 1.8.9-2, build 1601 at the time of writing.

Now you have to use PixelMath. Uncheck the "Use a single RGB/K expression" box and set the color space to "RGB Color". This is the clever bit. You do not want all the stray light that leaked in through your H-α filter to contaminate the image. There's a trick to make PixInsight disregard it.

Red: iif(H > 1.00*main_R,1.10*H,main_R),
Green: main_G,
Blue: main_B.

Let's unpack the red channel. The iif (indicative if) command has the following form.

$$iif\left(A \textit{ is true,do this if } A \textit{ actually is true,otherwise do this}\right)$$

The iif goes through every pixel in the images, pixel by pixel. If, for a given pixel, the H-a image is brighter than 1.00 times the main image, set that pixel equal to 1.10

times its value in the H-α image. Otherwise, just use the pixel from the main image. I experimented a little with the factors 1.00 and 1.10 to get the effect I wanted. The green and blue images are left unchanged.

Some people play with the blue channel to get the H-a to look a bit more crimson. I have not done this.

The new image generated is the one shown in Fig. 6.37.

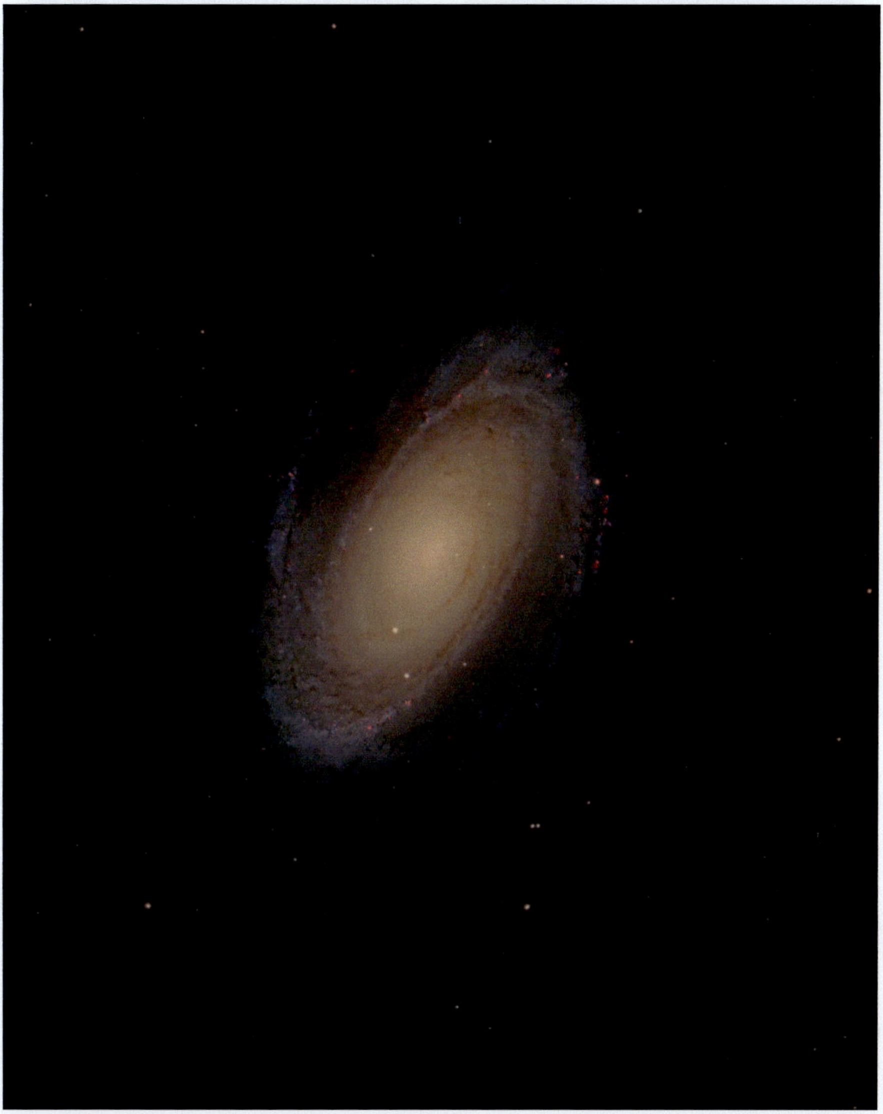

Fig. 6.37 M81 with H-α-rich regions shown. (Image by the Author)

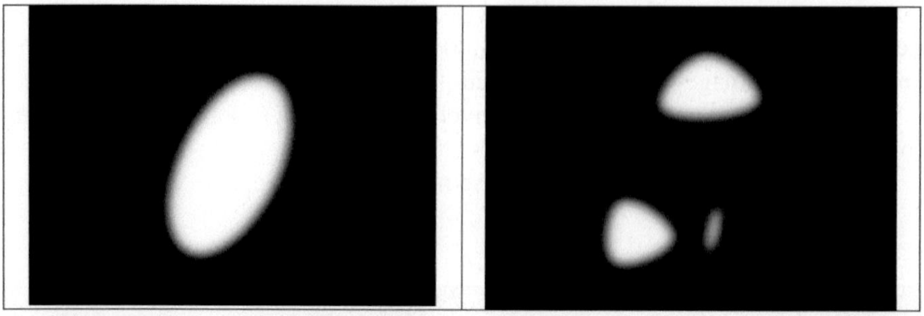

Fig. 6.38 The larger masks used to remove stars from the H-α image. (Image by the Author)

There is something I neglected to tell you. I did the H-a adding operation on images with the stars removed. I could do this automatically for the main image with StarXTerminator. Unfortunately, un the H-α image, StarXTerminator can't really tell stars from H-α regions. I removed the stars from my H image manually using masks. I did a bulk removal of all the faraway stars with one big mask, and some more with smaller masks, all generated as binary masks in GAME and made fuzzy in RangeSelection. These are shown in Fig. 6.38.

I also made a few masks to remove individual stars. These were ellipses in GAME, made very small and made using the Gradient Masks option. Of course, every time you think you have removed them all, you find others. But eventually you have gotten rid of them. Occasionally you have to make a judgement about whether an object is a sat or a H-α region. The clue is that H-α regions are not round. Figure 6.39 shows an example of such a mask.

Once I had done all this, I added the stars from the RGB image back in.

Why did I go to all this trouble? Because if you don't, you end up with pink stars.

One final tip is that the clever Pixel Math arithmetic described above can fail right at the galactic nucleus. Since we aren't going to want that part of the H-α image anyway, we might as well make a GAME gradient mask to remove it. After doing this, my H-a image is shown in Fig. 6.40.

Finally, it is not unknown for people combining separate LRGB images from a monochrome camera just to use the H-α channel instead of the red channel, and not bother to remove stuff they don't think is due to star-forming nebulae.

M81 Is a Groupie

The principal members of the M81 group are M81, M82 and NGC3077, a small galaxy that looks as if it might be elliptical, but it would be better described as irregular.

My method of producing the image in Fig. 6.41 is no different from the other images. It presented no special challenges. The only thing I changed was that I used

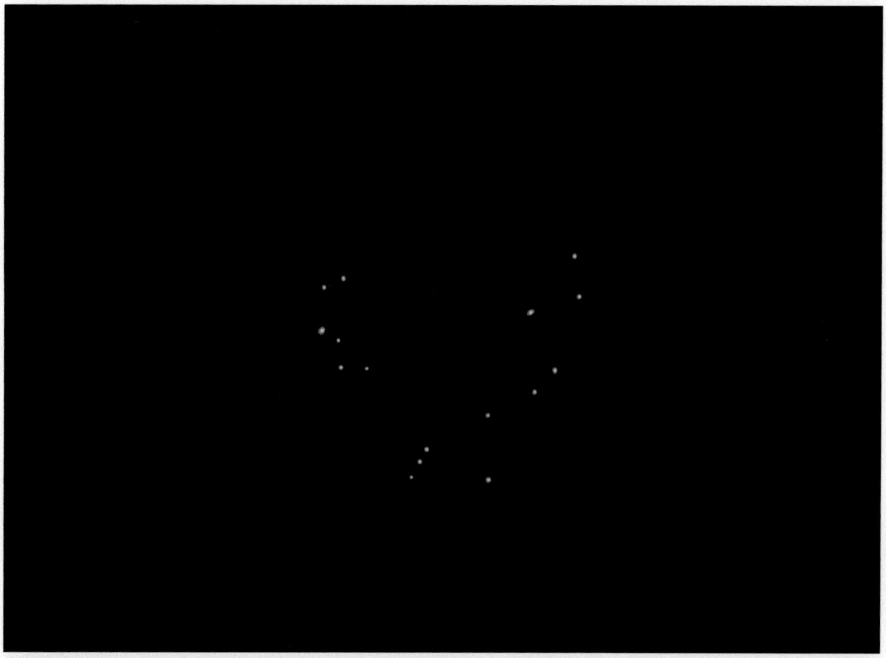

Fig. 6.39 A GAME mask to remove individual stars. (Image by the Author)

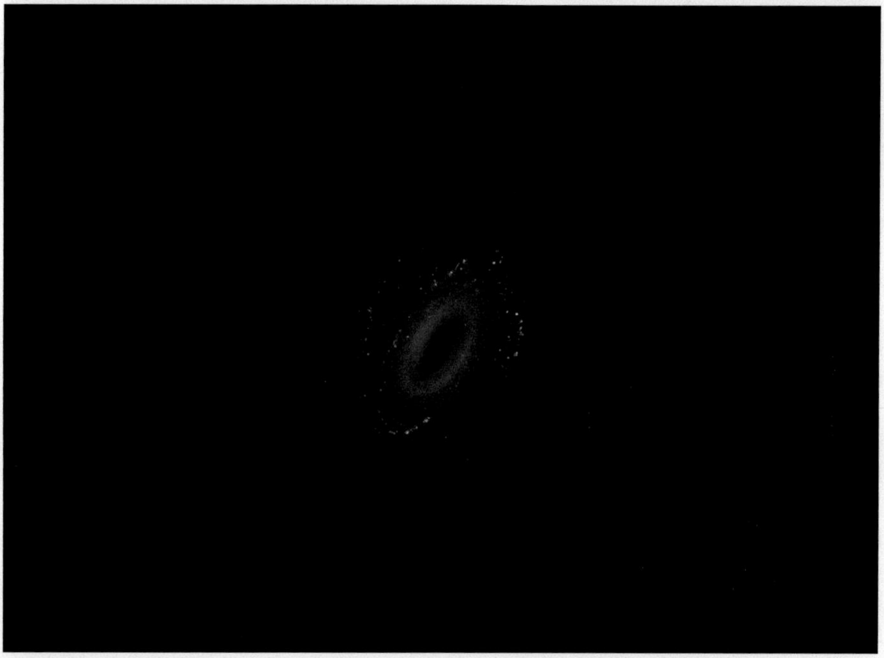

Fig. 6.40 The H-α image with stars and the nucleus removed. (Image by the Author)

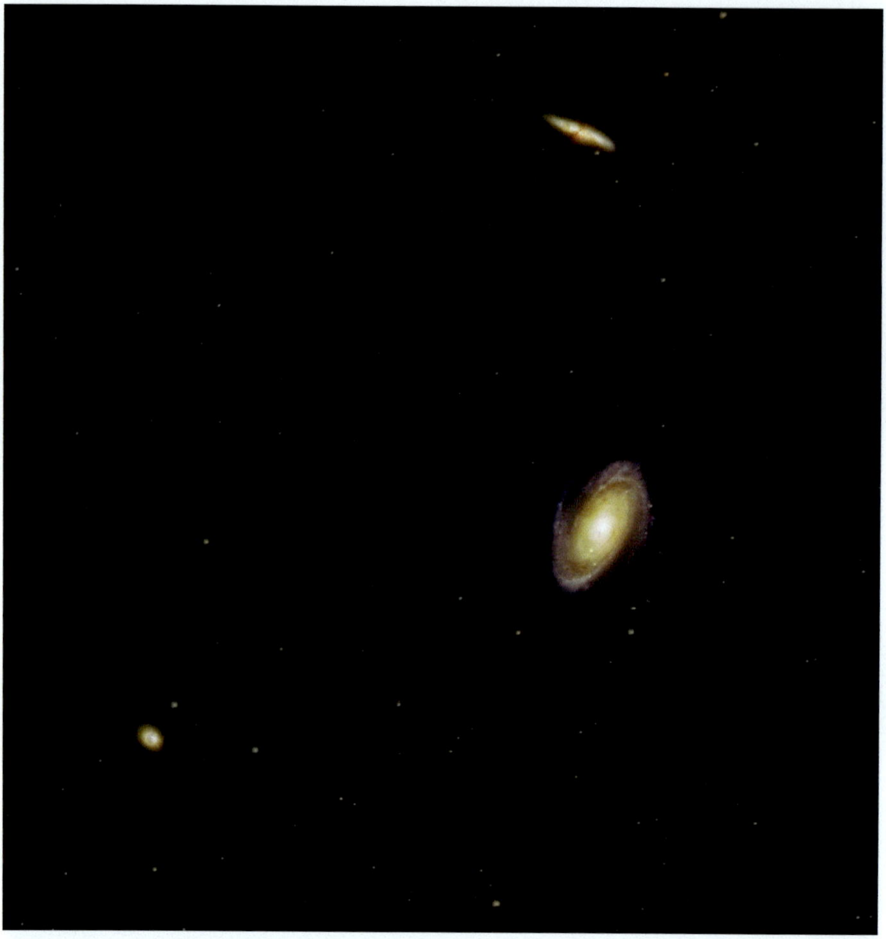

Fig. 6.41 The principal members of the M81 group: M81 at center right; M82 at the top; and NGC3077 at the bottom left. (Image by the Author, taken with a ZWO ASI2600MC camera attached to 73 mm apochromat on a Celestron CGX mount)

my William Optics 73 mm *f*/4.9 ZenithStar telescope. I piggyback it on my 11″ telescope, so it uses the same mount and guiding system.

These galaxies are close: M82 is only about three M81 diameters away from its neighbor, whereas M31 is about six Milky Way diameters from us. So are they interacting?

Smoking gun evidence comes from a radio image in a 1994 paper by Yun, Ho & Ko [50] in which they show monatomic hydrogen gas trails between these three galaxies. I've imposed their image onto my visible-light image in Fig. 6.42. It is hard to avoid the conclusion that the smaller two galaxies have both collided with M81. It is also interesting to note that the spiral arms of M81 extend way beyond those we can see in the visible-light image.

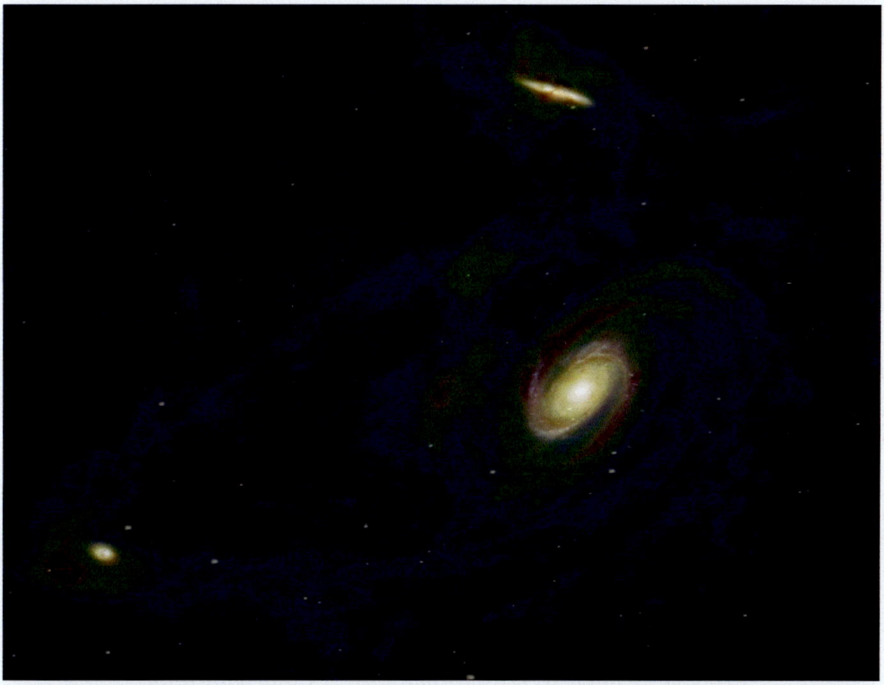

Fig. 6.42 Monatomic hydrogen image, by Min Yun, from a radio telescope, https://people.astro.umass.edu/~myun/m81hi.html, used with permission of the Author, superimposed on the Author's image in Fig. 6.41

One probable consequence of these collisions is that in the large nucleus of M81, star forming has stopped. There are no blue stars there. That makes M81 another example of a Green Valley galaxy of the type described in Fig. 5.1.

M82, the Poster Child Starburst Galaxy

This galaxy, on the other hand, came out of its interaction with M81 all guns blazing.

We see it almost, but not quite, edge-on. It is sometimes called the Cigar Galaxy. By the time I took the photographs shown below I had traded my Celestron C11 up for the 11-inch Edge HD version. These images are taken with that instrument. I changed nothing else, except that I replaced the *f*/6.3 focal reducer/corrector for the bespoke focal reducer for the Edge HD 1100, which makes the telescope an *f*/7 instrument.

The weather in April 2024, when I worked on this project, was not very cooperative. I had to grab my images when I could, in among cloudy weather.

I took images in white light (i.e. RGB), in H-α and in infrared.

After running through WBPP and dealing with color, blur & noise, I tried three methods to stretch my image. The tools I used were

Script → EZ Processing Suite → EZ Soft Stretch

Process → IntensityTransformations → ArcsinhStretch

Process → IntensityTransformations → GeneralisedHyperbolicStretch.

The first of these is a free add-on. If you want it, you should read the PixInsight users forum threads about this script. It is not quite as straightforward as adding most other add-ons. It's also quite an old script so it may be dropped from future releases. In this instance, the most promising approach proved to be the ArcsinhStretch. I continued with this, and parked the others.

I removed the stars and continued with several gentle applications of both

Process → IntensityTransformations → HistogramTransformation

and

Process → IntensityTransformations → CurvesTransformation,

Mostly the latter. At an intermediate point along the way, I used

Process → IntensityTransformations → ColorSaturation.

I also made a GAME Gradient Mask, shown in Fig. 6.43.

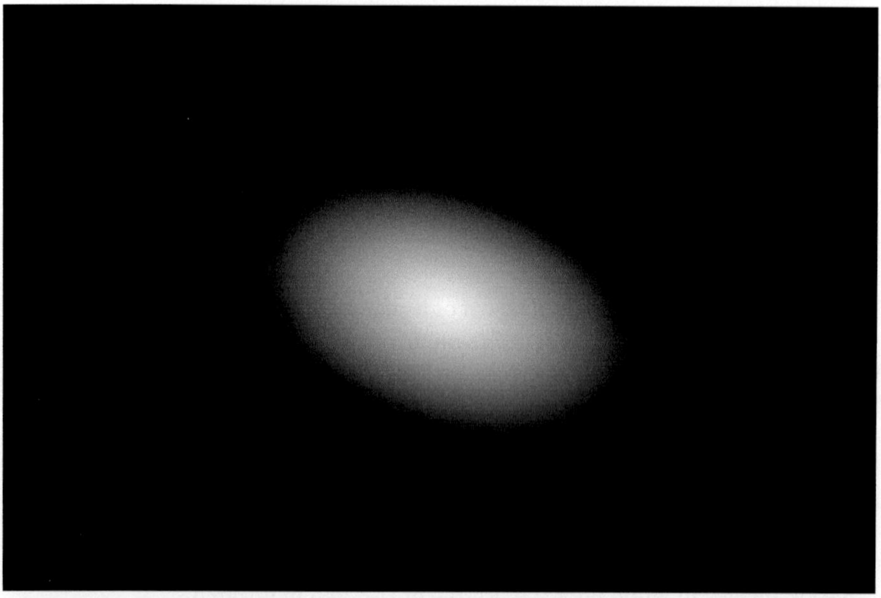

Fig. 6.43 The GAME Gradient Mask used to help suppress the background. (Image by the Author)

I toned the stars down with the Curves Transformation tool just described. Then I used Pixel Math to add them back in. My RGB image is shown in Fig. 6.44.

The total exposure time was four hours over several nights. In Fig. 6.44 I can see some H-α in red, some dust in orange, and hints, but no more than hints, of a galactic plane in blue towards the edges.

My H-α image was taken with a 3 nm filter, using my ZWO ASI2600MC color camera set to a gain of 350, and run at −15 °C. Of course, I ignored the green and blue channels in the image. I captured 420×60s frames, and was able to register and stack 386 of them, giving me 6 h 26 m of data.

Again, I tried EZ Soft, Generalized Hyperbolic and Arcsinh stretching. The last method was chosen to continue with. I was able to remove the stars with StarXTerminator, which saved me a lot of work. My stretched image looked like Fig. 6.45.

It may not show up well in a book, but there's a lot of background noise in Fig. 6.45. So I fed a tif version of the image to Topaz Denoise AI 3.7.2, an external application.

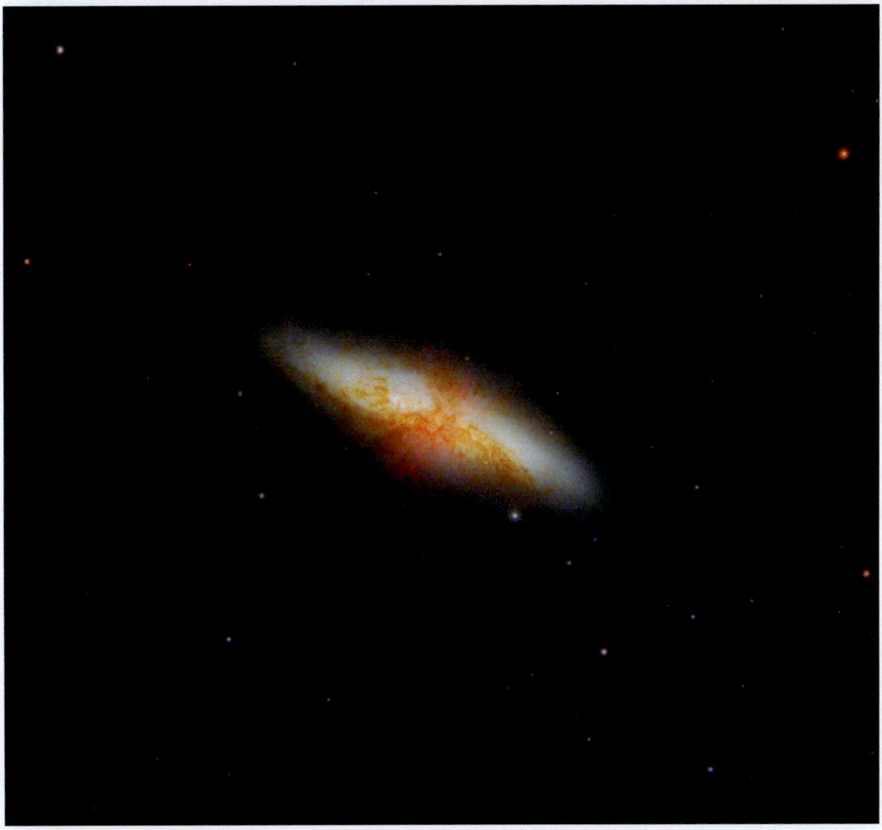

Fig. 6.44 My RGB image of M82. North is up. (Image by the Author)

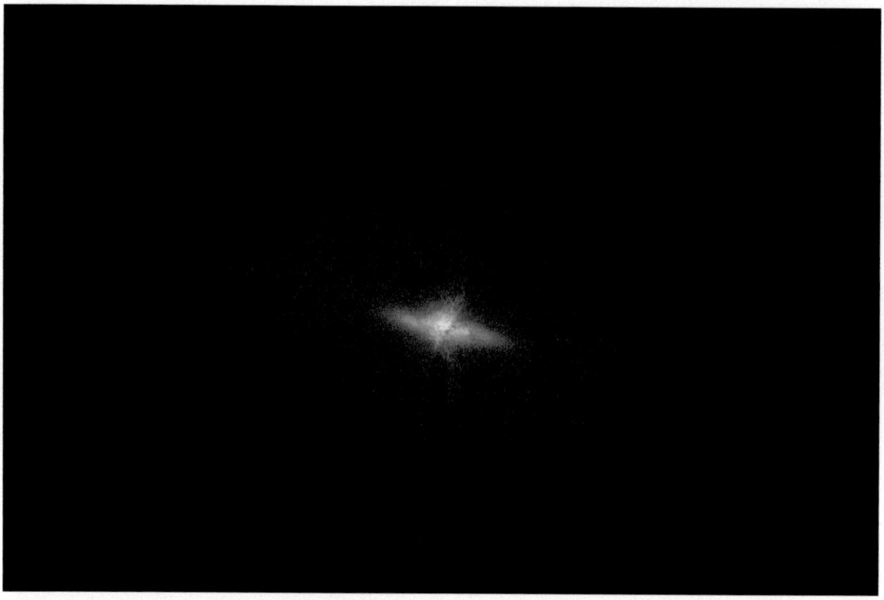

Fig. 6.45 An intermediate stage in stretching my H-α image. (Image by the Author)

Knowing that I was going to have to not pick up too much spurious light in the center of the galaxy, I toned this down somewhat in my next iteration of stretching. I used the Curves transformation shown in Fig. 6.46.

My H-α image is now as shown in Fig. 6.47.

This image was already aligned with the RGB image before I had removed the stars, so I was able to add the H-α t file, whose name was M82_Ha_asinh_R_9_DeNoi_Regd.xisf, to the existing RGB image, which I had called M82_4h_RGB_asinh_9.xisf, by the following Pixel Math.

Red: max(1.875*M82_Ha_asinh_R_9_DeNoi_Regd,1.0*M82_4h_RGB_asinh _9_R)
Green: M82_4h_RGB_asinh_9_G
Blue: M82_4h_RGB_asinh_9_B

The numbers 1.875 & 1.0 in the Red expression were arrived at by trial and error.

The RGBH-α image is shown in Fig. 6.48.

We can't call this galaxy passive. It is spectacularly unusual, and thoroughly merits its place in my 'Fab Four'.

I was inspired by a 1991 paper by Telesco and others [51], who found a bar in M82 by photographing it at several infrared wavelengths, to see if I could find it.

So I installed my ZWO ASI1600MM monochromatic camera onto the Edge HD with *f*/7 reducer, with a Baader-Bessell infrared filter in a filter wheel. To set

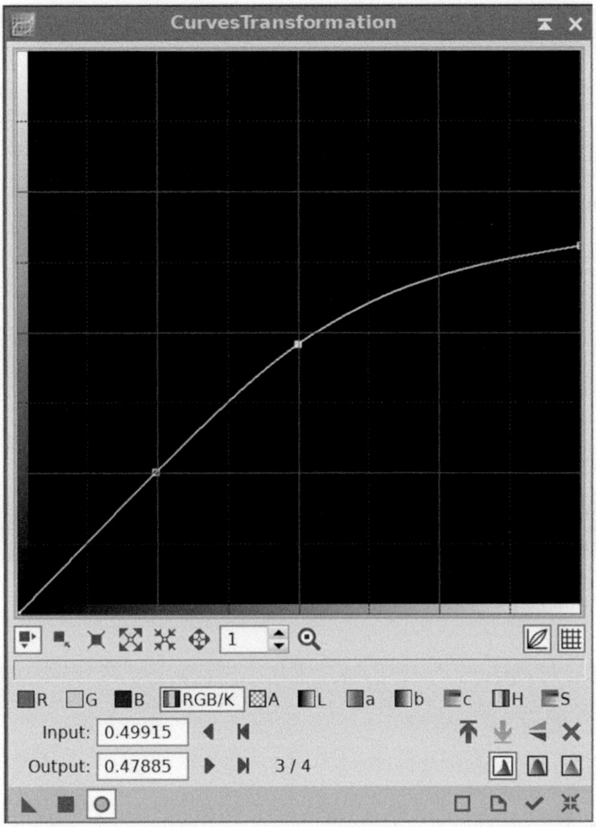

Fig. 6.46 The curves transformation used to suppress very bright regions. (Image by the Author)

everything up, I wheeled round to the Baader-Bessell V filter in visible light. I did, however, have the wit to know that I would have to re-focus in IR, which I did.

This was a total try-on. My IR filter transmits light between 750 and about 1100 nm, and the camera runs out of sensitivity at about 1000 nm, which is a much shorter wavelength than Telesco and collaborators used. It might or might not work. Let's see what I found with 4 hours of IR data.

I again tried three methods of stretching and settled on Arcsinh stretching. I was able to remove the stars with StarXTerminator, bit still used the GAME mask shown in Fig. 6.49 to remove the haloes of two bright stars. The tool

Process → BackgroundModelization → AutomaticBackgroundExtractor

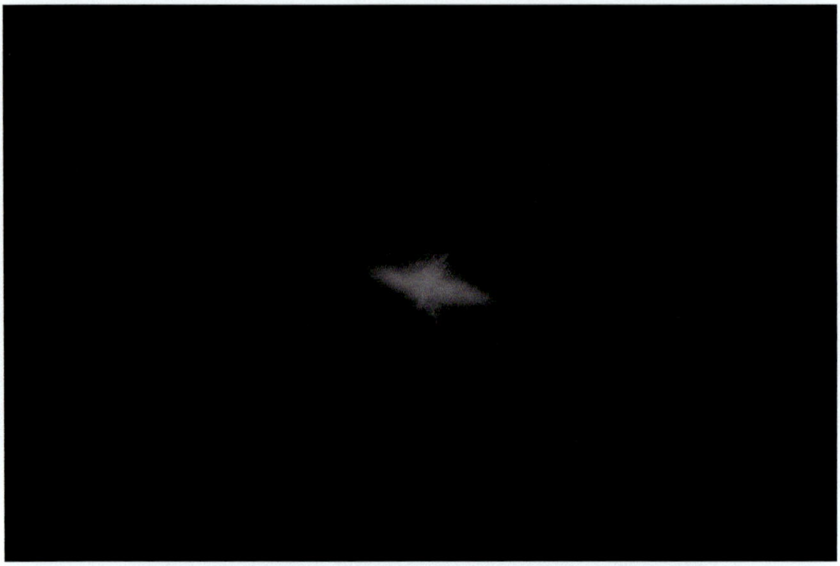

Fig. 6.47 My H-α image at this point. (Image by the Author)

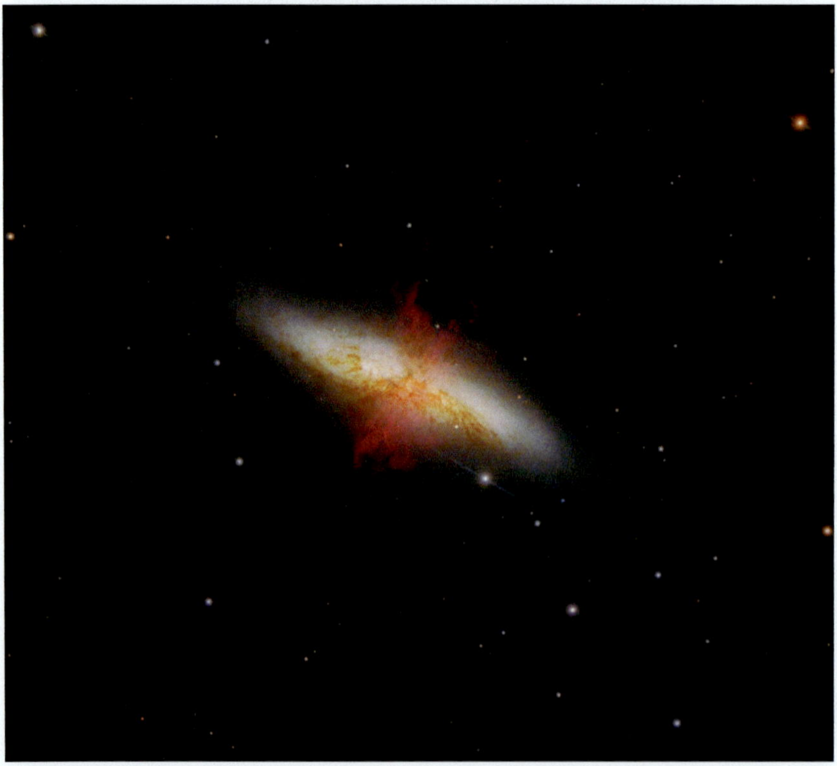

Fig. 6.48 The combined RGB and H-α image of M82. (Image by the Author)

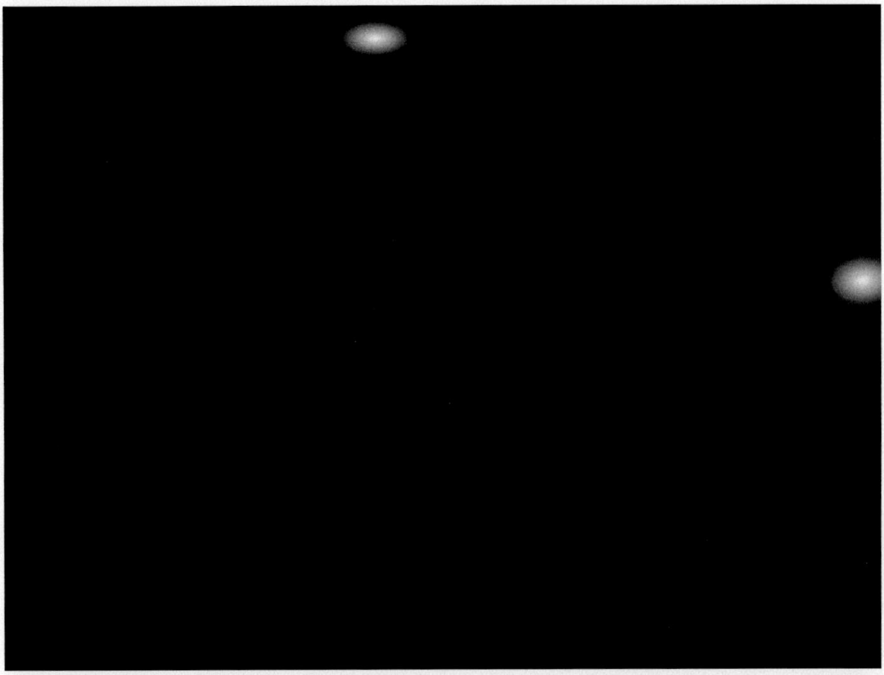

Fig. 6.49 The GAME mask used to remove the haloes from two bright stars. (Image by the Author)

Successfully removed some extraneous background light. Other than that, it was our old friend

Process → *IntensityTransformations* → *HistogramTransformation*

That handled the image stretching.

My final IR image is shown in Fig. 6.50. There is no reason to add false color. I don't know what the "color" of IR looks like any more than I know what the "colors" of X-rays look like.

The verdict? I think I can see a bar in this, which is almost, but not quite, parallel to the rest of the galaxy.

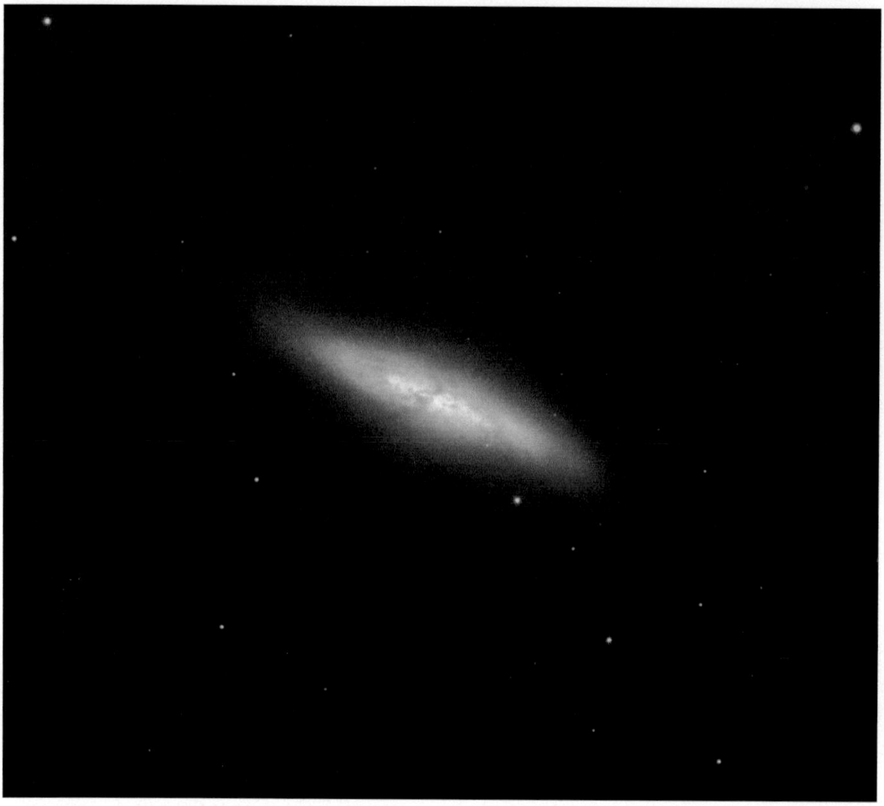

Fig. 6.50 My final IR image of M82. (Image by the Author)

Wrap-up

We have seen four very different galaxies, and on the way have investigated collisions in the M51 system and the M81 group. We've seen a supernova in M101. We've found fiery H-α and an infrared-visible bar in M82. We have also encountered lots of tips to produce stunning photographs of the Fabulous Four.

Chapter 7
Cluster Folk

We are cluster folk: we live on the outskirts of the Virgo cluster. Just as it was hard to appreciate the nature of the Milky Way galaxy from inside it, so it is easier to appreciate the nature of our cluster if we look at another.

The Coma Cluster

As it happens, this cluster is in a similar part of the sky to the Virgo Cluster. It is in the constellation of Coma Berenices[1], slightly to the north of the center of the Virgo Cluster. At 320 Mly away (about 100 Mpc) it is easily the most distant object I have set out to photograph. Its only numerical designation is in the Abell catalogue of galaxy clusters, first published by George O. Abell of the University of California in 1958. It is designated Abell 1656.

My frames were captured with my new Celestron 11″ Edge HD, on a CGX mount using a ZWO ASI2600MC color camera. After preprocessing, I had the image shown in Fig. 7.1, only of course it was in color. I temporarily stretched this image using the radioactive symbol button along the top of the PixInsight window. As you can see from the name I gave the image, I captured 235 × 60 s of usable frames, or 3 h 55 m of data.

It is immediately obvious that there is a diffraction spike coming from the brightest star. A Celestron Edge HD should not produce diffraction spikes. I made the decision to postpone this issue for now. I have to take it off my mount as soon as this manuscript will have been completed to do some maintenance on the mount, so I'm then going to trace the origin of the spike – presumably dirt. In the meantime, it actually helps me to distinguish stars from galaxies and galactic nuclei.

[1] This is Latin for "Berenices' comb". It is pronounced "KOH-ma Berra-NEE-chays".

J. Clark, *Photographing Galaxies from Light Polluted Skies*, The Patrick Moore Practical Astronomy Series, https://doi.org/10.1007/978-1-0716-4506-2_7

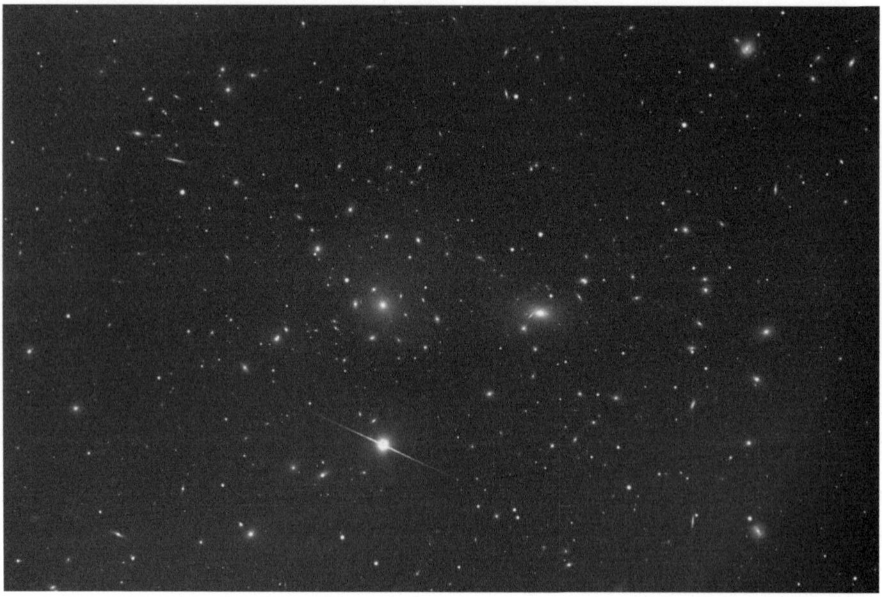

Fig. 7.1 The image from which I began postprocessing. (Image by the Author)

The main challenges are a blotchy background and a ring of vignetting around the edge of the image. I tried three methods of stretching: EZ Soft Stretch, Generalized Hyperbolic Stretch & Arcsinh stretching. The third of these was the most promising. It is found at

$$Process \rightarrow IntensityTransformations \rightarrow ArcsinhStretch.$$

Removing the stars gave a weird-looking main image, because StarXterminator got rather confused about which objects were stars. But I applied

$$Process \rightarrow \langle Etc \rangle \rightarrow GraXpert$$

To the "starless image", which reduced the blotchiness. After a little more stretching, I added the stars and removed galaxies back in, and then ran GraXpert again. The image really did not need stretching beyond the initial Arcsinh stretch. I tried to make a north-up image, but lost a lot of galaxies in the process, so I abandoned that effort, and contented myself with a 180° rotation to make the image north-ish up. Figure 7.2 shows my final image.

I did make an annotated version of this image, but there's no point in showing it to you because you wouldn't be able to read it. It labeled more than 180 galaxies.

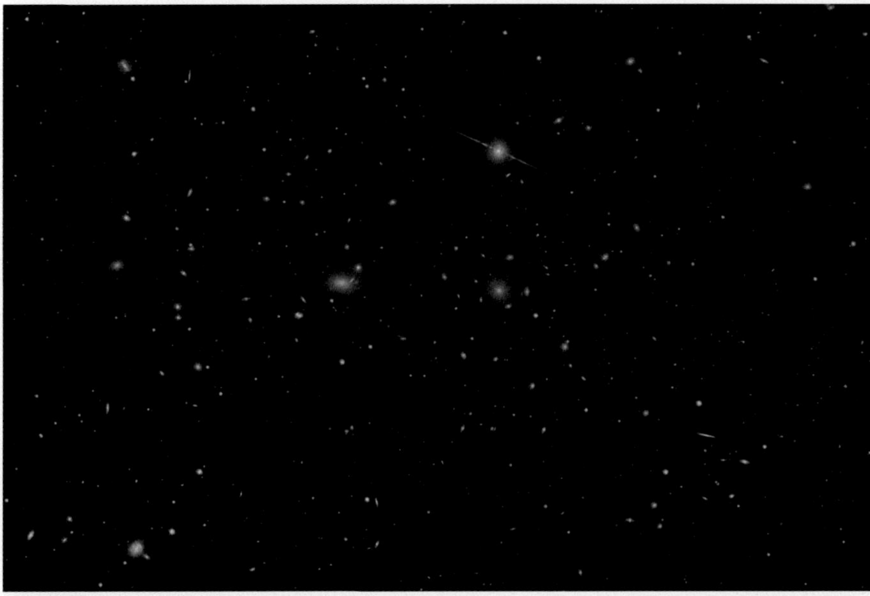

Fig. 7.2 The Coma Cluster. (Image by the Author)

Thus, this was not an especially challenging photograph technically. Yet, at 336 Mly or 103 Mpc away, it is the most distant object I have ever photographed.

But it is very illuminating about what a galaxy cluster looks like. I have a less good photo of the Perseus cluster I got late last January, when the weather turned bad again. It isn't exposed for long enough, and Perseus rapidly disappeared from my night sky, so I can't add more hours until late this coming summer – I'm writing in May – but I will show it to you anyway, in Fig. 7.3,

The moral of this image is that you need many more hours of exposure to obtain a good photo of this galaxy cluster. But I don't think it would be an exercise in eating soup with a fork. There's something there to work with. The central galaxy NGC1275, also known as Caldwell 24, has something interesting going on, which may repay closer study with my telescope using its full *f*/10 magnification. Now there's a challenge! In retrospect, the surprise is that I got away with just under four hours on the Coma Cluster.

One of the main things the clusters in Figs. 7.2 and 7.3 have in common is that the clusters center on one or more giant elliptical galaxies. The Perseus cluster and the Coma Cluster center on two. I will show you in due course that the Virgo Cluster centers on three giant elliptical galaxies. The galaxies immediately surrounding the elliptical giants are mostly smaller elliptical galaxies. Only on the

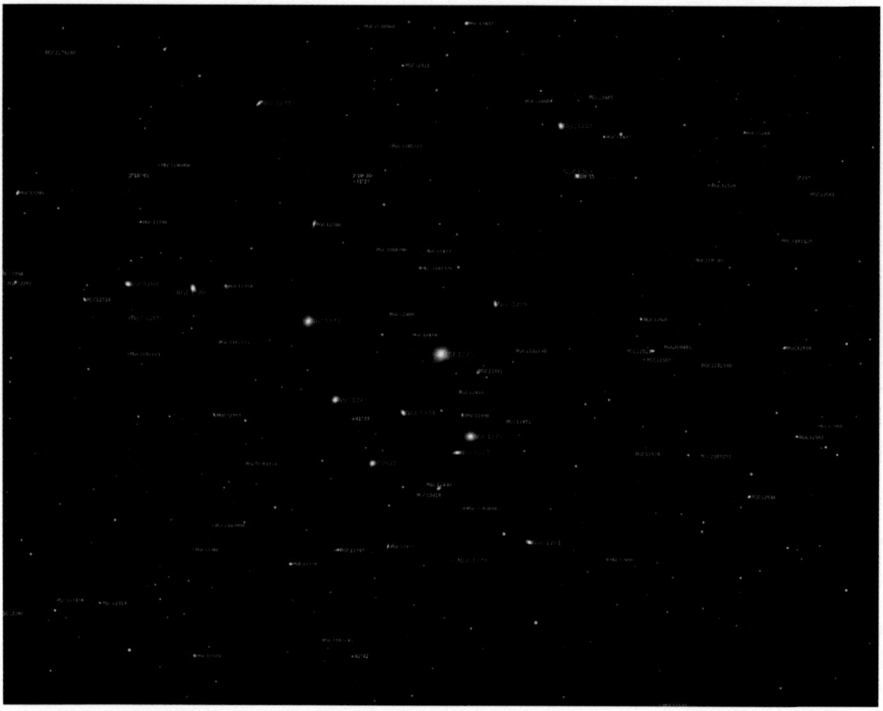

Fig. 7.3 A rather inadequate image of the Perseus cluster, Abell 426. 186 × 60 s, Celestron C11, otherwise usual equipment. (Image by the Author)

outskirts do spiral galaxies become common. Figure 7.2 shows this well. Figure 7.3 does show this if you stare hard, but you can see it more easily in the professional image Fig. 7.4.

This, then seems to be a general characteristic of galaxy clusters. They are gravitationally pulled together by giant elliptical galaxies. These central galaxies may themselves be the product of mergers.

Having discovered this, let's take a look at the center of the Virgo Cluster.

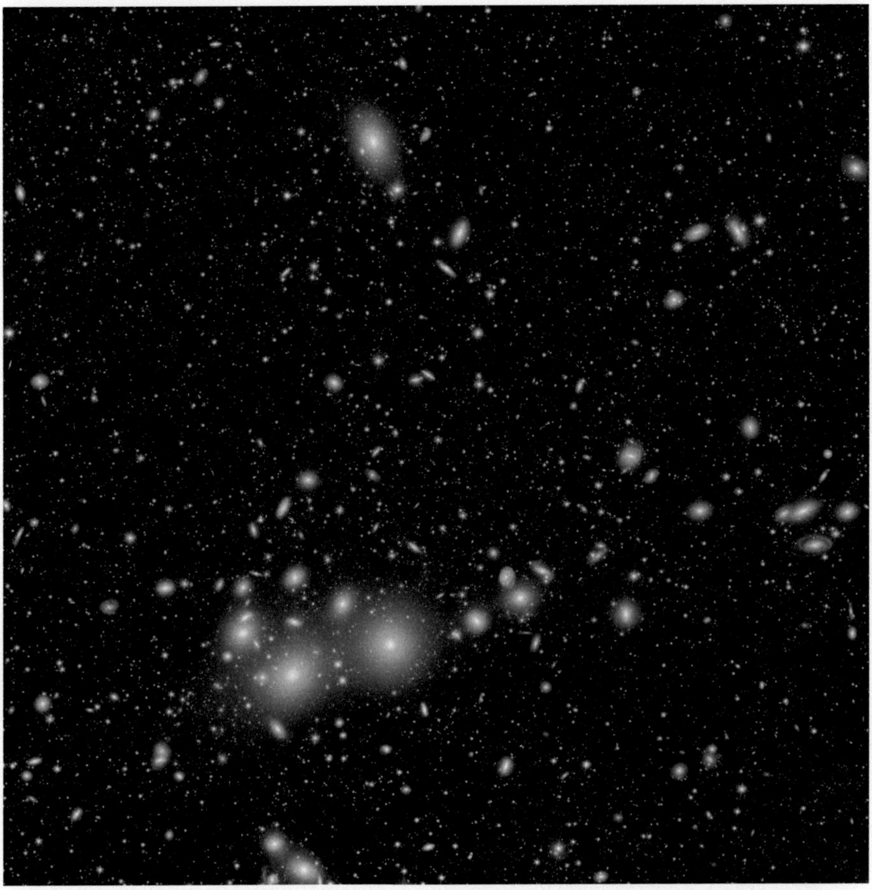

Fig. 7.4 A professional image of the Perseus Cluster. (Image Credit: ESA/Euclid/Euclid Consortium/NASA, image processing by J.-C. Cuillandre (CEA Paris-Saclay), G. Anselmi, CC BY-SA IGO 3.0)

M87 and Markarian's Chainu

These objects occupy enough sky that I used my 73 mm *f*/4.9 apochromatic refractor to capture them.

In the end WBPP accepted 388 × 60s of frames, or 6 h 28 m. I captured these over three nights.

A quick look at Fig. 7.5 shows the main processing challenge: a good deal of background light, which I needed to remove.

First, I removed the stars from the Image in my usual way with

Process → ObjectRecognition → StarXTerminator

and parked the stars. I later used

Fig. 7.5 A quick stretch of my image shows a lot of background light in various colors. The processing challenge is to remove it. (Image by the Author)

Process → IntensityTransformations → HistogramTransformation

To reduce their brightness. This was before the days of GraXpert in PixInsight, and I used

Process → BackgroundModelization → AutomaticBackgroundExtractor

To try to knock out the background. From there, I used Histogram Transformation gently in several steps, and interchanged this with using

Process → IntensityTransformations → CurvesTransformation.

I used

Process → ColorCalibration → AssistedColorCalibration

to intensify the colors. I went out to the external application Topaz Denoise AI to reduce the noise, and then used

Process → PixelMath → PixelMath

To add the stars back in, resulting in Fig. 7.6.

In Fig. 7.7 the galaxies are labeled. Not all the galaxies in the image are labeled: the 50+ that are labeled makes the image quite crowded enough. M86 near the center and M87 to the lower left are doubly-ringed.

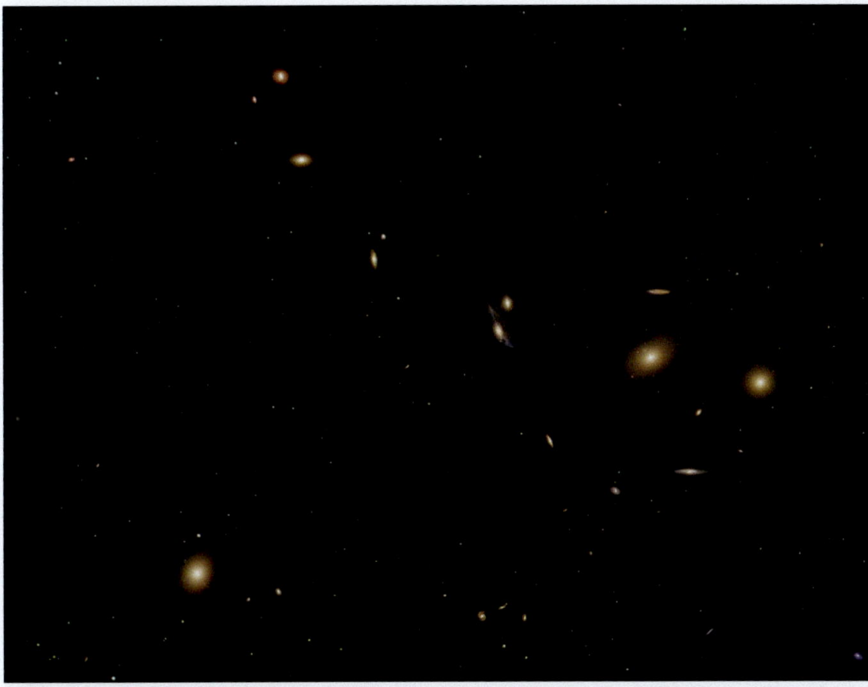

Fig. 7.6 Markarian's chain and M87. (Image by the Author)

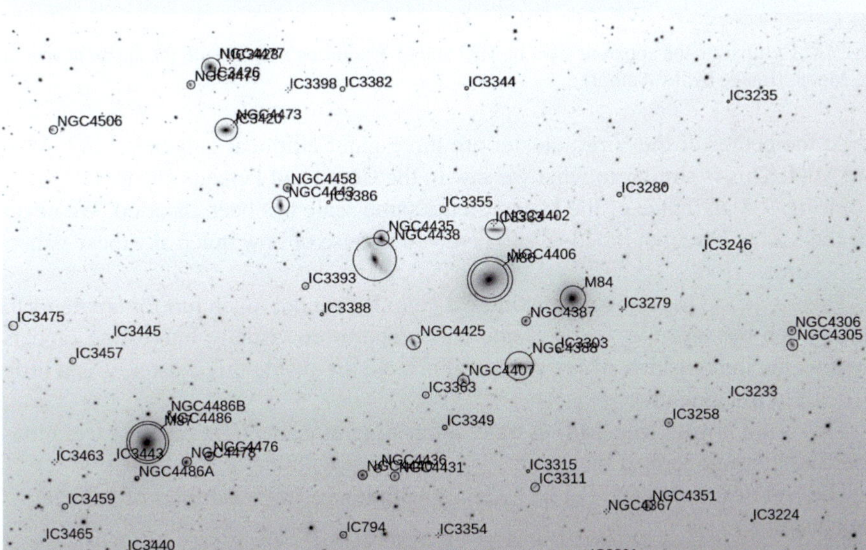

Fig. 7.7 The galaxies in Fig. 7.6 labelled. There are so many galaxies that it is sometimes hard to separate them out. In particular, the doubly-circled object in the Center is M86 and the doubly-circled object to the lower left is M87. (Image by the Author)

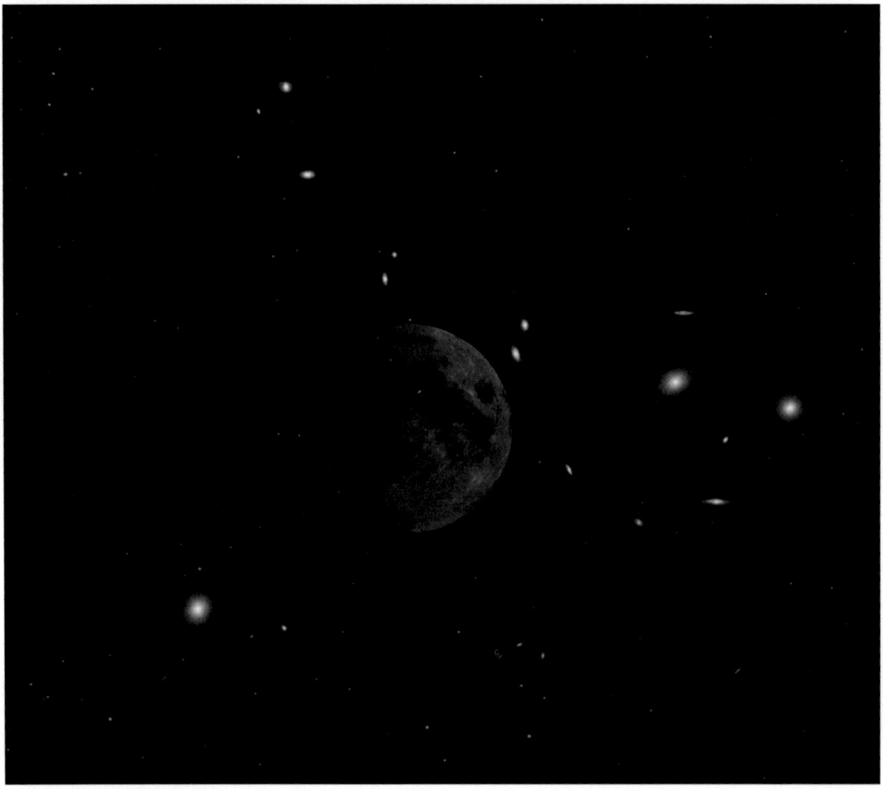

Fig. 7.8 Comparing the apparent sizes of M87 and of Markarian's chain with the apparent size of the Moon. (Image by the Author)

At the center of the Virgo cluster are three giant elliptical galaxies, M87, M86 and M84. This is similar to what we saw in the Coma and Perseus clusters.

In Fig. 7.8 an image of the Moon on the same scale has been inserted. We never see the Moon there, but the idea is to give you a sense of how much sky these galaxies occupy.

There's a case for saying that elliptical galaxies are not much fun for amateurs to photograph because they are so featureless. The more visually interesting objects are towards the outskirts of the cluster. Since we are inside this cluster, it is a little difficult to imagine this.

This is not always true. M87 is more interesting in visual light on close examination. We'll have a look at that soon.

The rest of the chapter will be taken up with interesting examples of galaxies in the cluster.

M87 and Its Active Nucleus

Figure 7.9 shows my starting point, temporarily stretched, after preprocessing, color calibration, blur and noise reduction. This was captured with my Celestron C11 and *f*/6.3 focal reducer/corrector. The horrible vignetting wasn't going to be a problem, because I was only interested in the center of the picture. The background, however, was quite unwanted.

I'm beginning to sound like a stuck record saying that the Arcsinh stretch worked the best. I can certainly think of counter-examples. Anyway, this time I ended up with Fig. 7.10.

In fact the image in Fig. 7.10 is over-stretched for what I want to show you. If you stretch the galaxy very gently and tenderly, you get the result in Fig. 7.11. The blue jet shooting out from the nucleus is now quite visible.

The blue jet is of a type of radiation emitted by particles travelling at close to the speed of light, called Cherenkov radiation. It is evidence that the nucleus of this galaxy is a violent place.

Indeed, it is well known that there is an unbelievably big black hole at the center of M87. Figure 7.12 shows the infrared image of the material surrounding the black hole using the Event Horizon Telescope, at an infrared wavelength of 1.3 mm. In 2019–20 I took a course in general relativity just as the first images of the event horizon (the point of no return) of this black hole were released. We were given the exercise of verifying that this event horizon is consistent with a multi-billion solar mass black hole. It is. The actual mass published in 2019 was 6.5 ± 0.7 billion solar masses [52].

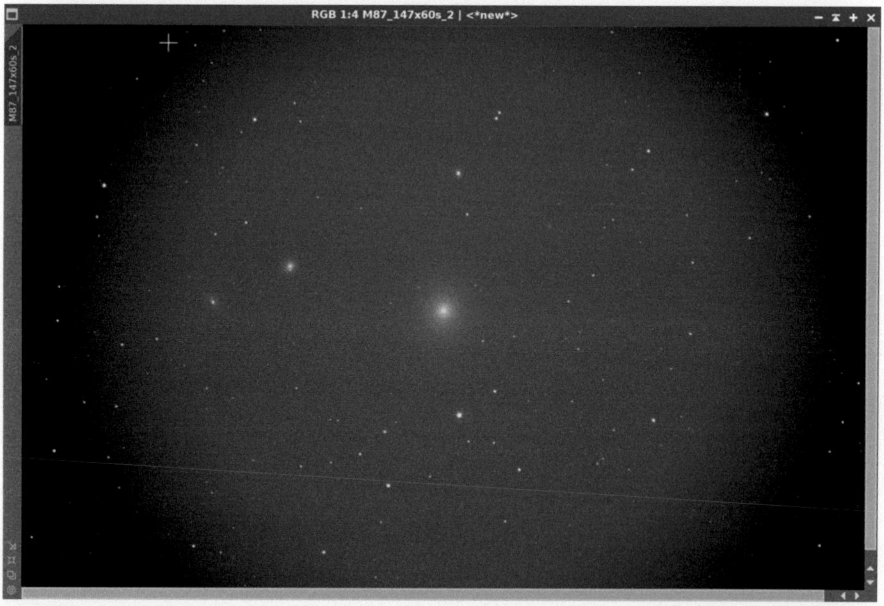

Fig. 7.9 My initial image after preprocessing, temporarily stretched. (Image by the Author)

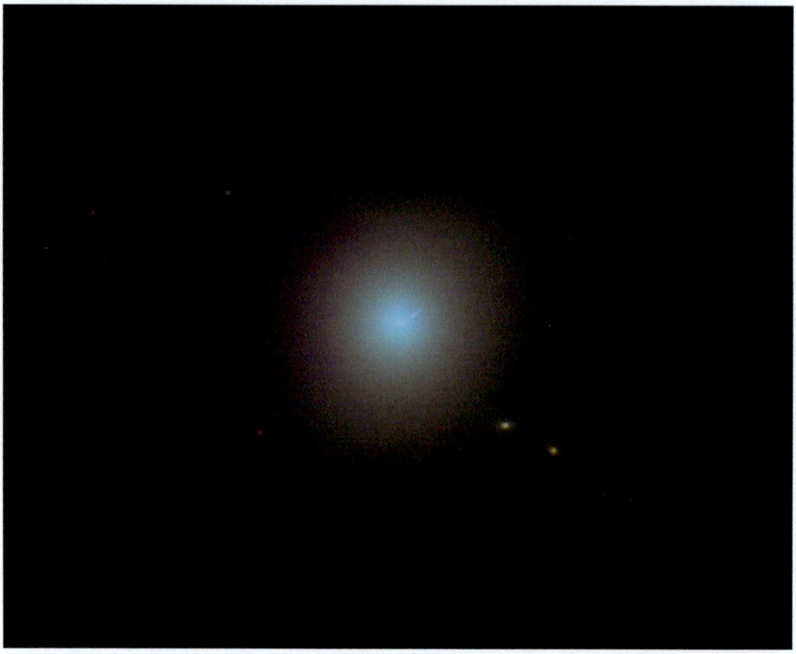

Fig. 7.10 My M87 image after Arcsinh stretching and cropping. (Image by the Author)

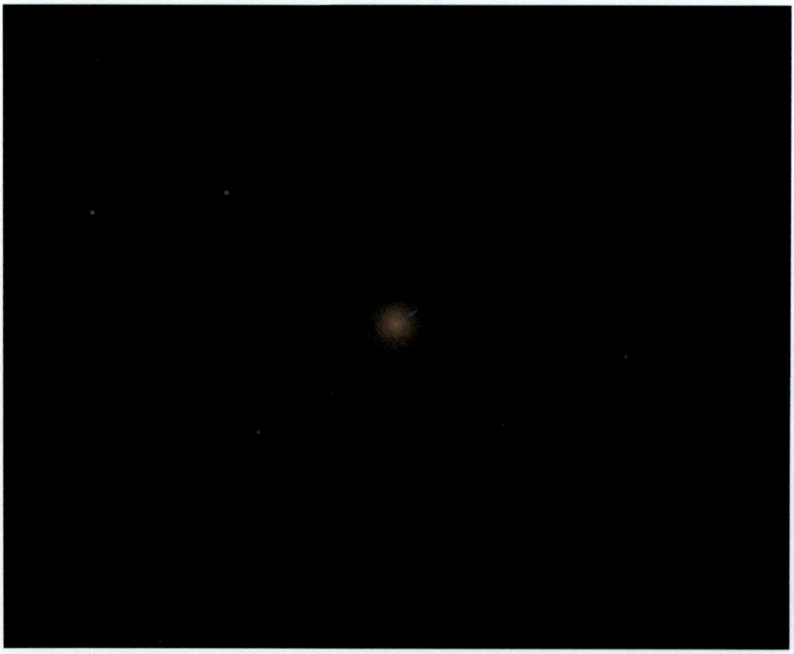

Fig. 7.11 M87, 143 × 60s (2 h 23 m) with very gentle stretching. You can now see the blue jet shooting out from the active nucleus. (Image by the Author)

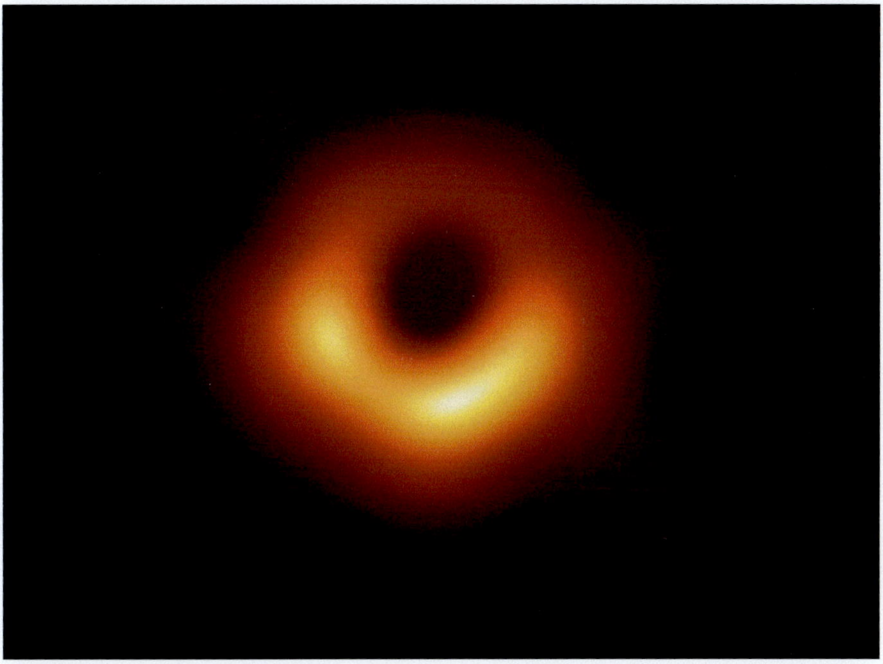

Fig. 7.12 The material surrounding the supermassive black hole at the nucleus of M87. (Image Credit: EHT Collaboration, CC BY 3.0)

The Leo Triplet

This is a cute group of three galaxies in Leo, about 17° west of M87. I could have photographed them all at once with my refractor, but I was more minded to see if I could make a collage, not least because I expected to get a better result if I were to give each of the three galaxies the processing they needed.

All three of the galaxies are spirals, but we are looking at one of them, NGC3628, edge-on, so we can only indirectly tell this: it doesn't look like any plausible other type of galaxy. The second, M65, is a tightly wound spiral with just a hint of a bar, and the third, M66, is a loosely wound, S-shaped barred spiral.

As it turned out, M65 & M66 required fairly similar processing. We'll go through M66 in detail to show you the general idea. NGC3628 required some similar features, but there is no very bright nucleus to control, or spiral arms that we can bring out.

First, I'll show you, in monochrome, what came out of WBPP, blur & noise reduction. In each case the images have been given a non-permanent stretch. You won't, of course, see that I had already done spectrophotometric color calibration.

Figures 7.13, 7.14, and 7.15 show some common features. The first is poor control of dust bunnies. The flat frames weren't good enough to remove them. I clearly needed to clean my optical train. Nowadays I have the correct equipment to do this. You can buy lens cleaning kits & chip cleaning kits online. I also take fresh flat

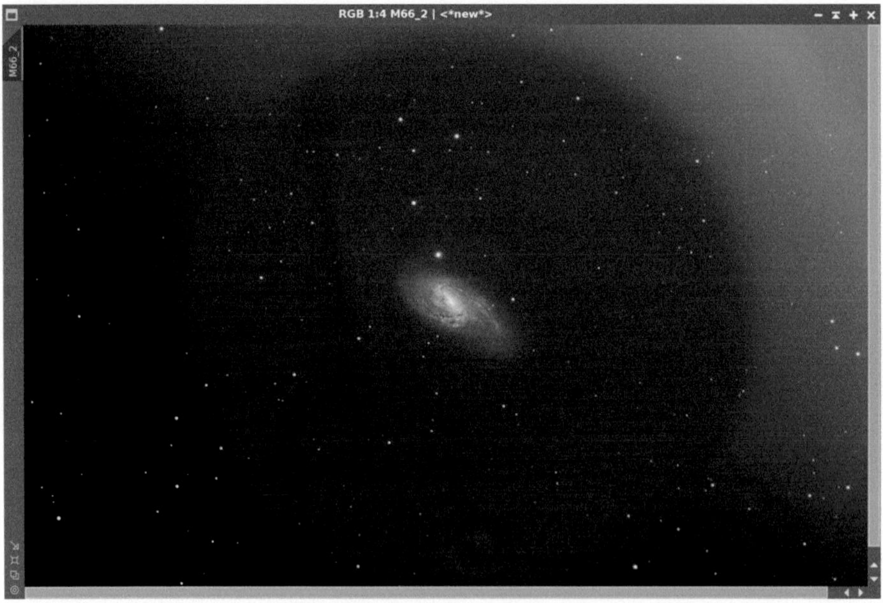

Fig. 7.13 M66 preprocessed. (Image by the Author)

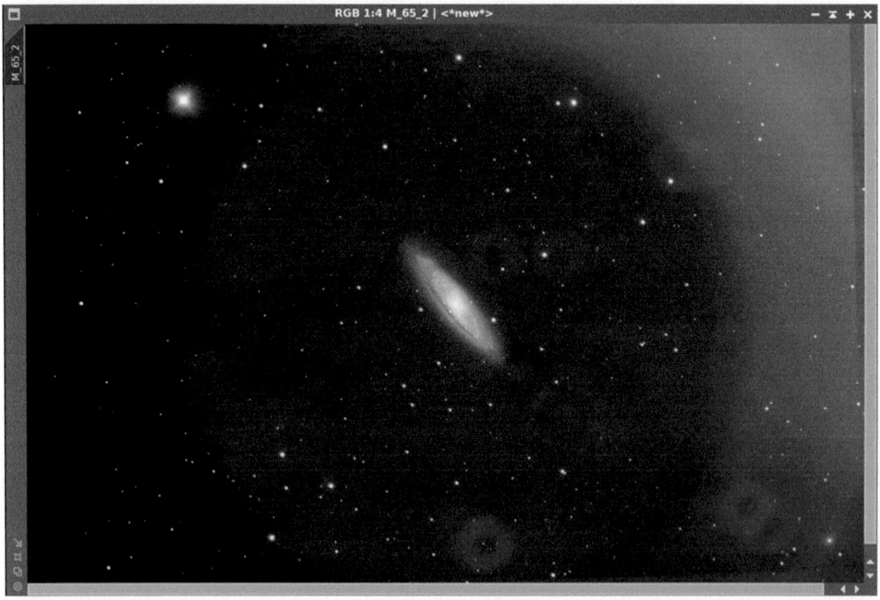

Fig. 7.14 M65 preprocessed. (Image by the Author)

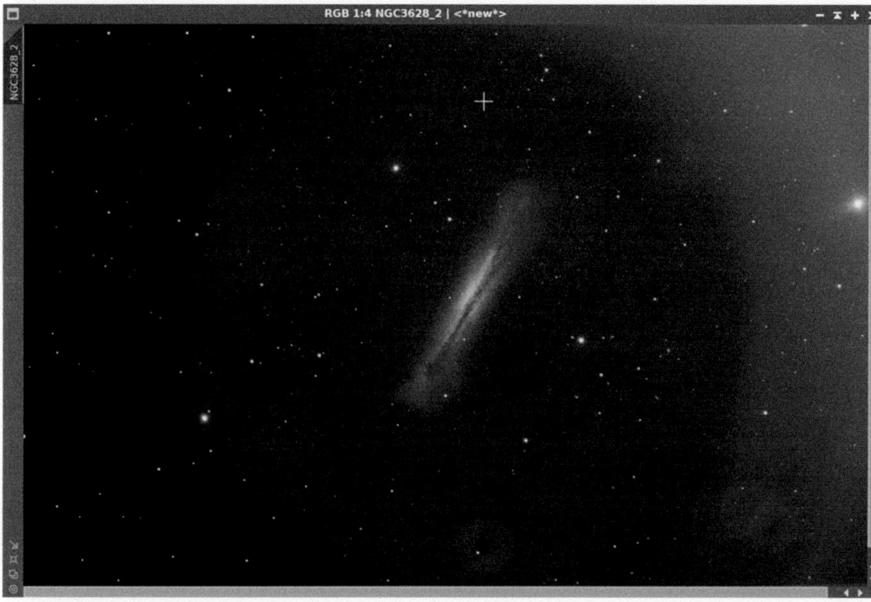

Fig. 7.15 NGC 3628 preprocessed. (Image by the Author)

frames every single time I remove a camera. However, there is now a technique in PixInsight to deal with the dust bunnies. It wasn't there fifteen months ago.

The second issue is, I think, basically light pollution. The same pattern of circular fogging and fogging to the top right is common to these images. I captured these images some time before I wrote this, and I reprocessed them to write this section. The first time around, the images weren't too bad, so I knew that none of these problems were going to make me fail.

Going to M66, I found that Generalized Hyperbolic Stretch,

$$Process \rightarrow IntensityTransformations \rightarrow GeneralizedHyperbolicStretch,$$

Was the most promising way to stretch the image.

In Fig. 7.16, the image already looks a lot better.

Next, I applied the GraXpert tool, found at

$$Process \rightarrow \langle Etc \rangle \rightarrow GraXpert.$$

The GraXpert window is shown in Fig. 7.17. This is a new piece of software, which is rapidly growing its capability, so it may not be the same by the time you read this. Anyway, I needed background extraction, and chose the Division option to mimic a flat frame. (The alternative is to use it to subtract.) This is a really clever little AI-based tool that can sometimes appear to work miracles. It sometimes fails miserably, but this time it performed very well.

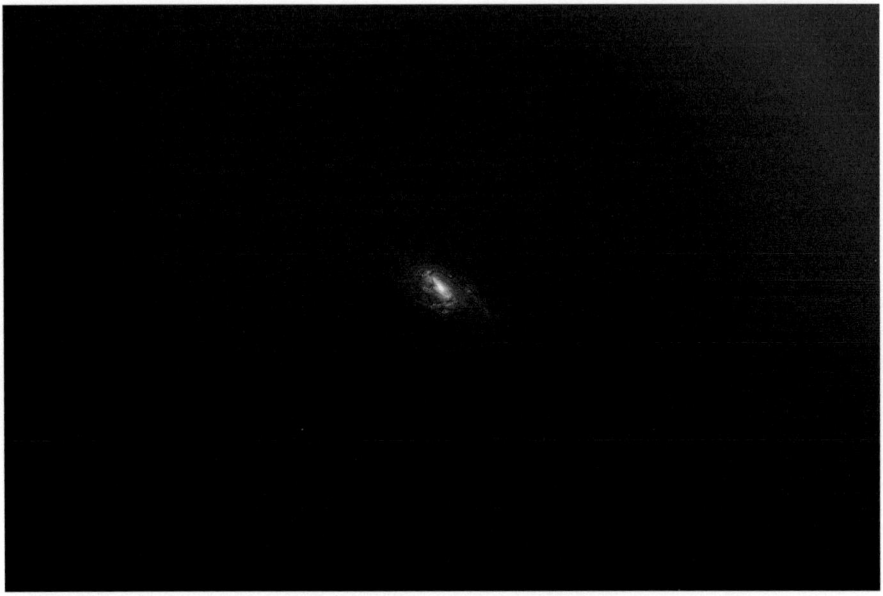

Fig. 7.16 M66 after application of Generalized Hyperbolic Stretch and star removal. (Image by the Author)

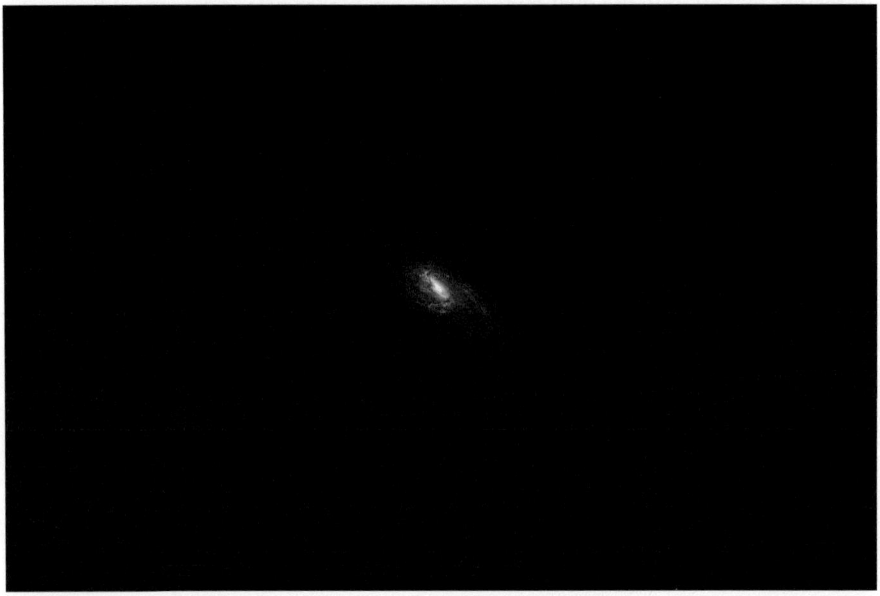

Fig. 7.17 Next, I applied the GraXpert tool. (Image by the Author)

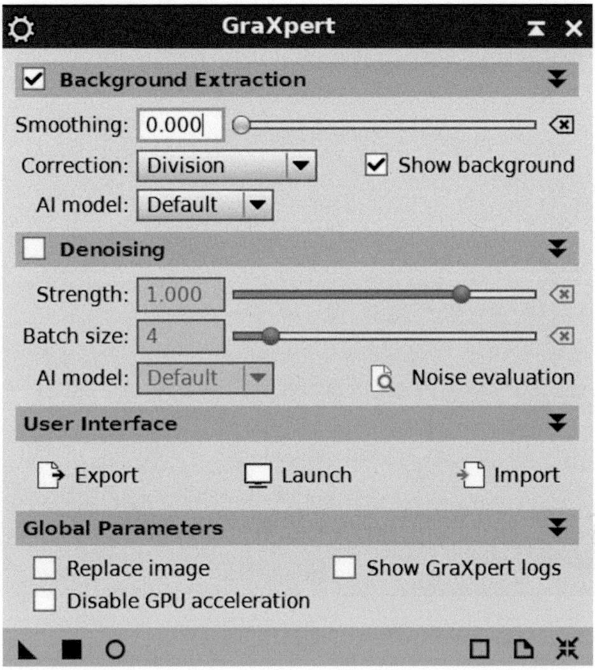

Fig. 7.18 The GraXpert tool, set to divide the background away. (Image by the Author)

In Fig. 7.18, we can see that the light pollution problem has gone away. Some of the dust bunnies are still visible around the edges, but we will deal with those. Since I will want to stitch three images together, I don't want to use my time-honored method of cropping the image to throw trouble away. Next, I tried an Arcsinh stretch to make the galaxy look more brilliant. Arcsinh stretches also bring out color.

Process → IntensityTransformations → ArcsinhStretch,

We have more galaxy, but also more dust bunnies. The result is shown in Fig. 7.19.

I got rid of these by making a mask, so that I could darken the edges of the image without affecting the galaxy. I don't like making GAME masks (Script → Utilities → GAME) that are similarly sized to the galaxy. It's then very difficult to make an image where I can see the scar of the mask staring at me. I much prefer, if possible, to make the mask much bigger than the galaxy. This I did, making a binary mask, which I then fed to the Range Selection tool (Process → MaskGeneration → RangeSelection) to make the edge fuzzy (Fig. 7.20). I then used

Process → IntensityTransformations → HistogramTransformation

to kill off the dust bunnies in the unmasked area.

Next I went to work on the over-bright nucleus in Fig. 7.19. First you have to make a GAME script Gradient Mask that just covers the nucleus (Fig. 7.21). You

Fig. 7.19 The result of an Arcsinh stretch. (Image by the Author)

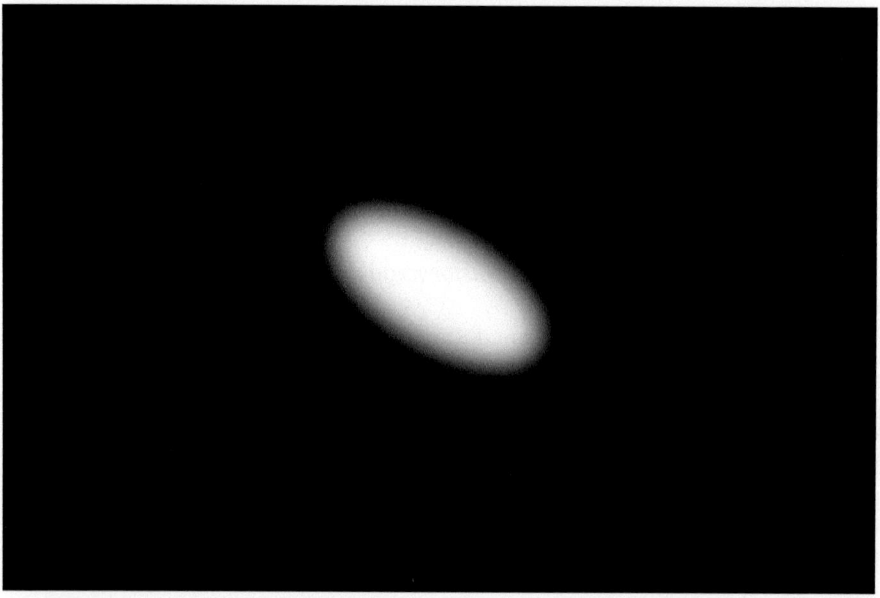

Fig. 7.20 The mask I made to remove the dust bunnies. (Image by the Author)

only want to operate on the are underneath the white in Fig. 7.21. If necessary, go to Mask → Invert mask.

Then you use a tool

Process → MultiscaleProcessing → HDRMultiscaleTransform

To tone down the nucleus of the galaxy. I usually stick with default values in Fig. 7.22, but if, as occasionally happens, that overdoes the reduction, I have dropped to three layers.

The next thing I did was to use a tool to bring out the spiral arms. The tool is

Process → IntensityTransformations → LocalHistogramEqualization.

The default values are shown in Fig. 7.23. You use the hollow circle at the bottom left to open a live processing window, and slide the Kernel Radius slider to taste. Once you have achieved the desired effect, the Amount slider adds to transformed image to the untransformed one, and the Amount is the weighting. 1.0 means you only keep the transformed image. 0.5 means you average them. 0.75 means three quarters transformed, one quarter untransformed, and so on. You slide this button until you like the result. I think of it as an aesthetic effect rather than a scientific one.

My galaxy now looks as in Fig. 7.24.

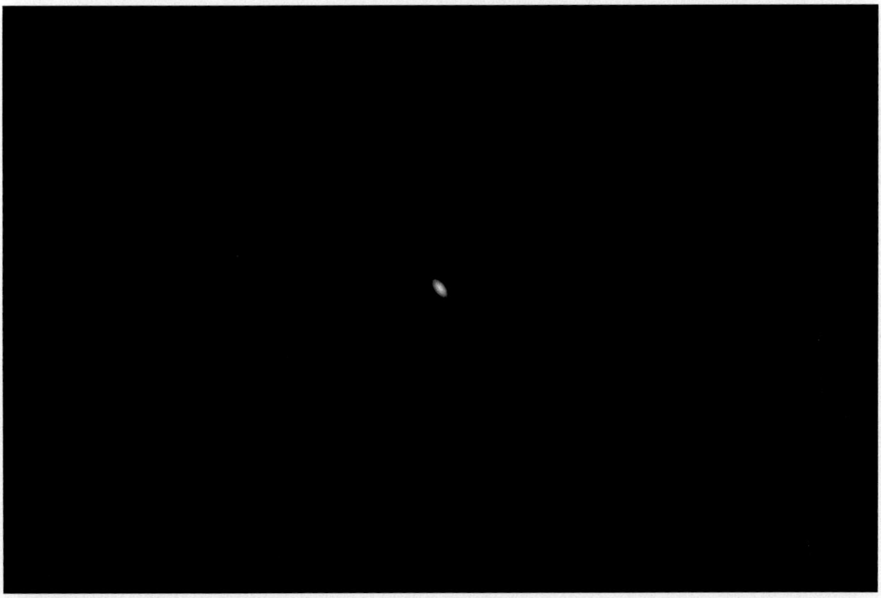

Fig. 7.21 A teeny oval mask to cover the nucleus of the barred galaxy. (Image by the Author)

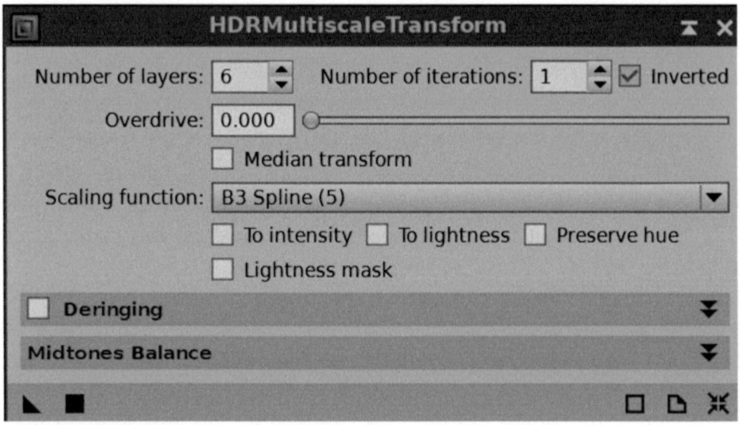

Fig. 7.22 The HDR Multiscale Transform tool. (Image by the Author)

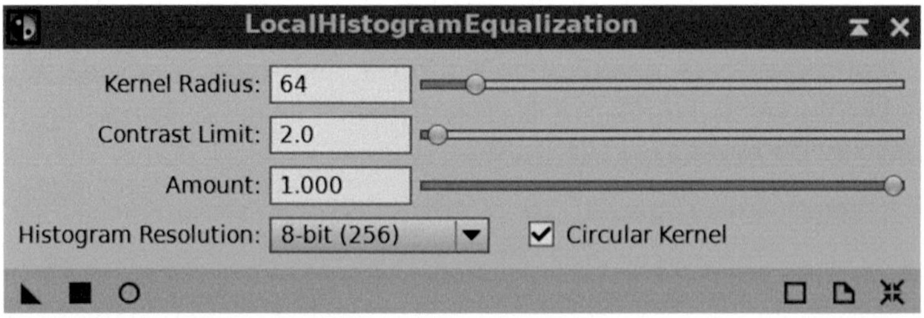

Fig. 7.23 The default settings of the Local Histogram Equalization tool. (Image by the Author)

Now all I have to do is to use Pixel Math to add the stars back in. I did not tone them down, because I wanted to stitch three images together; and the plate solver requires lots of good stars to solve the images to be joined.

My final galaxy is shown in Fig. 7.25. My very similarly processed M65 is shown in Fig. 7.26.

NCG3628 required a slightly different approach. First, I used Arcsinh stretching (Process → IntensityTransformations → ArcsinhStretch). It worked better than Generalised Hyperbolic stretching. But I came to the conclusion that I had over-stretched the image shown in Fig. 7.27. I didn't really believe the outline shown. It was almost as if GraXpert had used a mask. I have no idea if it did, but that's what it looked like.

I used

Process → IntensityTransformations → CurvesTransformation

to tone down the grey area surrounding the galaxy. I also wanted to make the nucleus look a bit fierier, for aesthetic not scientific reasons. So the curve I applied looked like the one in Fig. 7.28.

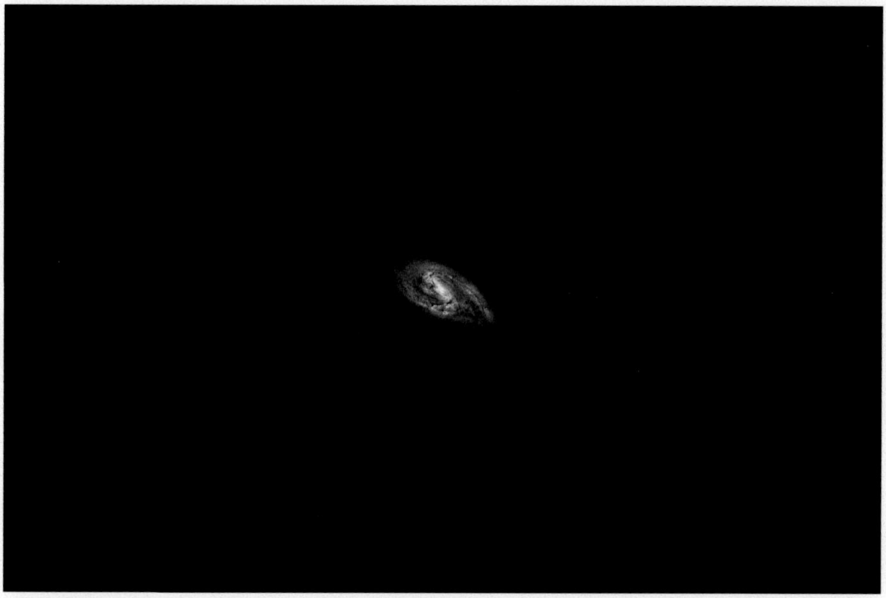

Fig. 7.24 M66 with the nucleus tamed and the spiral arms brought out. (Image by the Author)

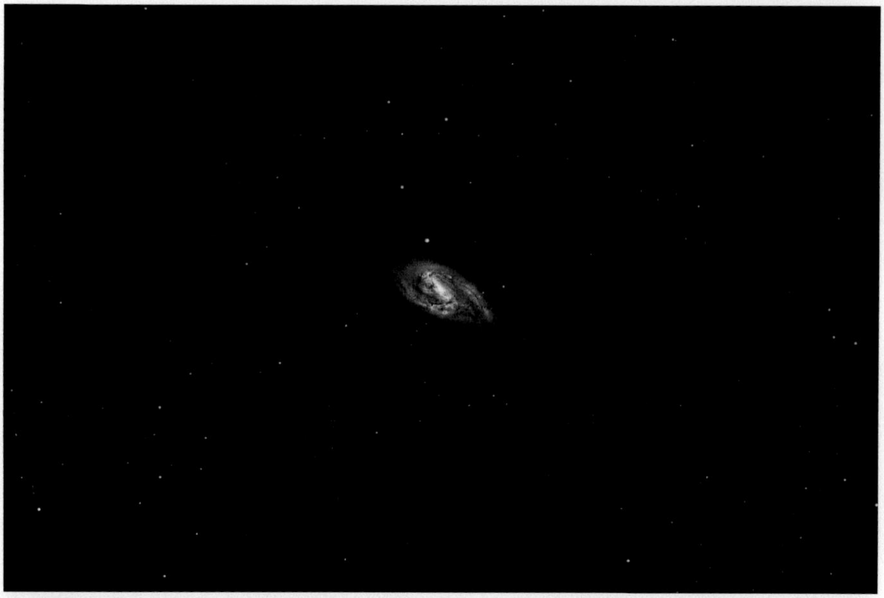

Fig. 7.25 My final M66. (Image by the Author)

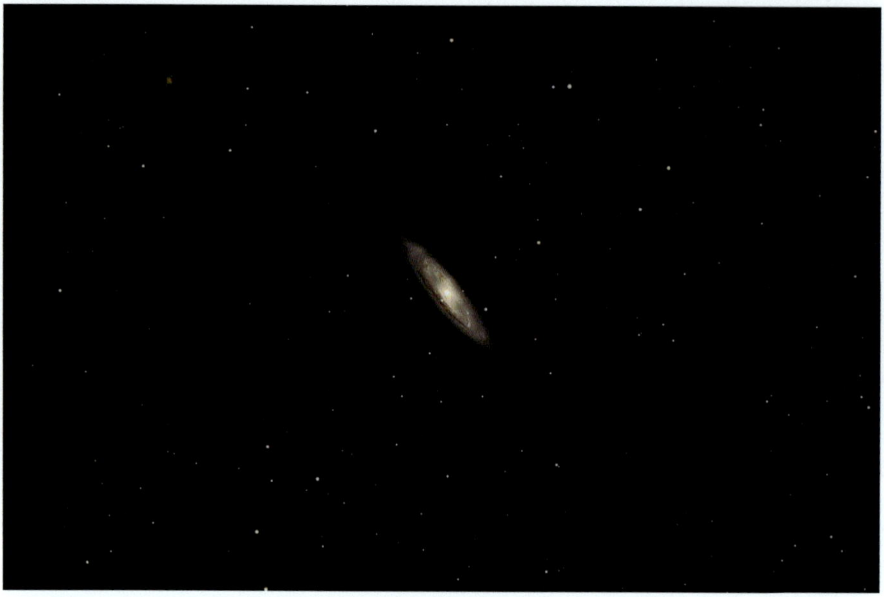

Fig. 7.26 My very similarly processed M65. (Image by the Author)

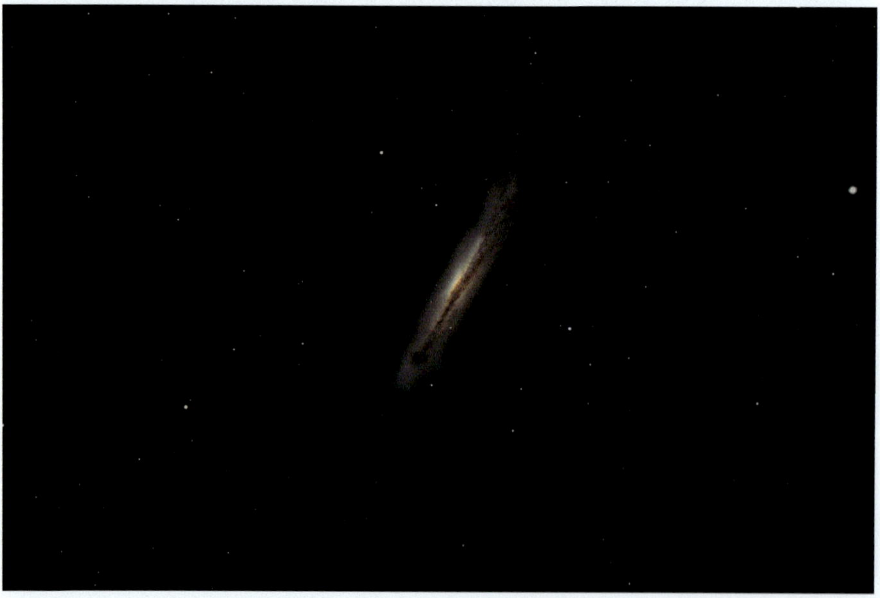

Fig. 7.27 An intermediate stage in NG3628. I wasn't sure there was anything really wrong with this: I did, however, think it looked dull. (Image by the Author)

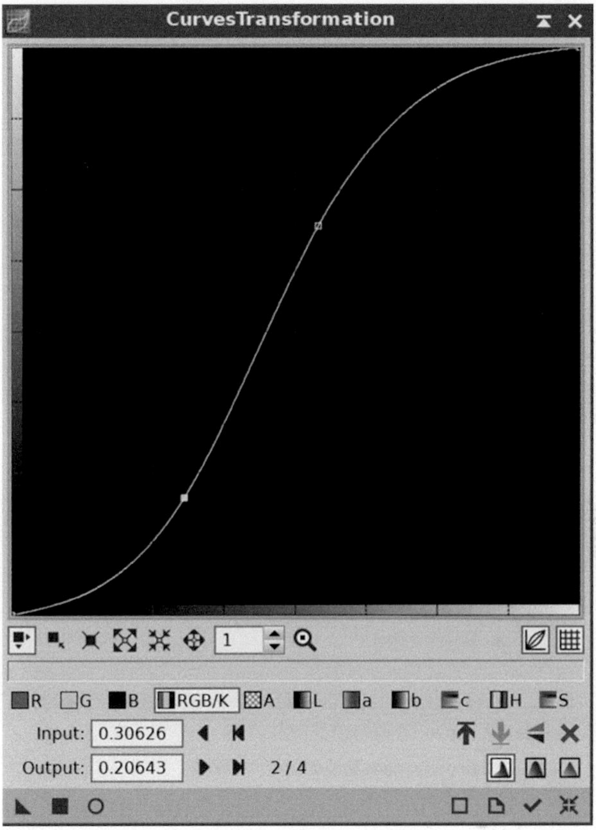

Fig. 7.28 An "S-curve" stretch to increase the contrast in an image, darkening the dark parts and brightening the bright parts. (Image by the Author)

At an earlier stage I suppressed the nose in the outer areas of the image by taking a cunning approach with the Curves Transformation tool to deal with the dust bunnies.

In Fig. 7.29, you can see that I pinned the curve at three points to prevent changes to brighter parts of the image. Then I was able to lover the brightness of only the lowest regions. That got rid of the dust bunnies.

Now I had an image of NGC3628 with which I was happy. I added the stars. The result is shown in Fig. 7.30.

The next stage, the panoramic stitching, requires that the three galaxy images be plate-solved. For that reason, I saved them in PixInsight's native file format, xisf.

PixInsight does not call Panoramic stitching by the name most photographic processing apps use. It calls it "Mosaic Merging". This is a three-stage process. The first stage is to load the images to be merged into PixInsight, and make sure that they are plate solved.

Stretched images sometimes refuse to plate solve using

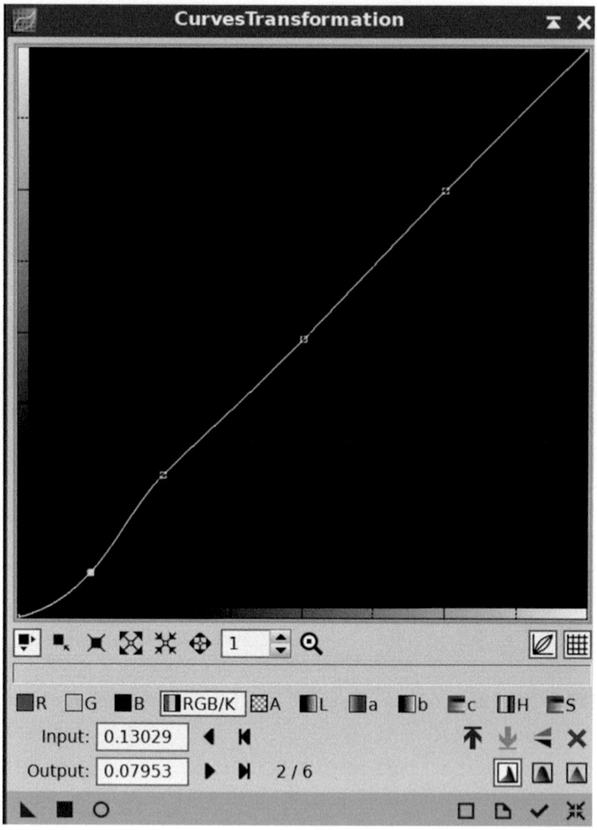

Fig. 7.29 Making use of Curves Transformation tool to suppress unwanted stuff in dark areas. (Image by the Author)

Script → Image Analysis → ImageSolver.

If that happens to you, as long as you have not cropped your image, you can use

Script → Utilities → CopyAstrometricSolution

To copy the solution from your preprocessed image into your final one.

The next step is to preprocess the images by running the script

Script → Mosaic → MosaicByCordinates,

Which is shown in Fig. 7.31. After preprocessing the images, you save them into a folder. The default values for this script are fine, but don't forget to choose an output directory.

Now comes the actual joining. This is done using the tool script

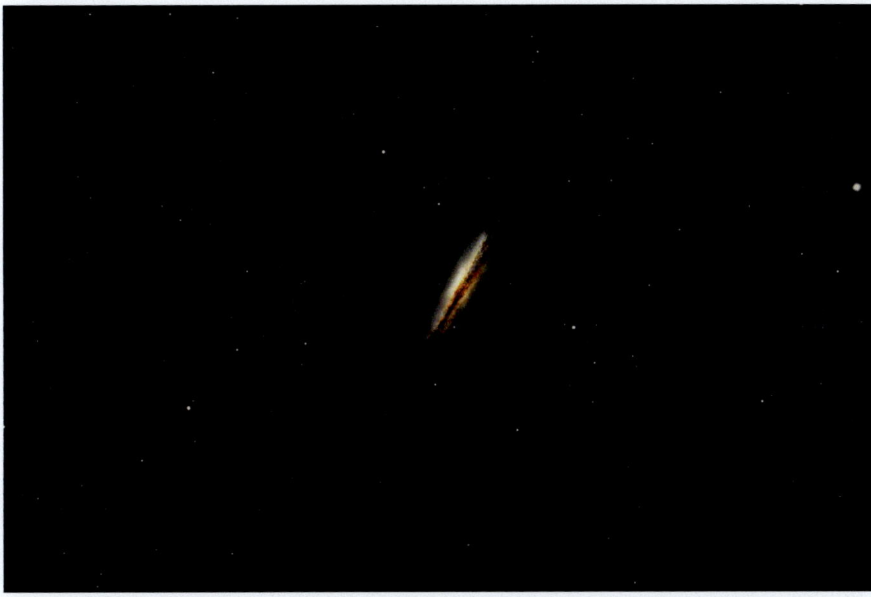

Fig. 7.30 My final image of NGC3628. (Image by the Author)

Process → GradientDomain → GradientMergeMosaic,

Shown in Fig. 7.32. With this tool, I changed from the default values to those shown in Fig. 7.32, and picked up the files saved by the Mosaic by Coordinates script. You press the solid circle at the bottom left of the tool and the images should merge. Each step of this process may take a few minutes to run, and the merged file may be large.

Finally, I used

Process → Geometry → DynamicCrop

To give me a rectangular image, since that is what people seem to like. You can rotate the rectangle you choose with this tool. After some experimentation, I chose the jaunty angle shown in Fig. 7.33.

The psychology of image composition can sometimes be unexpected. I also made a north-up version of the image, as scientific tradition seems to favor (Fig. 7.34). In this orientation, the galaxies create quite a different feeling in me.

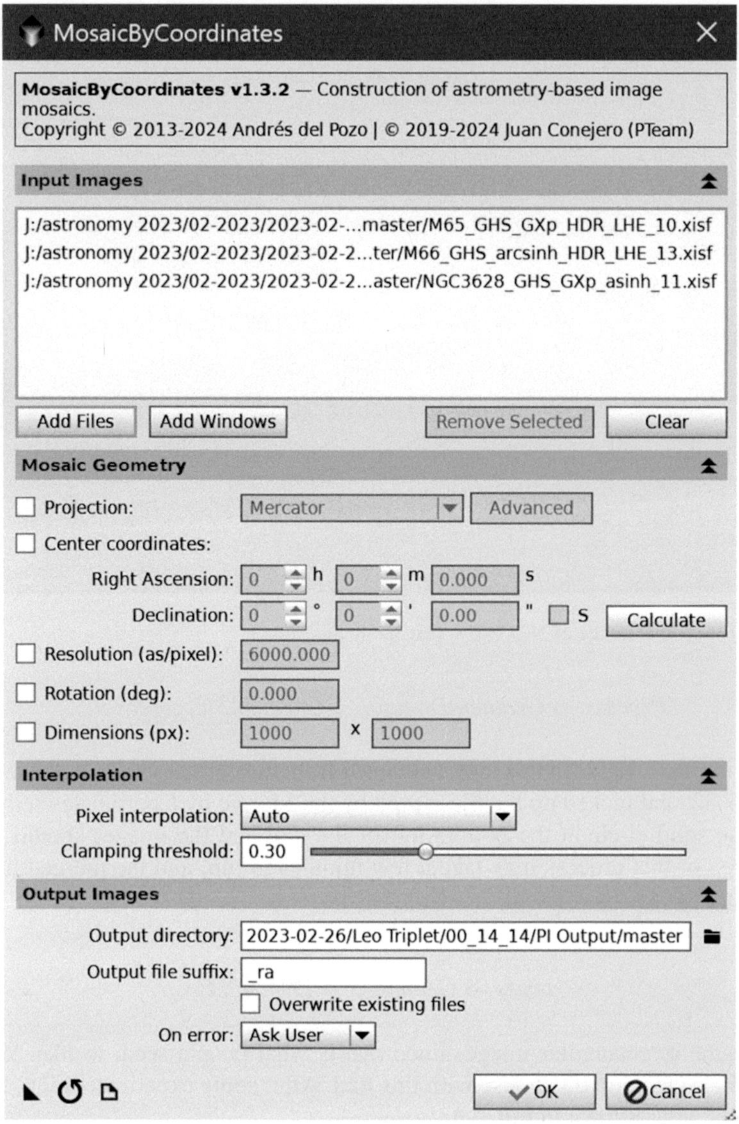

Fig. 7.31 The Mosaic by Coordinates script. (Image by the Author)

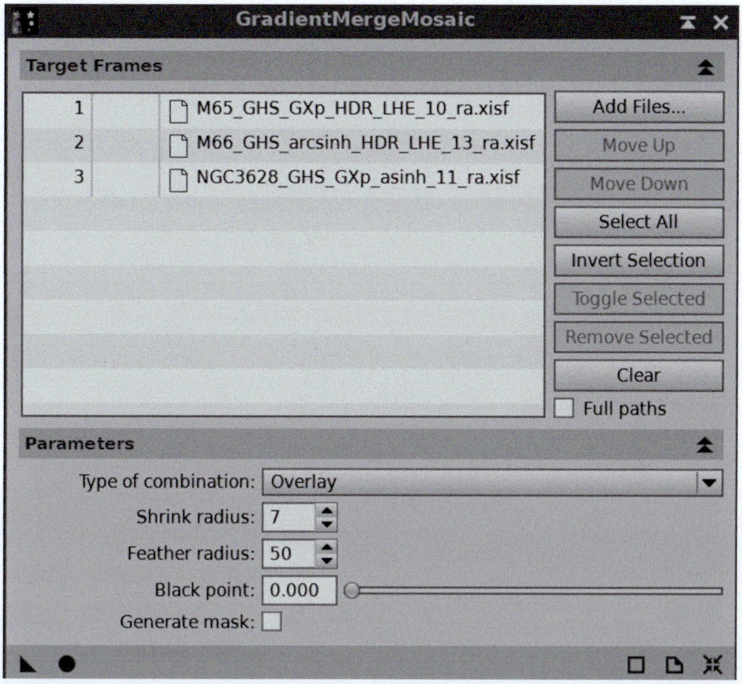

Fig. 7.32 The Gradient Merge Mosaic tool, with the values I used in this case. (Image by the Author)

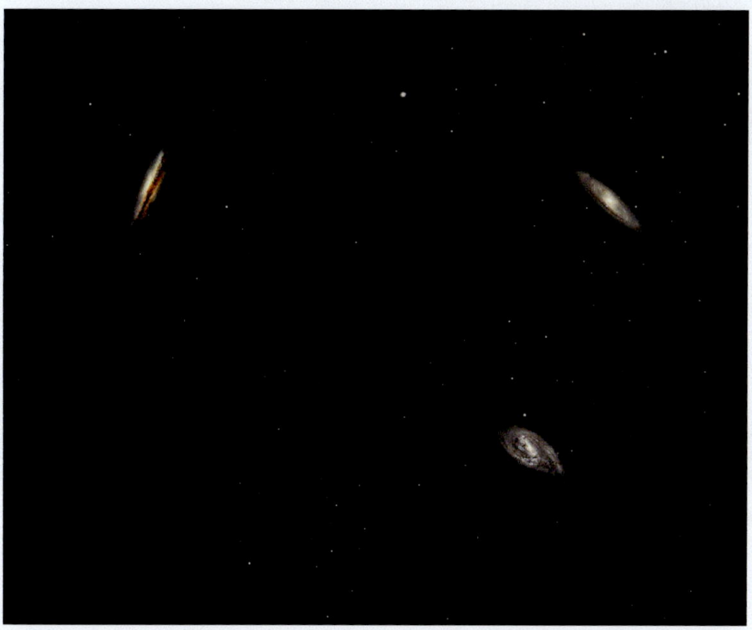

Fig. 7.33 The final, merged image of the Leo Triplet. (Image by the Author)

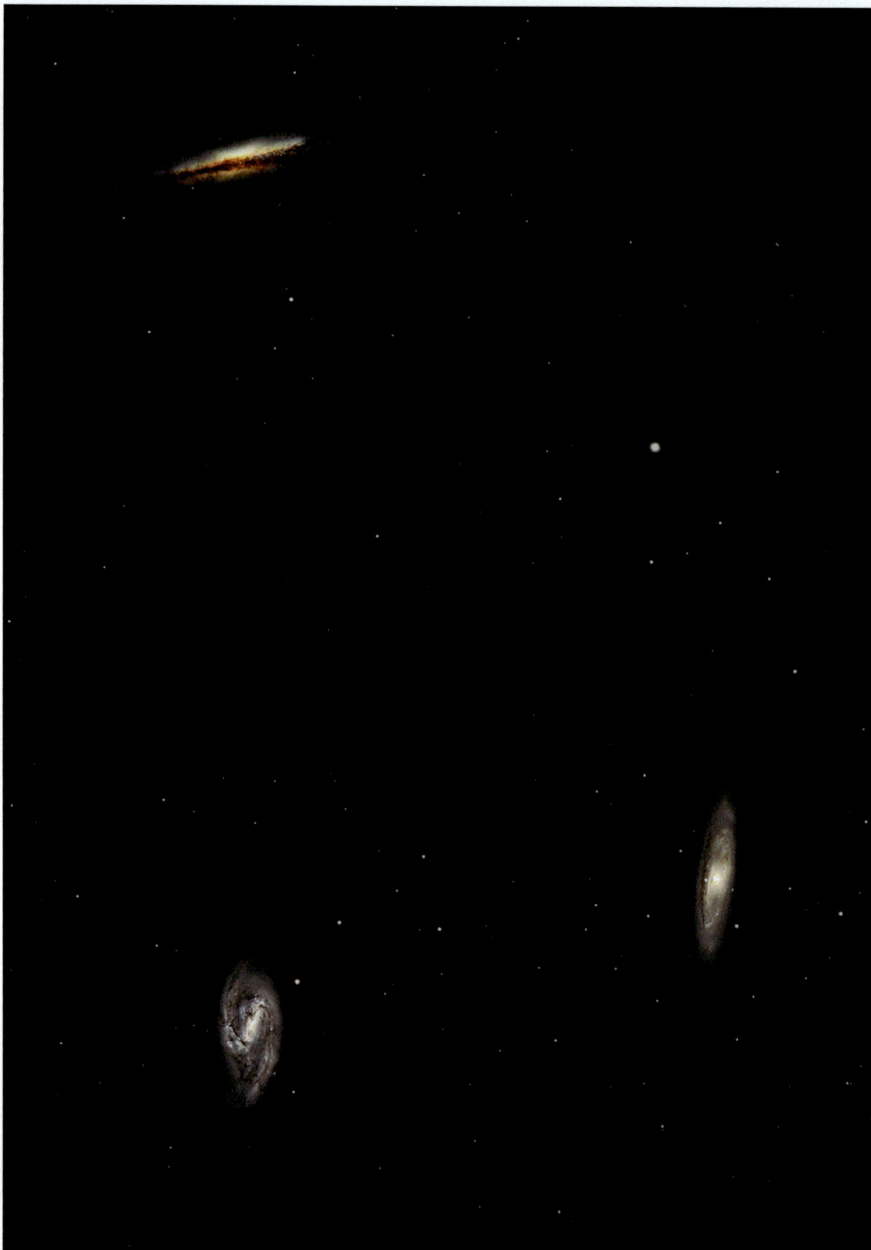

Fig. 7.34 The Leo Triplet, shown north-up. (Image by the Author)

M63, the Sunflower Galaxy Hides a Spiral

Photographing M63, or NGC5055, in white (RGB) light presents no especially interesting challenges, except that it is possible to make it come out in almost any color. A quick Google search shows images in all manner of colors. My earlier efforts at discovering the color of this galaxy were not credible: all over the place would not have been an unfair description.

I used the tool

Process → Color Calibration → SpectrophotometricColorCalibration

And, although I enhanced the color saturation, I did not deviate from the recommendation of this tool. My image is shown in Fig. 7.35.

This galaxy is what is called a *flocculent* galaxy. I have heard it claimed that "flocculent" means fluffy and flaky. It actually comes from the Latin "floccus", meaning "a tuft of wool". For a galaxy, what this really means is that you see a lumpy structure instead of spiral arms. Yet the galaxy is clearly spiral. Flocculent galaxies are not rare. M63 is the best-known example.

I found a 1996 paper [53] entitled "Uncovering Spiral Structure in Flocculent Galaxies" in which the author uses IR light of wavelength 2100 nm, i.e. about three times the wavelength of red light, to photograph flocculent spirals. I have no capability at that wavelength, but between my Baader-Bessell IR filter and my ZWO ASI1600MM monochrome camera, I can photograph at wavelengths between about 750 and 1000 nm.

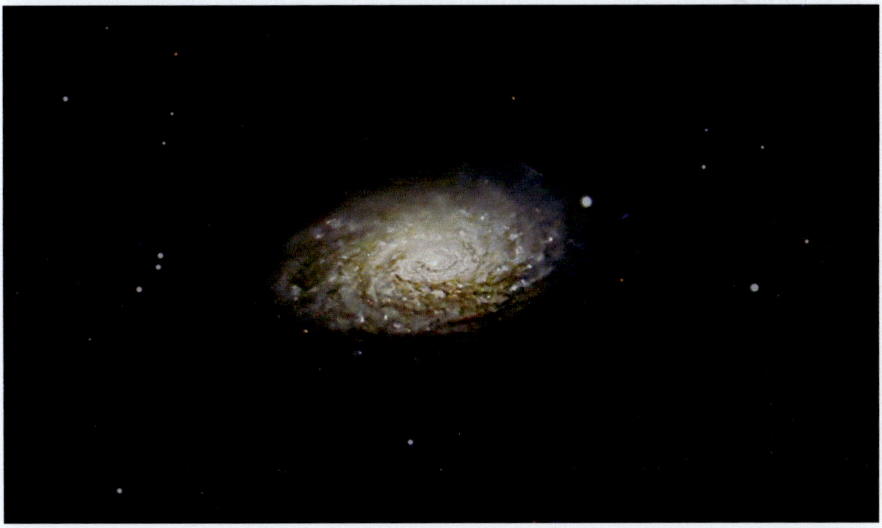

Fig. 7.35 The Sunflower Galaxy M63 in Canes Venatici. (Image by the Author)

What did I have to lose by having a go to look for the spiral in M63 in the very near infrared? The worst I could do was to fail.

My technique was the same as that in Chap. 6 when I photographed M82 in infrared. Line everything up, attain rough focus, and collimate, in visible light, then focus fully in IR on a bright star. I knew from adolescent experience that cameras of the 1970s and before often marked a focus spot on the lens with a symbol meaning that IR would be in focus at infinity at that point. It was not quite the same focus point as infinity for visible light. In the days of black-and-white film photography, using IR-sensitive film was a well-known way for landscape photographers to penetrate haze. You could recognize these photos because the sky was black with white clouds, but most members of the public were none the wiser. I never actually tried this, though I was a keen photographer in my teens, but I knew to re-focus for IR.

When this is done, find the galaxy, and off you go. The preprocessed image is shown in Fig. 7.36. It is typical of my few infrared efforts in that there's a lot of background noise.

The process I followed was:

$$Process \rightarrow IntensityTransformations \rightarrow GeneralizedHyperbolicStretch,$$

$$Process \rightarrow ObjectRecognition \rightarrow StarXTerminator,$$

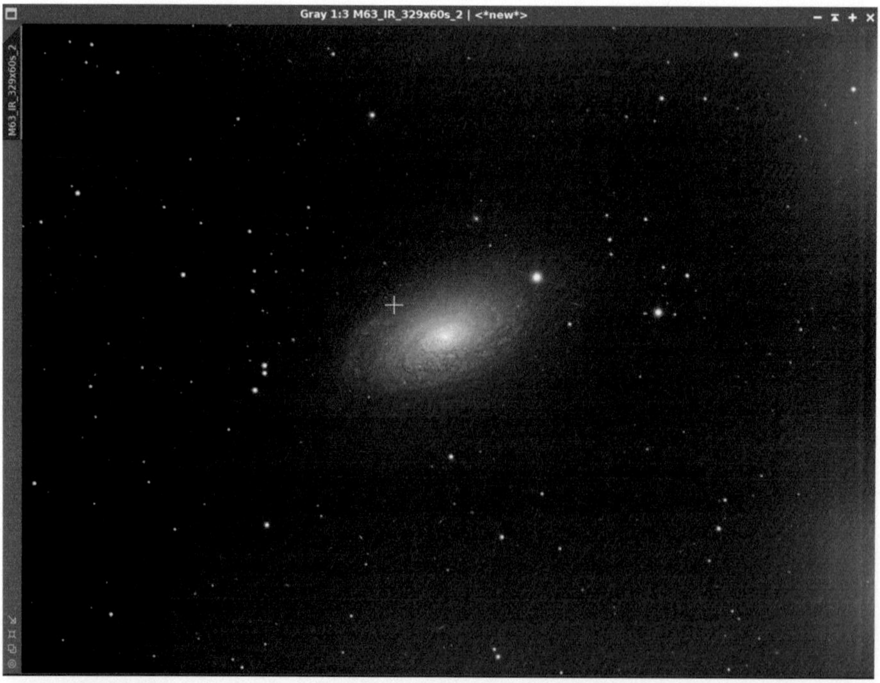

Fig. 7.36 M63 in infrared: my starting point. (Image by the Author)

$$Process \rightarrow \langle etc \rangle \rightarrow GraXpert\left(\text{in subtract mode}\right),$$

$$Process \rightarrow IntensityTransformations \rightarrow LocalHistogramEqualization,$$

$$Process \rightarrow IntensityTransformations \rightarrow CurvesTransformation,$$

$$Process \rightarrow PixelMath \rightarrow PixelMath\left(\text{to add the stars back in}\right),$$

$$Process \rightarrow IntensityTransformations \rightarrow Invert.$$

Figure 7.37 shows my final image. I can make out two spiral arms in this, so I regard my experiment as a success.

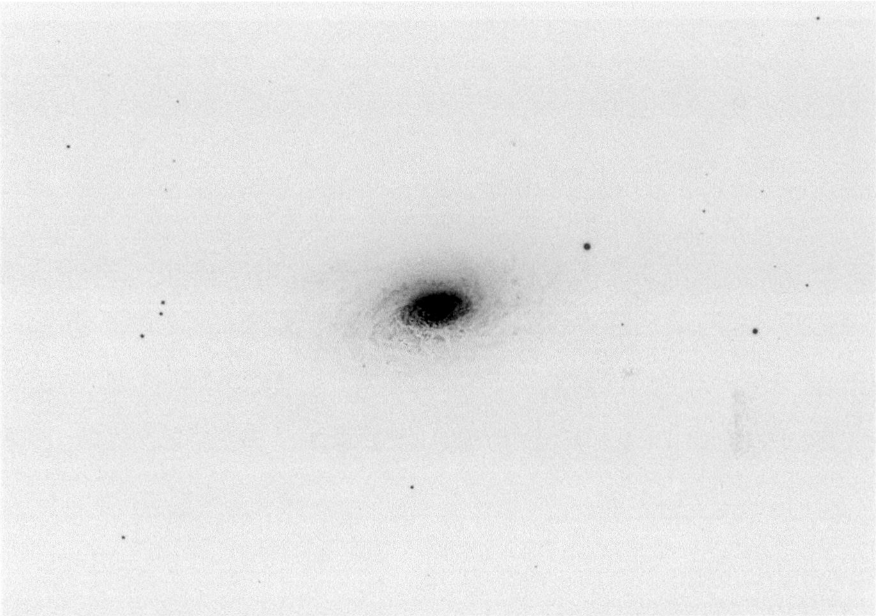

Fig. 7.37 The final infrared image of M63. I can now see two spiral arms. (Image by the Author)

M64: Black Eye or Sleeping Beauty?

Vera Rubin once wrote a paper [54] whose title was "Kinematics of NGC 4826: a Sleeping Beauty Galaxy, not an Evil Eye". As I'm sure you know, the "evil eye" is a potent symbol in folk Judaism, so it isn't an especially flattering name for a galaxy. As was noted in Chap. 1 the 'black eye' feature of M64, or NGC4826, was first seen by William Herschel.

Otherwise, M64 is a very subtle galaxy. The most notable feature to me was that the old Photometric Color Calibration tool completely failed, and gave me quite implausible colors.

This was the galaxy that forced me to download the Gaia database to enable me to use the Spectrophotometric Color Calibration. The result (Fig. 7.38) was now much more believable.

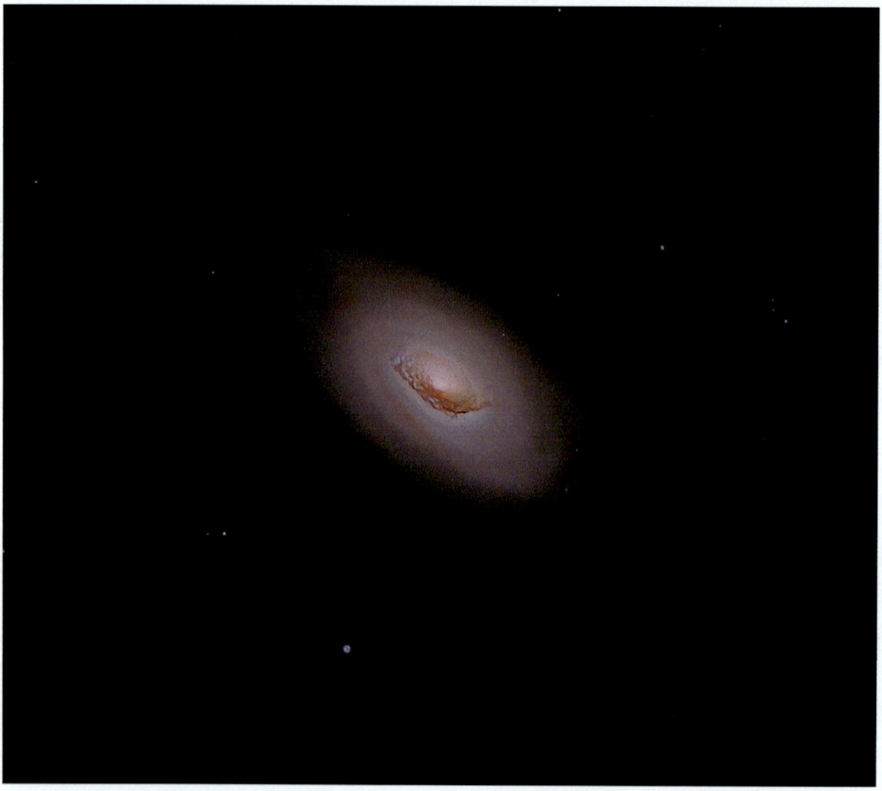

Fig. 7.38 M64 color calibrated with Spectrophotometric Color Calibration. (Image by the Author)

I also sent the image to the external application Topaz Sharpen AI, Version 1.4.0, to sharpen the dust feature that gives the 'black eye' effect.

The spiral arms are very subdued. I was quite surprised that the Local Histogram Equalization tool did absolutely nothing to enhance the arms. But that's the "sleeping beauty" aspect of this galaxy.

It is easier to imagine the back eye feature being towards us, which I imagine it is, if it is shown below the nucleus, as in Fig. 7.38, than if it is shown above the nucleus. That's another example of how it pays to think about the psychology of image composition.

Talking of Eyes

The 'Eyes' galaxies are a noticeable feature midway along Markarian's chain.

Figure 7.39 shows my preprocessed image. There is some weird stuff going on at the right-hand edge of the frame. The Eyes galaxies themselves, NGC4435 and NGC4438, are in the center-left of the frame, so I can afford not to worry about the strange stuff on the right. Their smallness does present its own challenge, though.

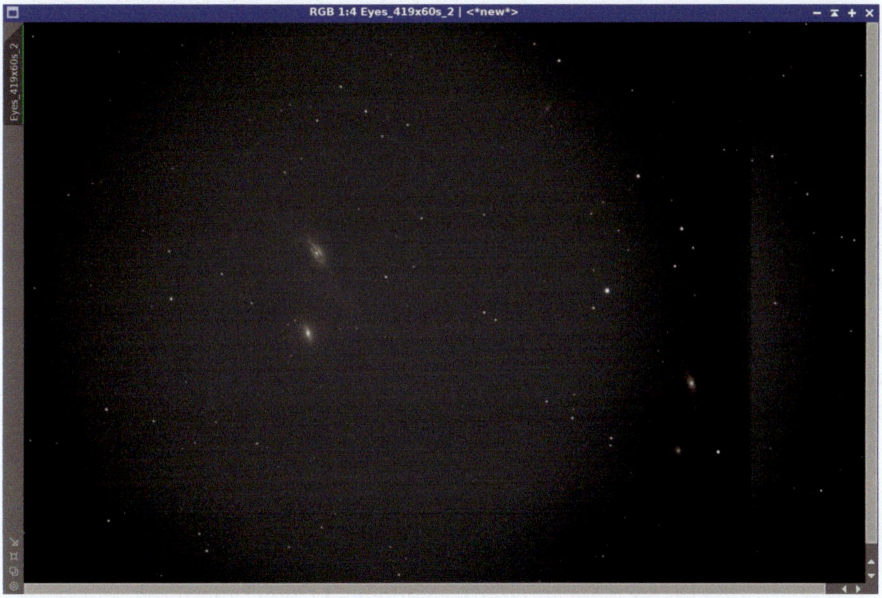

Fig. 7.39 My preprocessed image of the Eyes galaxies. (Image by the Author)

They are a long way away, 52 Mly or about 16 Mpc away. You can make a case that I should not have used the *f*/6.3 reducer-corrector on my Celestron C11. I may revisit these galaxies next year now that I have an Edge HD 1100, and see if I can do better at higher magnification, taking the 'hit' on the longer exposure time required at *f*/10; and hoping that my guiding is good enough. I think it should be. One reason for keeping my frames to 60 s is that the guiding does not need to be so good as for 5-minute frames.

I found this image difficult to process. There are 30 images in the PixInsight project file, most of which were failed attempts at processing. That's well above the average. I tried three ways of stretching:

$$Script \rightarrow EZ\ Processing\ Suite \rightarrow EZ\ Soft\ Stretch,$$

$$Process \rightarrow IntensityTransformations \rightarrow GeneralizedHyperbolicStretch, \&$$

$$Process \rightarrow IntensityTransformations \rightarrow ArcsinhStretch,$$

I soon abandoned the first of these, went some way with Generalized Hyperbolic Stretch before abandoning it, and eventually settled on Arcsinh Stretch.

As usual I removed the stars, and tried removing the background with

$$Process \rightarrow BackgroundModelization \rightarrow AutomaticBackgroundExtractor.$$

This tool failed to remove much background. Next I tried

$$Process \rightarrow \langle Etc \rangle \rightarrow GraXpert,$$

And did some stretching. Now I had an image I could work with: the stuff I needed to jettison was not mixed in with my target. This stage is shown in monochrome in Fig. 7.40.

I made a circular mask so that I could attack the ring of light, making first a binary masks with the GAME script, then making its edges fuzzy with the Range mask tool (Fig. 7.41).

I did a few steps with

$$Process \rightarrow IntensityTransformations \rightarrow Histogram\ Transformation.$$

And then enhanced the color with

$$Process \rightarrow IntensityTransformations \rightarrow ColorSaturation.$$

Then some more use of the histogram Transformation tool, and I was beginning to gat an image I liked.

I manually enhanced the colors with

$$Process \rightarrow ColorCalibration \rightarrow AssistedColorCalibration$$

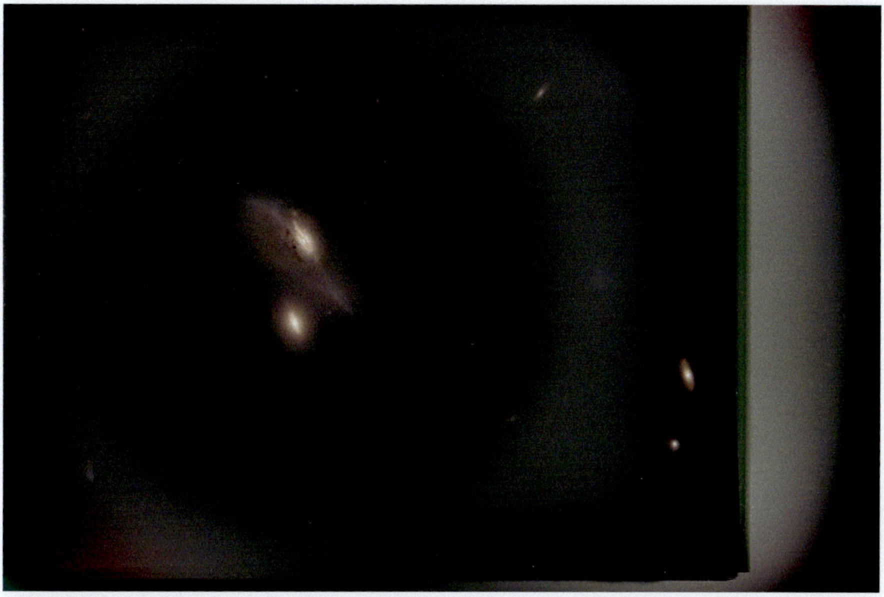

Fig. 7.40 An intermediate step in processing the Eyes galaxies. (Image by the Author)

Fig. 7.41 The mask I used to protect my Eyes galaxies image while I suppressed unwanted background. (Image by the Author)

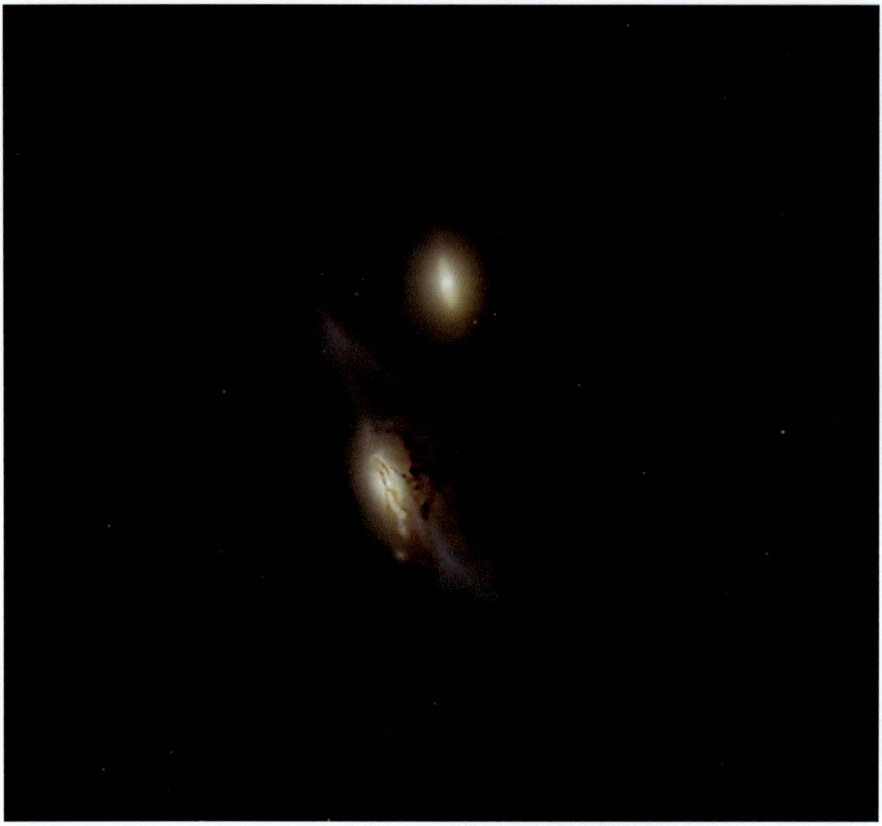

Fig. 7.42 My final image of the Eyes Galaxies, NGC4435 & NGC4438. (Image by the Author)

because I thought the image looked too red and not blue enough.

I now rotated the image to north-up, added the stars, cropped the image and sent it to Topaz Denoise AI to sharpen the dust – this app sharpens as well as de-noises. I would not now do that: I would de-noise before adding the stars, as the Topaz application sometimes messes stars up.

The final result is shown in Fig. 7.42.

I have many examples of galaxies around the Virgo cluster, and hope to image quite a few more when circumstances allow. I hope I have given you some ideas of what to photograph in this cluster and how to photograph them.

Chapter 8
Going South

For Southern Hemisphere dwellers, these are your 'low-hanging fruit galaxies. For a northerner like me, most of them are completely out of view, though I have included a few that are so low on my horizon that they require several nights to obtain a decent image because they rise late and set early. With the British Climate… .'nuff said.

I subscribed to a remote telescope service to satisfy my curiosity about what is out there.

Telescope Live

I'm sure that Telescope Live isn't the only company to offer the services it does, but it is the one I went with. It can be found at https://telescope.live/home.

I've been paying a monthly subscription There are several levels of subscription. I went for the lowest level that got me access to the tutorials. There's overabundance of tutorials there now. It is almost worth the subscription for these, but you certainly need the basic ones.

The files you buy using the credits from your subscription are unprocessed, except that they have been calibrated with darks, flats and bias frames.

The deal is basically that you own the rights to any processing that you do, but you must credit them with either of two sets of wording, which they specify:

<Your Name> / Telescope Live, or
Image processed by <Your name>, original data from Telescope Live.

If you are a Southern hemisphere dweller, and want some northern galaxies to play with, you can do much worse than to check out the ones listed earlier in this book.

J. Clark, *Photographing Galaxies from Light Polluted Skies*, The Patrick Moore Practical Astronomy Series, https://doi.org/10.1007/978-1-0716-4506-2_8

Table 8.1 The twenty brightest galaxies with southern declinations

Galaxy	Blue magnitude	Constellation	RA (H)	Ra (min)	RA (sec)	Dec (°)	Dec (min)
LMC	0.63	Dorado	5	24	0	−69	48
SMC	2.79	Tucana	0	51	0	−73	6
NGC 5128	7.89	Centaurus	13	22	32	−42	45.5
NGC 253	8.13	Sculptor	0	45	8	−25	33.7
NGC 55	8.22	Sculptor	0	12	24	−39	28
NGC 5236	8.51	Hydra	13	34	10	−29	36.8
NGC 300	8.7	Sculptor	0	52	31	−37	57.4
NGC 6744	9.24	Pavo	19	5	2	−63	56.3
NGC 4594	9.28	Virgo	12	37	23	−11	21
NGC 6822	9.35	Sagittarius	19	42	7	−14	55.7
NGC 1313	9.37	Cetus	3	17	39	−66	40.7
NGC 1291	9.42	Eridanus	3	15	29	−41	18.5
NGC 247	9.51	Cetus	0	44	40	−21	2
NGC 1316	9.6	Fornax	3	20	47	−37	23.1
NGC 4945	9.6	Centaurus	13	2	32	−49	12
NGC 7793	9.65	Sculptor	23	55	15	−32	52.1
NGC 3115	9.98	Sextans	10	2	44	−7	28.5
NGC 3621	10.03	Hydra	11	15	50	−32	32.4
NGC 4697	10.11	Virgo	12	46	1	−5	31.7
NGC 1097	10.16	Fornax	2	44	11	−30	29.1

Source: Reference [36]

If you are a Northern Hemisphere dweller, Table 8.1: The twenty brightest galaxies with southern declinations lists the twenty brightest galaxies in the Southern Celestial Sphere. This list will give you a good start.

How to Use WBPP to Preprocess Separate LRGB Images

The procedure may be less automatic in other preprocessors but it won't be very different.

The images pick up information from the headers in FITS files, including exposure length and the filter used, provided of course that your image capture software writes this information to the FITS file. It also picks up Right Ascension and Declination information, and focal length information.

As Fig. 8.1 shows, the light frames are arranged by color filter and, although this is not shown, by exposure time. It will appropriately process these files, and stack the L, R, G & B images correctly, and separately.

It is noticeable that there are focal lengths missing in the table of frames in Fig. 8.1. Experience with my own images tells me that the image solution in WBPP can fail if incorrect focal lengths are given. In the instance shown above, the reference image the software chose must have been a correct one, so I got away with it

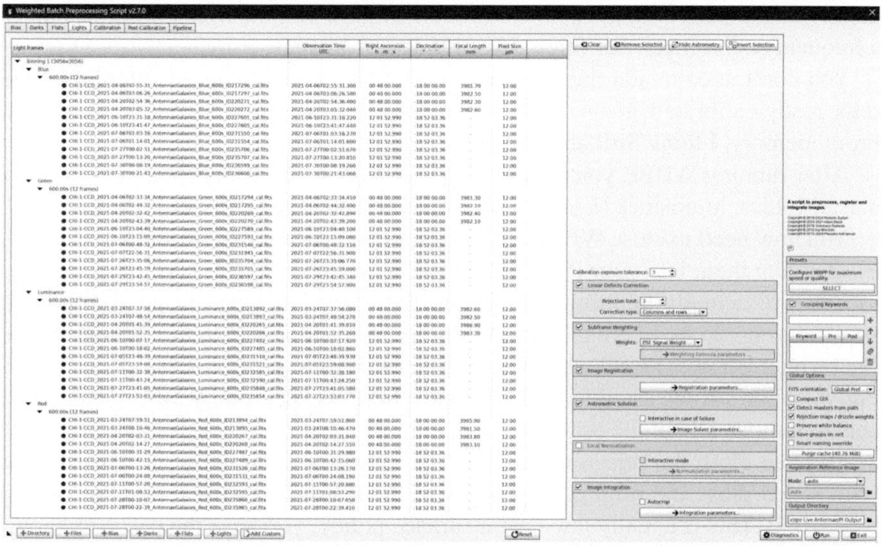

Fig. 8.1 The lights tab in WBPP for LRGB images. (Image by the Author)

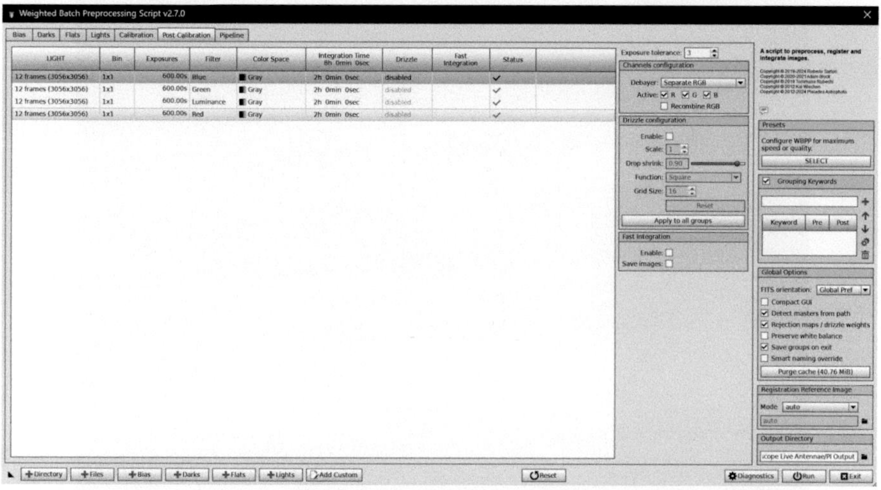

Fig. 8.2 The post-calibration window settings to process separate channels. (Image by the Author)

this time. But actually, this is no more than a minor irritant. You can re-plate-solve later.

Don't forget to add the correct output directory.

In the post calibration tab, note (Fig. 8.2) that I have checked the separate channels option in the Channels configuration. Not also that if you have a mix of 300 s and 600 s exposure times, you need to make the number "Exposure tolerance" greater than 300 in *both* the post-calibration window and the lights window. This is a frequent occurrence with Telescope Live sets of Images.

You don't need to add flat, dark and bias frames to calibrate, as Telescope Live sells ready-calibrated images. WBPP accepts this situation, but does warn you. From memory, I think Siril also issues a warning.

After running WBPP, you then end up with four separate master images in your master folder: luminance (L), red (R), green (G) and blue (B).

You now need to close WBPP and import these four master files into PixInsight, by dragging them into your PixInsight window.

You will actually find three images in PixInsight for every image you thought you had dragged in. There is also a high-rejection and low-rejection image. These can be interesting to reveal what defects have been removed. The example shown in Fig. 8.3 shows some satellite trails that have been weeded out. WBPP seems to be able to remove all but the brightest satellite trails.

You can close the rejection images. My advice is to duplicate the actual light frames and close the originals, to preserve them from accidental damage during processing. This is a really low-cost insurance policy: you have the peace of knowing that the worst you may have to do is start gain.

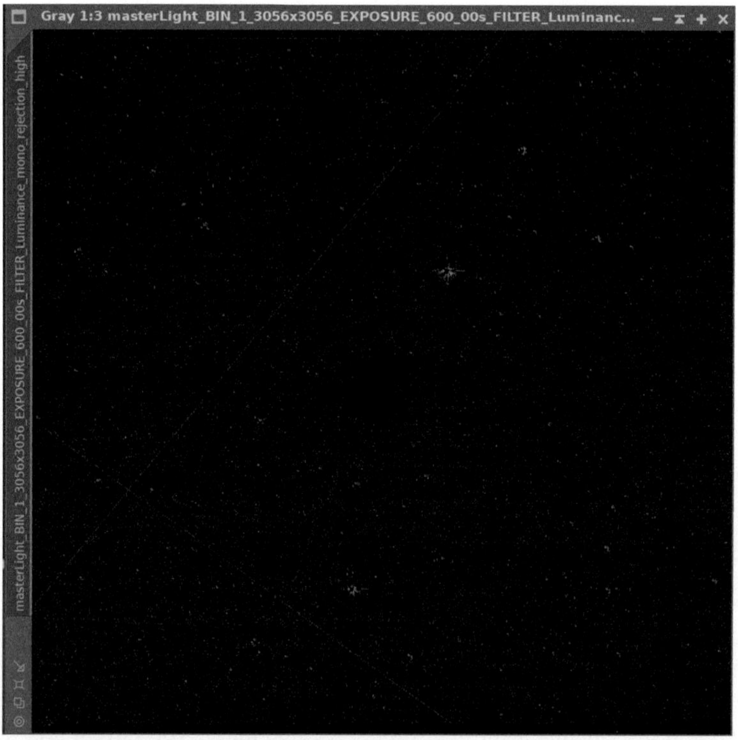

Fig. 8.3 A high-rejection image, temporarily stretched, showing satellite trails that have been eliminated. (Image by the Author/Telescope Live)

Now we need to combine the L, R, G & B frames. There are tools to do this in PixInsight. One is

Process → ChannelManagement → ChannelCombination.

However, I have not used any of the built-in tools. I did the channel combination in Pixel Math. This is because I have found in the past that the channel combination tool puts too much emphasis on the L channel, which overwhelms the color channels.

I have tried adding the luminance channel to each of the R, G & B channels, and I have tried multiplying the R, G & B channels by the L channel. Multiplication gives much more detail. Figure 8.4 shows the Pixel Math settings, with the green channel shown. I have added a factor of 1.00 into my multiplication. Multiplying by one changes nothing, of course, but I have the option to experiment with changing that number.

So my recommended formulae are

Red: R × (L × weighting factor)
Green: G × (L × weighting factor)
Blue: B × (L × weighting factor)

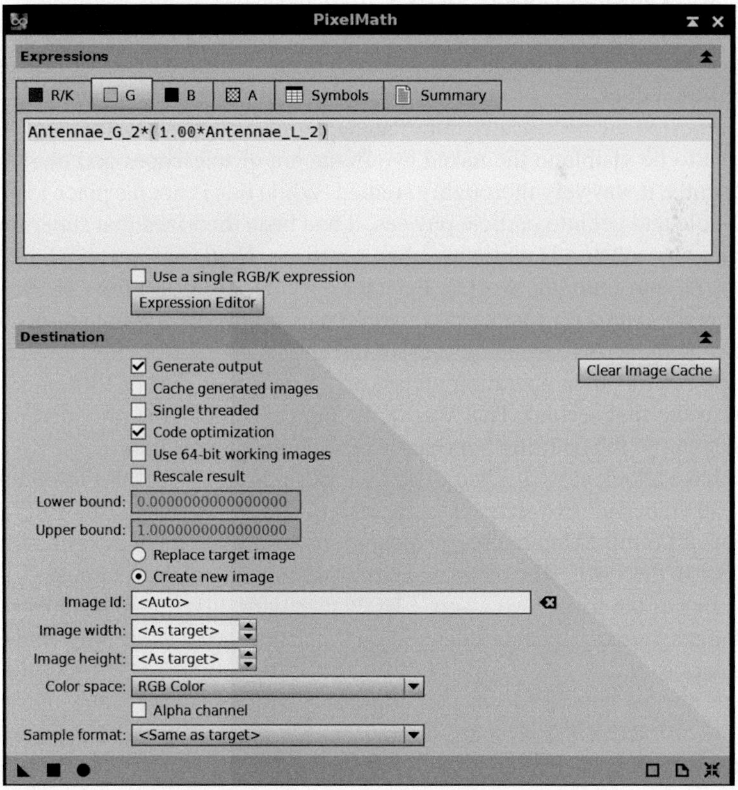

Fig. 8.4 The G window in Pixel math when combining the channels. (Image by the Author)

By clicking on the square button at the bottom of the Pixel Math window, you will create an LRGB image.

From here, you can proceed to process in the same way as for a one-shot color camera.

The Large Magellanic Cloud (LMC)

This galaxy is a satellite of the Milky way, about which most of us in the north have heard, but know little. That certainly applied to me; and it was the first object that came into my mind when I wanted to satisfy my curiosity about the galaxies of the south.

Therefore, this item topped my list of objects to process from Telescope live; and I was not disappointed. I learnt a lot about this galaxy. It helped that I had to give a talk in a hurry to Cardiff astronomical Society because there was an unplanned gap in their schedule; and I offered to learn what I could about the LMC in a few days.

It contains the brightest nebula in the Local Group of Galaxies, the Tarantula Nebula, also known as Dorado 30, NGC2070 and C103. If this were the same distance from us as the Orion Nebula, it would be visible in daylight, and occupy a quarter of the sky. That would have an even worse effect on galaxy photography than the Moon does.

It also hosted the historically important supernova SN1987A. This was the only supernova to be visible to the naked eye in the era of telescopes and photography. Consequently, it was very thoroughly studied. While this is not the place to go down a rabbit-hole and get into particle physics, it had been theorized that supernovae are blown apart by subatomic particles called neutrinos. Neutrinos are very hard indeed to detect. At that time, the world's detectors, located deep in mines to avoid contaminating the data with cosmic rays, would pick up maybe a couple of neutrinos a day between them. A few minutes before the optical signal from SN1987A arrived, the three detectors then operating picked up 25 neutrinos during 13 seconds, most of them in the first second. That was quite unprecedented, and provided powerful reinforcement of the neutrino hypothesis of supernova explosion.

Let's have a look at what I had to do to tease an image out of the Telescope Live data. I downloaded two sets of data taken with their Takahashi FSQ-106ED (106 mm) *f*/3.6 refracting telescope located in Australia. One was centered on the central bar in the LMC. The other was centered on the Tarantula Nebula.

The ones of the main LMC were LRGB files with differing exposures of 300 s and 600 s per frame. The ones centered on the nebula were RGB only frames with varying exposure times per frame of 300 s and 450 s.

I have always wondered whether luminance frames added value. Early color televisions worked with one sharp luminance (i.e. black & white) image and three superimposed R, G & B images at much lower definition. The idea was that this would look reasonable at a cost people could afford. Early color televisions were expensive. It also relied on the fact that the human eye has about 30 luminance sensors (rods) for every RGB sensor (cones). On the forums like Cloudy Nights and Stargazers' Lounge, much ink seems to be spilled on this subject. In theory, because the eye works more on luminance than color, L frames ought to help. But we have an opportunity to test the theory here.

I preprocessed all the frames in the usual way, with one run through WBPP for each image.

For the main LMC I used Pixel Math to make an LRGB image using the following settings for the frames:

Red: R × (L × 1.00),
Green: G × (L × 1.00),
Blue: B × (L × 1.00).

I then stretched the LRGB image using

Process → IntensityTransformations → ArcsinhStretch.

I actually did two arcsinh stretches, the first one stretching as far as possible; and the second one stretching only a little more. Then I removed the stars using

Process → ObjectRecognition → StarXTerminator.

At this point a little problem made itself known.

Red defects were clearly going to ruin the image (Fig. 8.5). Could I make masks to get rid of them? No. I tried that. Would using Generalized Hyperbolic Stretching help? Nope. My most ingenious idea was to try to modify the Pixel Math used to create the image:

Red: iif(LMC_R_2<0.975,LMC_R_2*(1.00*LMC_L_2),0),
Green: G × (L × 1.00),
Blue: B × (L × 1.00).

In other words, I tried to kill off any very bright red pixels by replacing them with black, i.e. zero brightness. Brilliant though this was, it did not work at all. I even tuned the 0.975 number to try & make the method work. The failure remained just as miserable. My best ideas are lost on this world.

Now what? I tried omitting stages. First the noise extermination.

No dice.

Then the blur extermination and the noise extermination.

That worked!!

The BlurXTerminator was causing the defects. I added the noise extermination back in, but omitted the blur extermination. That also worked. Repeat the arcsinh stretching and I was back on track, but the image was a bit blurry. I doubt if I would even have noticed that a year ago, but I notice now.

Everything I have ever read or heard about noise and blur extermination says that you should do it before you do any stretching. Well, tough. That hadn't worked.

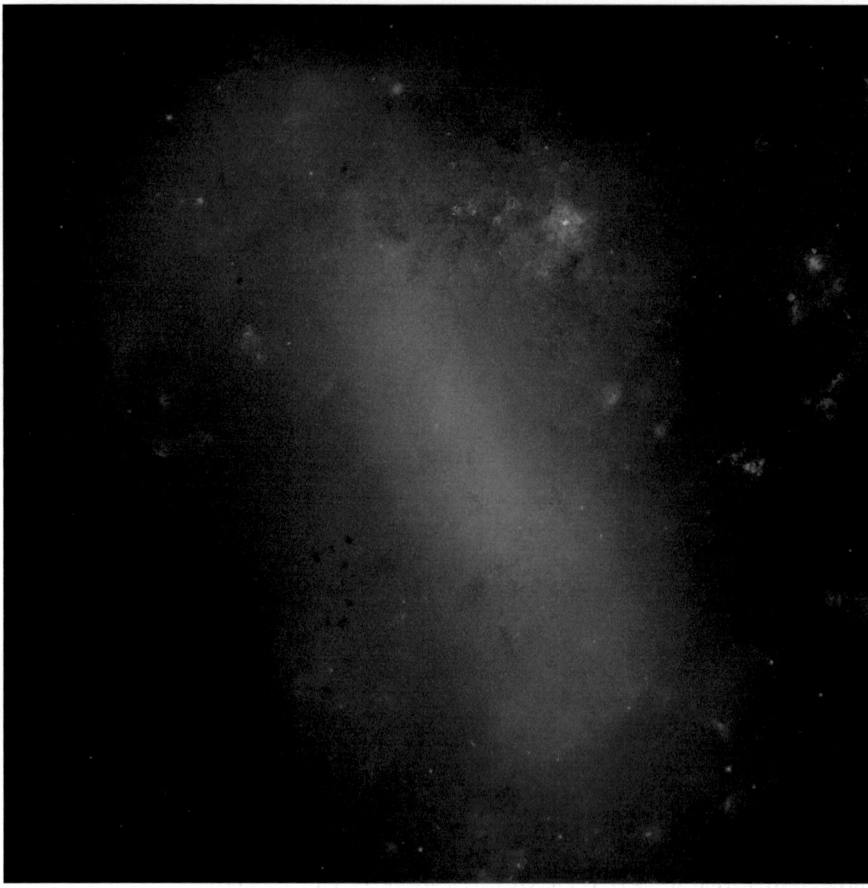

Fig. 8.5 This otherwise good image is marred by red defects. There are several. The most visible is at the top left. (Image by the Author/Telescope Live)

What did I have to lose by doing a blur extermination now, with the default settings? Nothing, so I tried it, and it worked, perhaps not quite as well as it would have on the linear image, but at least the stars looked respectably sharp. So I kept the deblur-while-stretched image.

The brightness of the Tarantula nebula required attention. It was whited out. The solution seemed to be to make a mask. I made the Range Mask in Fig. 8.6 using

$$Process \rightarrow MaskGeneration \rightarrow RangeSelection.$$

I tried the default settings in

$$Process \rightarrow MultiscaleProcessing \rightarrow HDRMultiscaleTransform.$$

Fig. 8.6 A range mask to operate on only the brightest parts of the image. (Image by the Author/Telescope Live)

This worked very nicely, so I added the stars back in with Pixel Math, and obtained the image shown in Fig. 8.7.

Next, I repeated the process for the second image, centered on the Tarantula nebula. This was taken with the same telescope and the same CCD camera, a FLI PL16083 with 4096 × 4096 9 μm pixels.

This time, there was no L image to add in, so I used Pixel math to make an RGB image using

Red: R
Green: G
Blue: B

This time round, the kinds of defects shown in Fig. 8.5 did not put in an appearance. I held the image of Fig. 8.7 in front of me in the PixInsight window, to make

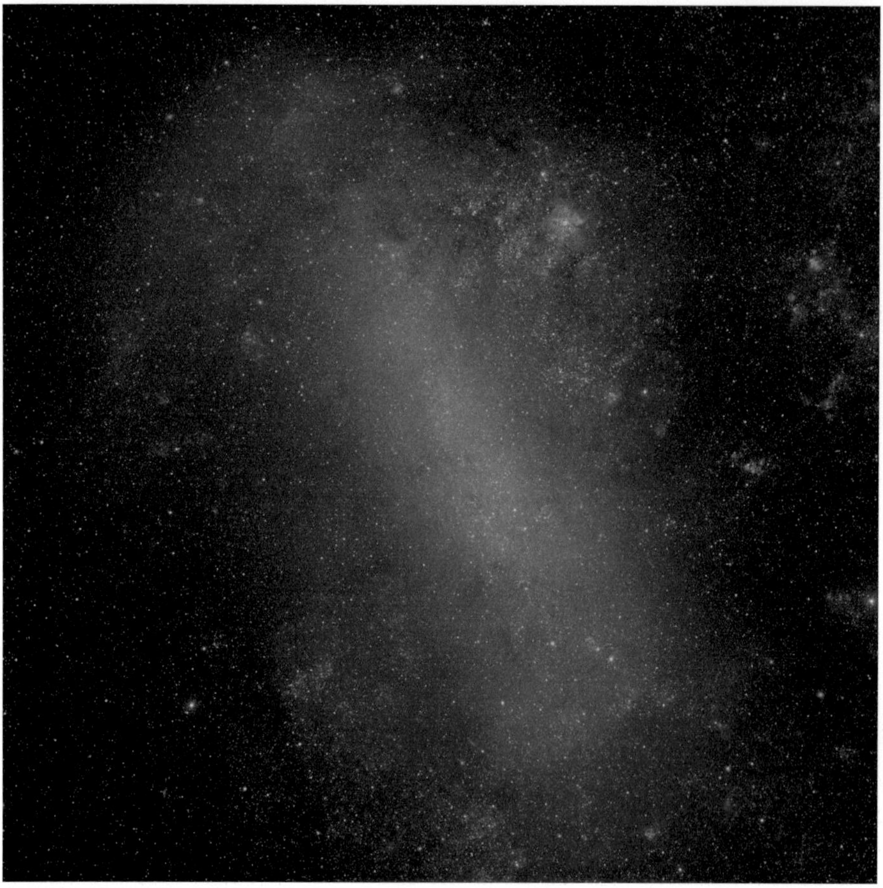

Fig. 8.7 The central part of the LMC. Note that the Tarantula nebula and the other nebulas are not whited out. (Image by the Author/Telescope Live)

the Arcsinh stretch look as much like this as possible, so that I would be able to stitch the images together. The final image is shown in Fig. 8.8.

So, despite my effort to make the frames similar, has the luminance image added value? I think it has. It does seem to have brought out fainter detail.

To stitch the two images together, you first load the files you want to merge into the Mosaic by Coordinates tool (Fig. 8.9). This makes images the size of the final mosaic, and registers the two images. The images have to have been plate solved to make all this work. I inverted the image in Fig. 8.8.

They are actually merged by the Gradient Merge Mosaic tool, which is at

Process → GradientDomain → GradientMergeMosaic.

Fig. 8.8 My final image of the central part of the LMC. (Image by the Author/Telescope Live)

I had to try different values of the parameters Shrink radius and Feather radius to prevent the join from showing. I also found that the Type of combination worked better as Average than as Overlay. In the end I used the parameters shown in Fig. 8.10.

There was one wrinkle left – an aesthetic wrinkle more than anything. My mosaic was not rectangular, and could not be made so. I thought that the default black background looked a little bit stark. I therefore made a background which was a similar color to a typical background of the actual image, only darkened. You can get PixInsight to display the RGB combination at any point in your image. My background colors were:

Red: 0.06
Green: 0.07
Blue: 0.05

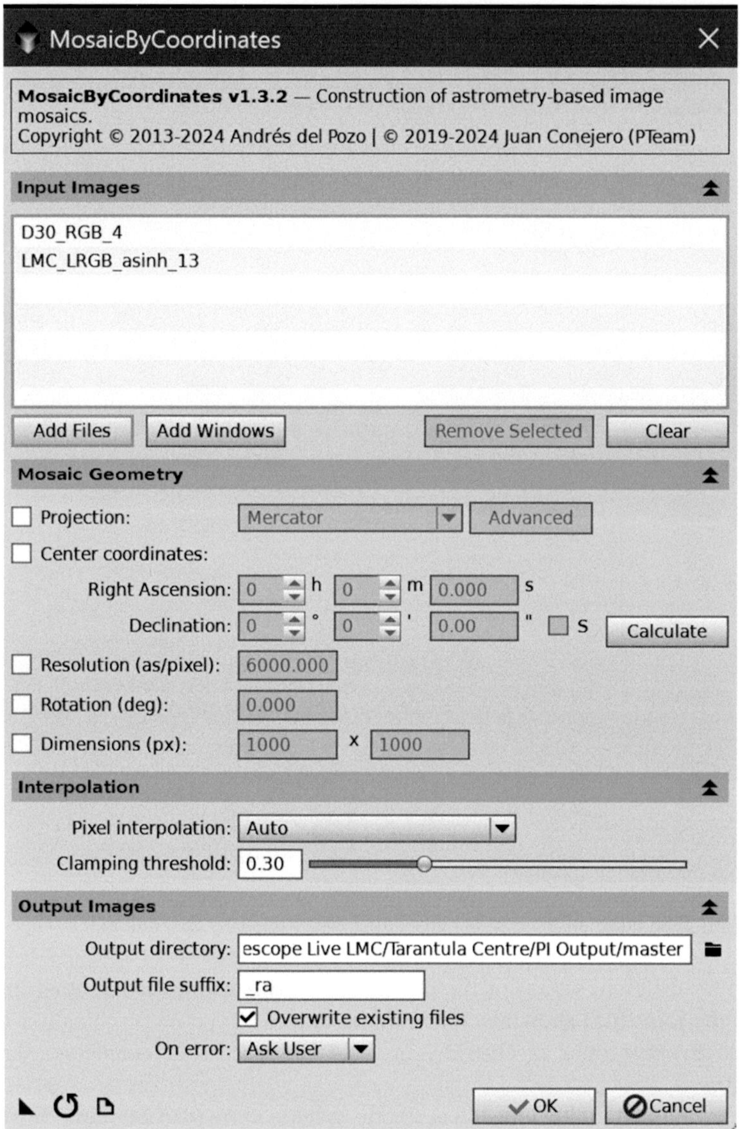

Fig. 8.9 The Mosaic by Coordinates script, under Script → Mosaic. (Image by the Author)

You make the actual image your target and add these colors as numbers in Pixel Math, making sure to un-check the "Use a single RGB/K expression" and to make the color space RGB. Pixel Math will then make an image of uniform color of the same size as the target image. I called it Bkgrd_Grey_4. I then went back in Pixel

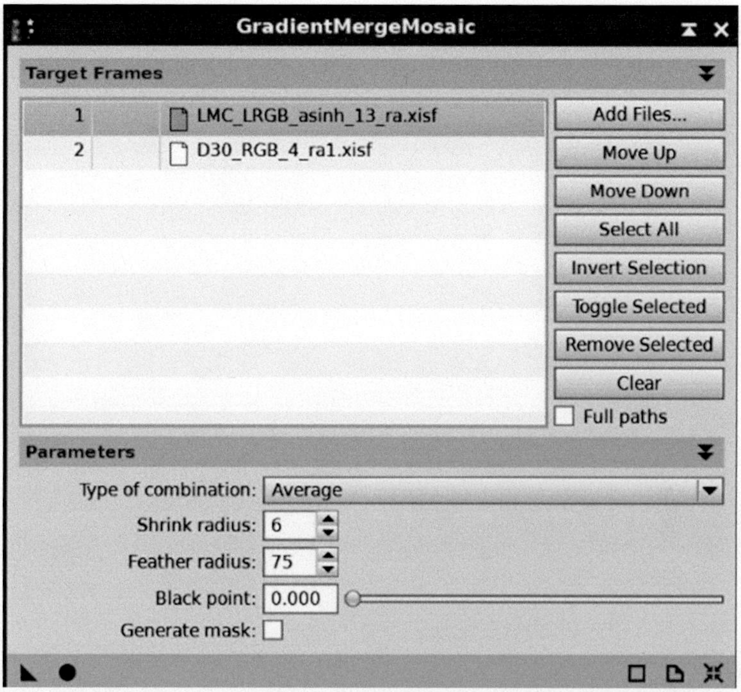

Fig. 8.10 These are the final values I found for the Gradient merge Mosaic tool. (Image by the Author)

Math to using a single RGB/K expression, and making the color space the same as the target; and used the expression "max(LMC_merged_4,Bkgrd_Grey_4)" to change the color of the missing parts of the image.

Figure 8.11 shows my finished image of the LMC.

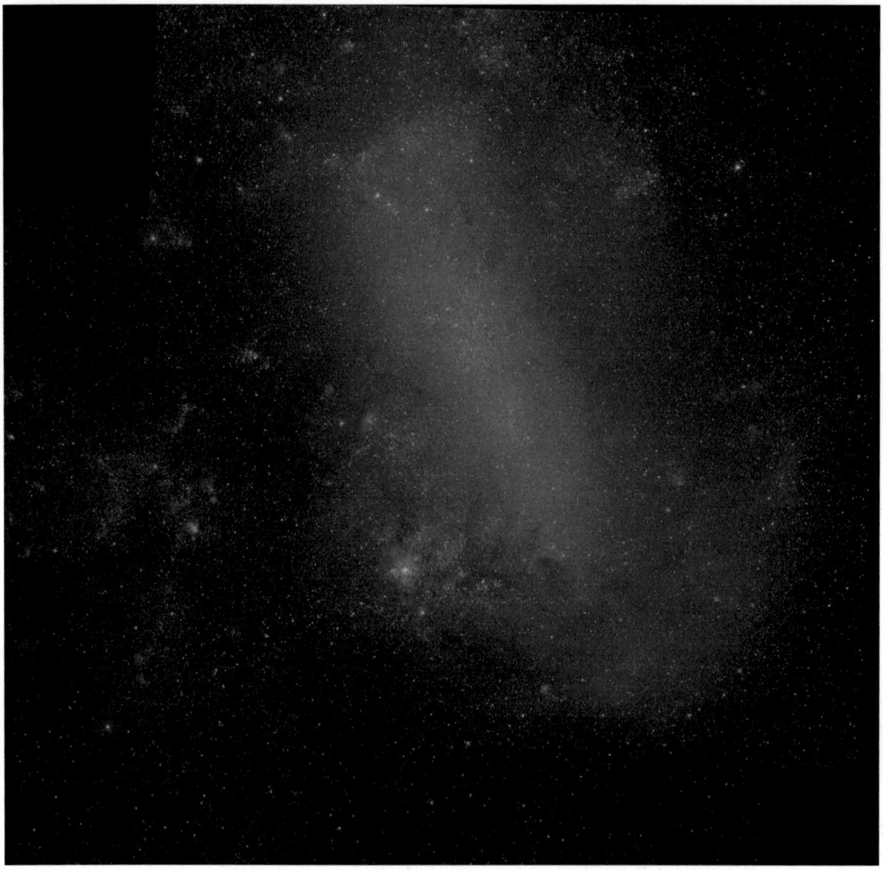

Fig. 8.11 My finished image of the LMC. (Image by the Author/Telescope Live)

Linear Defect Removal

This is a particular difficulty with CCD cameras, in that they can leave vertical or horizontal lines of noise in the image. WBPP knows all about this, and runs linear defect correction process to deal with it. Most satellite trails get removed at the same time.

When, afterwards, you load a master light image into PixInsight for postprocessing, you get three images: the light frame itself, and two other images showing low and high rejection material respectively. I always check these to see if anything interesting was removed.

Figure 8.12 shows an example. Both vertical camera defects and randomly-angled satellite trails have been removed from the master light frame.

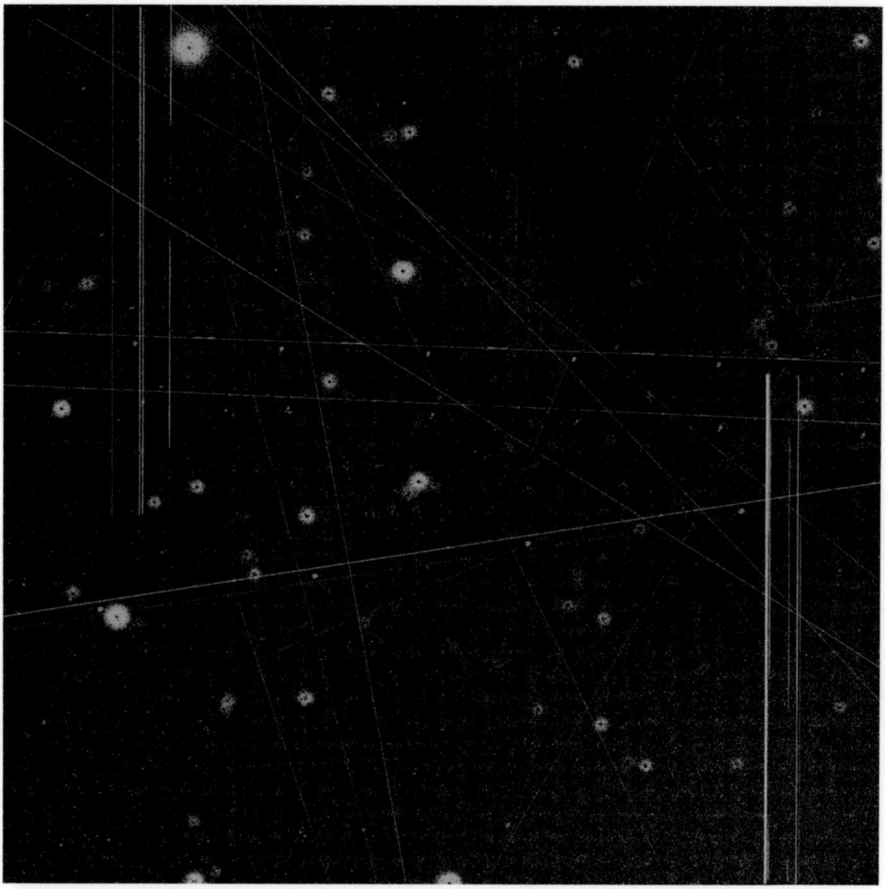

Fig. 8.12 An example of a rejection-high image. Both satellite trails and camera defects have been removed by WBPP. (Image by the Author/Telescope Live)

But what if WBPP fails to remove a camera defect? This happened to me with a master blue frame from the Small Magellanic Cloud (SMC), shown in Fig. 8.13.

It took me a while to figure out what to do about this. There are horizontal and vertical defect removing tools in PixInsight, which is what the support line at Telescope Live advised me to try. This was sensible advice, only it didn't work.

Therefore, I had to try to find a way to overwrite the defect manually. I reasoned, and indeed discovered the hard way, that if I overwrote the linear defect with a uniform color, no matter how carefully chosen, it would still show.

What I needed to do was to overwrite the defect with a noisy rectangle. With any luck, that would not show.

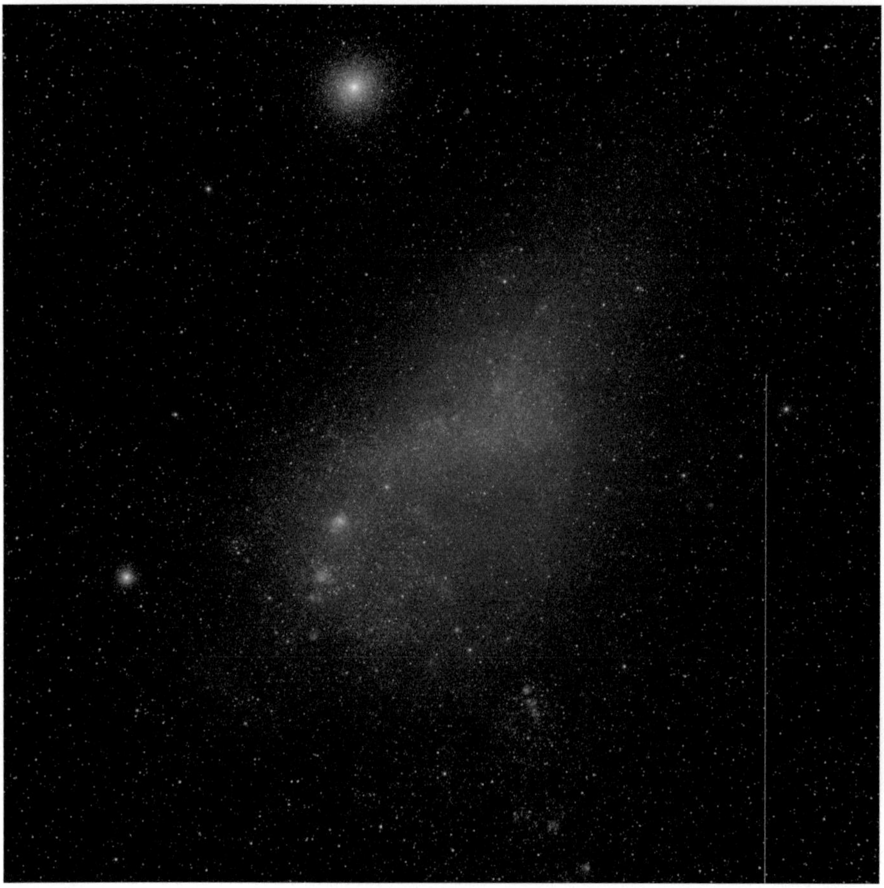

Fig. 8.13 A defective master frame from an image of the SMC. There is a glaring vertical defect. (Image by the Author/Telescope Live)

I discovered by running the mouse around the image that the horizontal coordinates of the defect went from x = 3559 to x = 3564. Since coordinates are measured from the top left of the image, the overwritten region only had to run for values of y greater than 1722.

By trial and error, I found that I could create a rectangle in the relevant area by randomly varying its brightness about the mean brightness of the target value. I created a variable called qexp1 to do this. I then used it inside an if statement. In Pixel Math, $T always means the target image, i.e. the one you last clicked on.

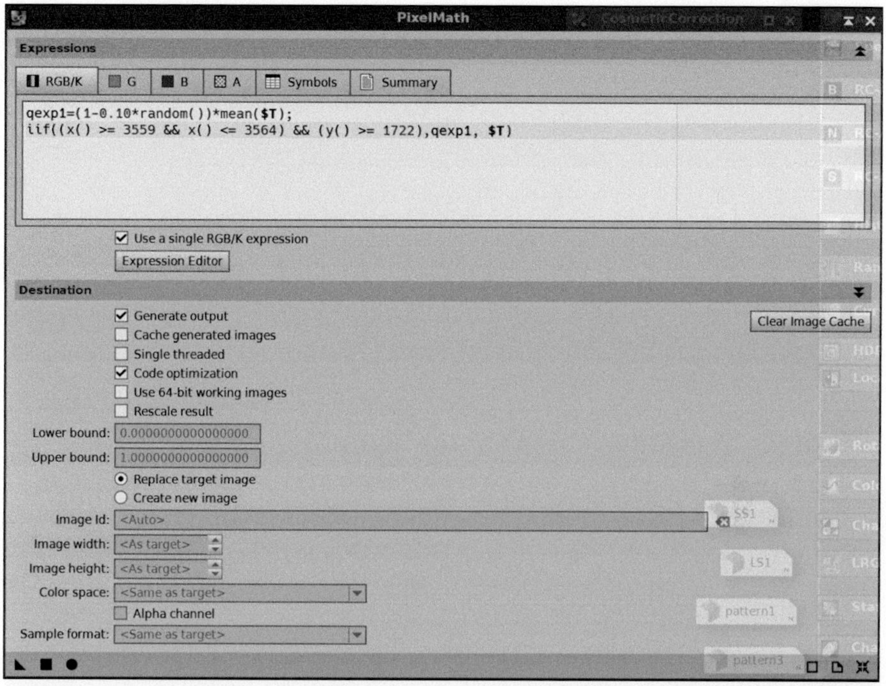

Fig. 8.14 The Pixel math I used to overwrite the linear defect shown in Fig. 8.13. Note the semicolon at the end of the first line. (Image by the Author/Telescope Live)

Figure 8.14 shows the Pixel Math I used. The idea was that the overwritten area might have a value randomly anywhere between the mean of the target and 90% of the mean of the target. You must also list the variable “qexp1” in the symbols tab.

Did it work? Judge for yourself in Fig. 8.15.

This is an example of just what a powerful tool Pixel Math is. I’m well aware that many, if not most, people are violently allergic to mathematics. But I hope to plant a tiny little seed in your mind that mathematics has its uses. Pixel Math does most of the hard work for you.

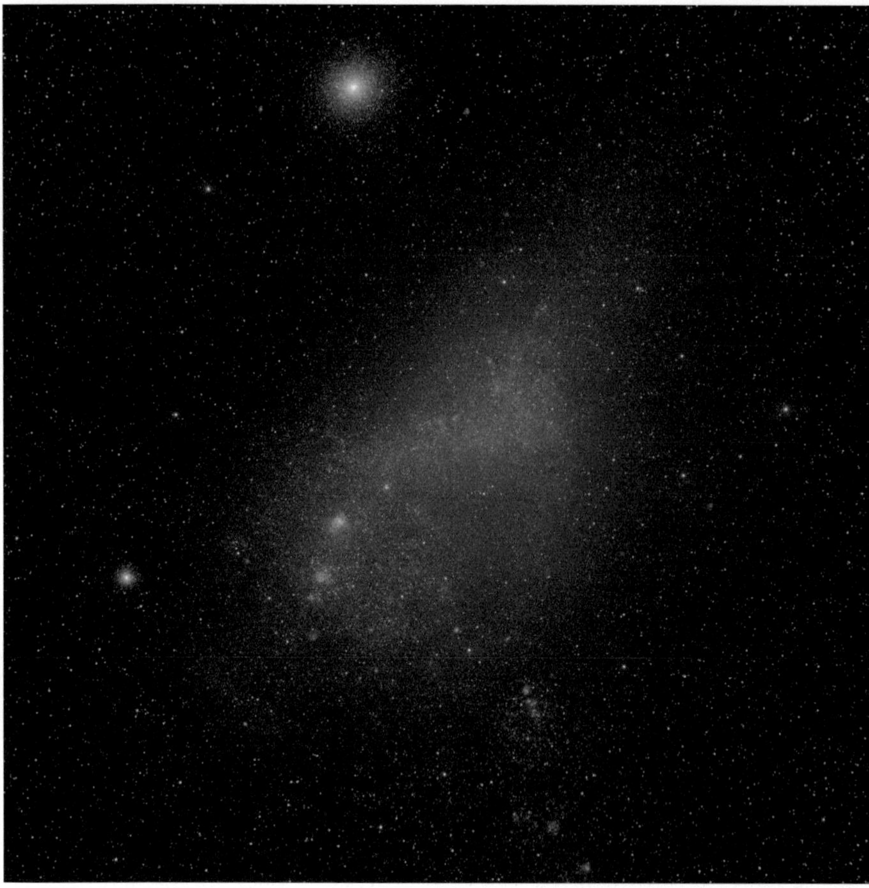

Fig. 8.15 The Small Magellanic Clod with the obtrusive white line suppressed. (Image by the Author/Telescope Live)

The Antennae

This is the nickname for the pair of colliding galaxies more formally known as NGC4038 and NGC4039; and Arp 244.

They may not be the brightest galaxies, but they are unquestionably two of the best galaxies of the Southern sky. They do in fact peep over my horizon at 51.6°N, but never become high enough to for me capture a decent photograph.

Figure 8.16 shows my starting point, temporarily stretched.

For me as a suburban dweller, the lack of spurious reflections in Fig. 8.16 caused by light pollution is striking.

I tried several processing steps, which have been described earlier in the book. The image I liked best went through the following stages.

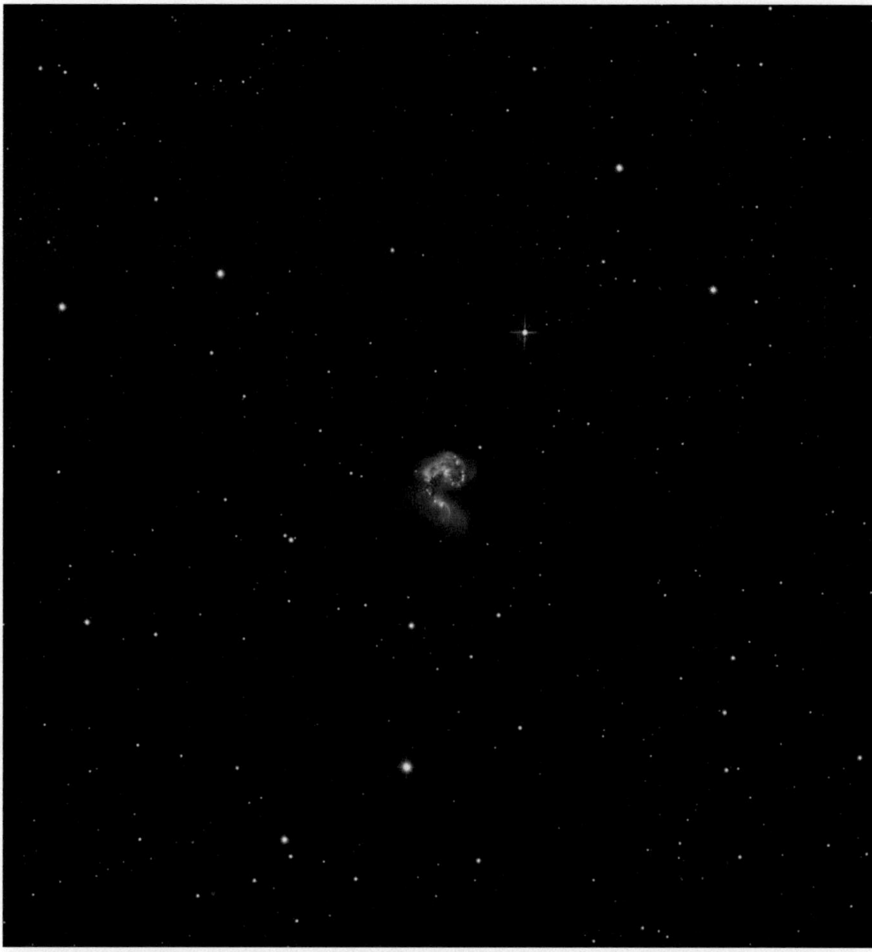

Fig. 8.16 My temporarily stretched starting point. (Image by the Author/Telescope Live)

1. Multiply R, G, B channels by L channel & combine in Process→ PixelMath → PixelMath
2. First stretch in Process → IntensityTransformations → ArcsinhStretch. Do this twice.
3. Remove the stars with Process → ObjectRecognition → StarXTerminator.
4. Make the mask shown in Fig. 8.17 to enable me to stretch everything except the nucleus.
5. Make a separate mask shown in Fig. 8.18 to enable me to suppress the halo around the brightest star in the image.
6. With the first mask, use Process → IntensityTransformations → CurvesTransformation to stretch the stuff not in the nucleus

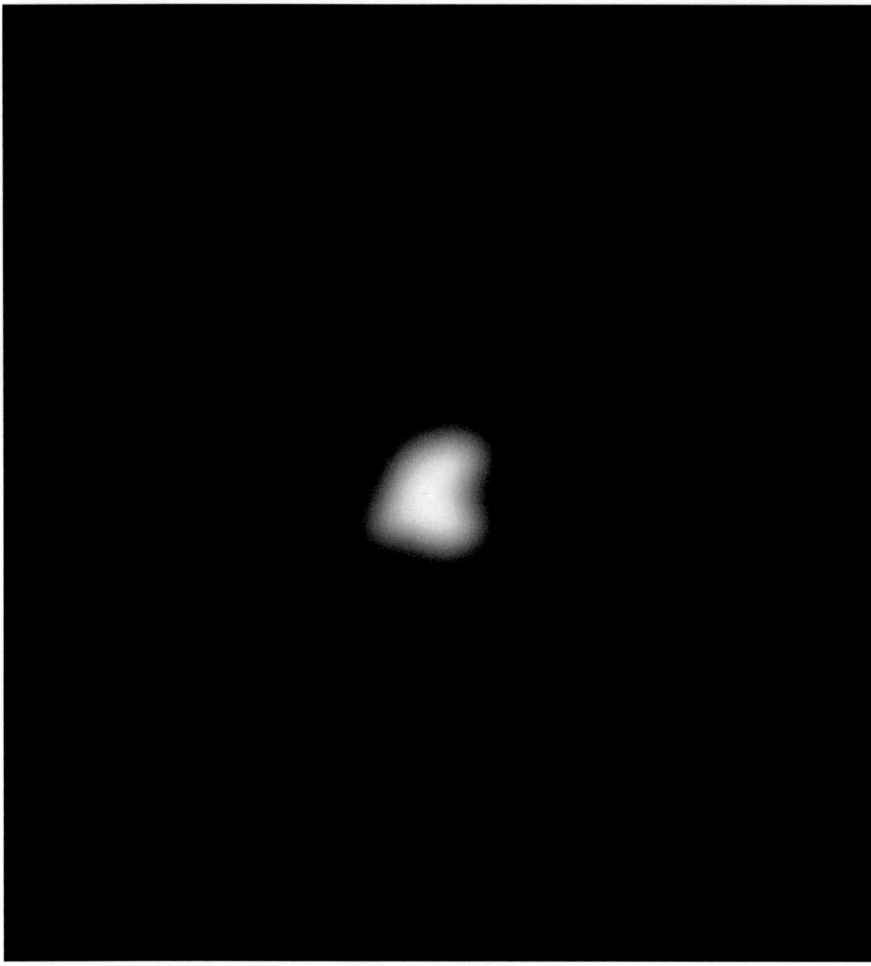

Fig. 8.17 Mask made by drawing three binary ellipses in the GAME script, then feeding to Range Selection to soften the edges. (Image by the Author)

7. Use Process → ColorCalibration → AssistedColorCalibration to add some blue to the galaxy tails
8. Remove the first mask, then add the second mask and use Process → IntensityTransformations → HistogramTransformation to suppress the halo around the brightest star.
9. Feed the starless image to the external app Topaz Sharpen AI and sharpen to bring out features in the nuclei. Save the result as a 16-bit tiff file and drag and drop it into PixInsight.
10. Add the stars back in, but after some experimentation, use main image + (0.9 × stars) to stop the stars overwhelming the image. My aim was to achieve a balanced composition.

Fig. 8.18 Mask made to suppress background light around the brightest star. (Image by the Author)

11. Crop and rotate to taste.

The image I ended up with was the one in Fig. 8.19.

Unusually for a photo of a galaxy, the surrounding stars are significant actors in the composition of this image. I experimented quite a bit to find a crop that showed the galaxies in their neighborhood but where the eye was not led away from the galaxies, and the effect was pleasing. One of the things I picked up by reading and by talking to people in camera clubs, and to an art teacher at school, was that there is an art to leading the viewer's eye around pictures; and that it is easy to get this hopelessly wrong. I hope I didn't.

There's also a second story to tell about these colliding galaxies. They were one of the earliest galaxies to have their collision analyzed to figure out how they ended

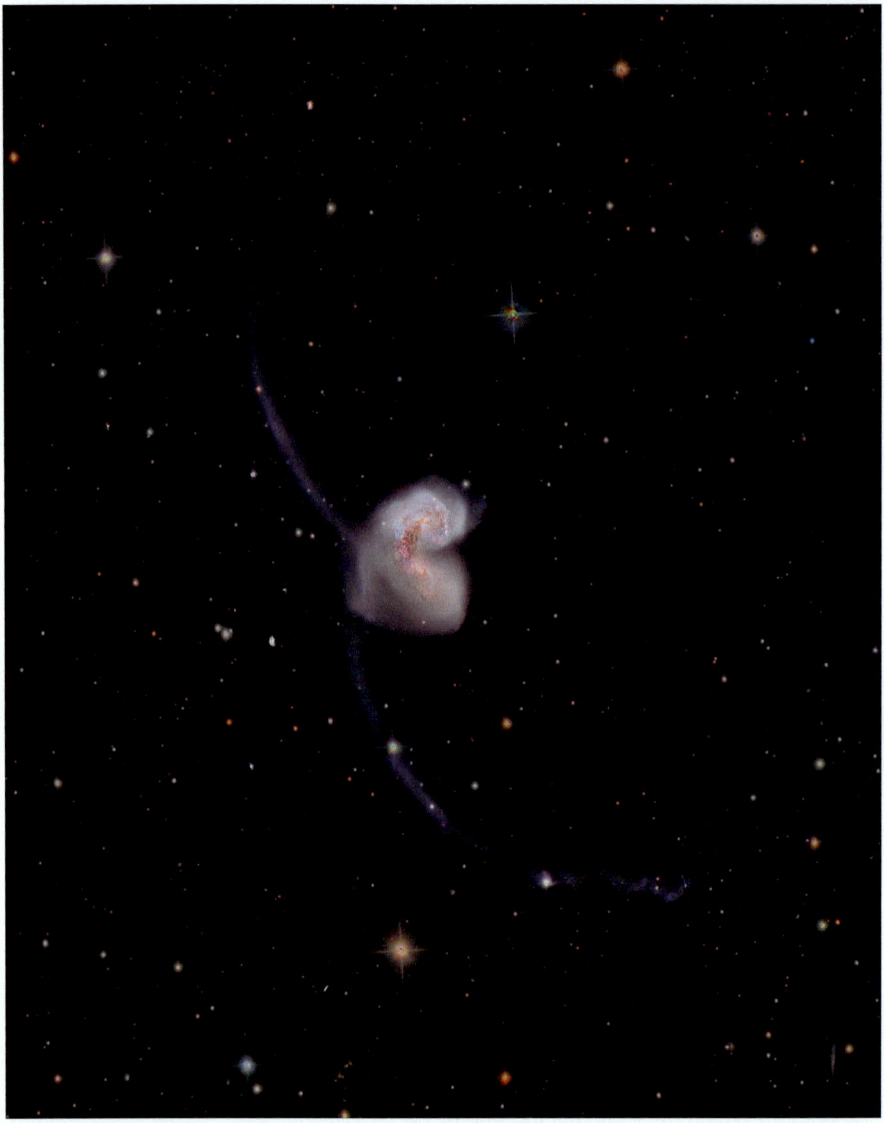

Fig. 8.19 My processed image of the Antennae Galaxies. (Image by the Author/Telescope Live)

up the way we see them. A close look at Fig. 8.19 will reveal to you that the upper tail belongs to the lower nucleus and vice-versa. In the mid-1970s the Toomre brothers analyzed this galaxy [55] and showed that its behavior con be predicted from simple physics, having an overwhelmingly dominant mass in each of the nuclei, and allowing the stars around the nucleus to feel the gravitational attraction of the nuclei but not of each other.

The computer simulation is illustrated in Fig. 8.20.

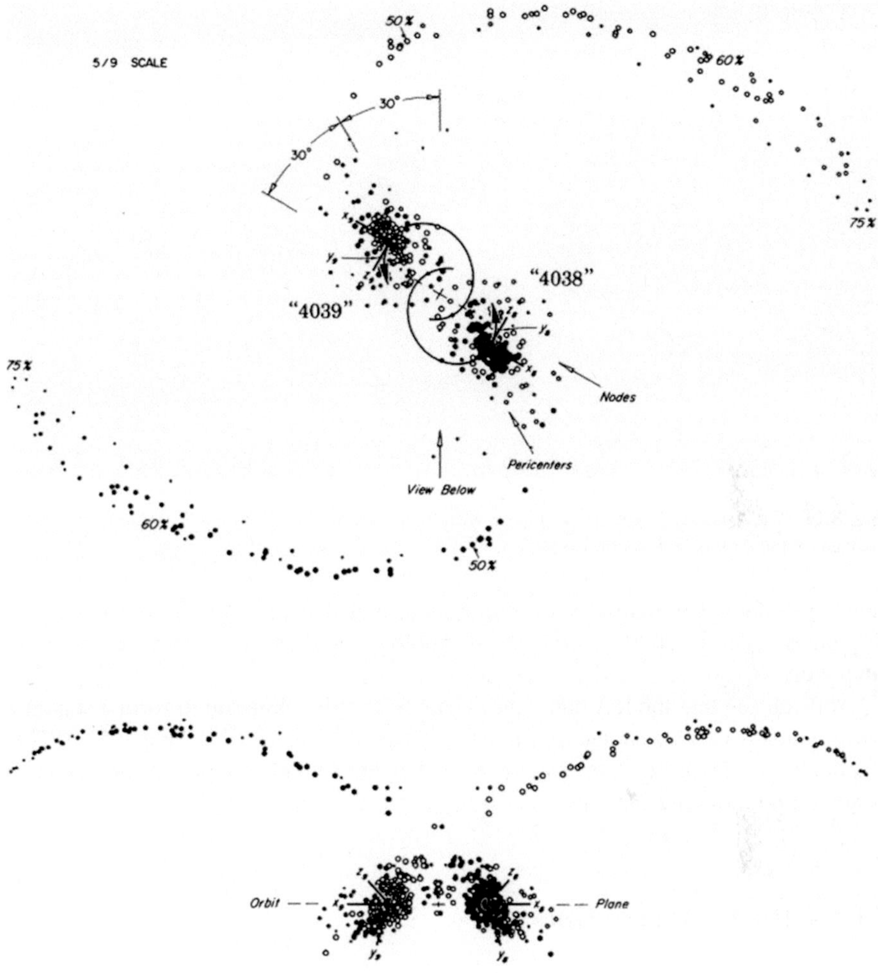

Fig. 8.20 The model of the collision of the Antennae galaxies by Toomre & Toomre. The e ≈ 0.6 mentioned means that the path of the galaxies around each other initially followed an ellipse of eccentricity 0.6. The stars originally from NGC4039 are shown as open circles, and the stars of NGC4038 are shown as solid circles. The upper image shows the galaxies flying past one another. The lower image shows the view from the Earth. The fact that the tails appear to have swapped galaxies is an illusion of perspective. The time unit is 100 million years. Image from Reference [55], by permission of the authors. (Image © American Astronomical Society)

The simulations of M51 system and of the Antennae collisions were made into movies by Toomre & Toomre using 1972 technologies. These have since been updated and, at the time of writing, are available on YouTube at https://www.youtube.com/watch?v=zvXLWEd2ZGU.

I rotated and re-cropped my image to show in Fig. 8.21 how the actual antennae compare with the Toomre & Toomre model. In their paper, these authors freely

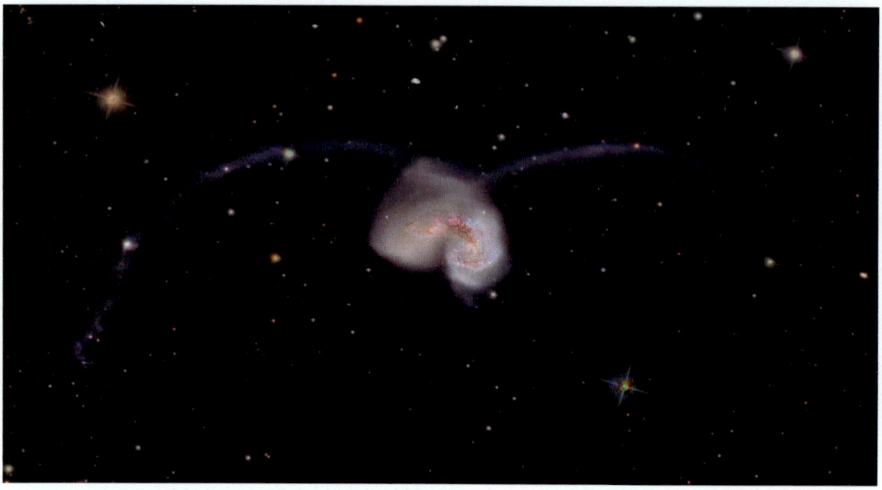

Fig. 8.21 The Antennae rotated to compare with the prediction of Toomre & Toomre in Fig. 8.20. (Image by the Author/Telescope Live)

admit that their simulation is oversimplified, and that the galaxies are not really symmetric, but given the computer technology of their time, they did remarkably well.

You can see that the left-hand tail in Fig. 8.21 is breaking up to form a separate dwarf galaxy. This was noted by Mirabel, Dottori & Lutz in a 1992 paper [56]. This phenomenon has been cited as evidence that the hypothesis that dwarf galaxies require a pool of dark matter in which to form [57] is not always correct.

M104, the Sombrero Galaxy

This galaxy is low down in my horizon, so I decided to try an image taken from Telescope Live's observatory near Santiago, Chile. It is a rather subtle target, because of the delicate halo.

The most awkward challenge with this image was that the calibrated light frames supplied did not remove all the dust bunnies. In a way I was relieved by this: if people who do astrophotography for a living can produce flat frames that don't fully work, I don't need to feel quite such an idiot when I do the same.

Figure 8.22 shows the worst dust bunnies, in the top left of the luminance master. I could always crop out the dust bunnies well away from the galaxy, but one in particular needed to be removed, towards the top left.

It turned out to be very easy to request the flat frames at no cost once you have purchased the original images. This can be done on the Telescope Live web site directly from "My Advanced Requests". The website says: *Simply click on the three-dots action menu for the given request, and click on "Download calibration*

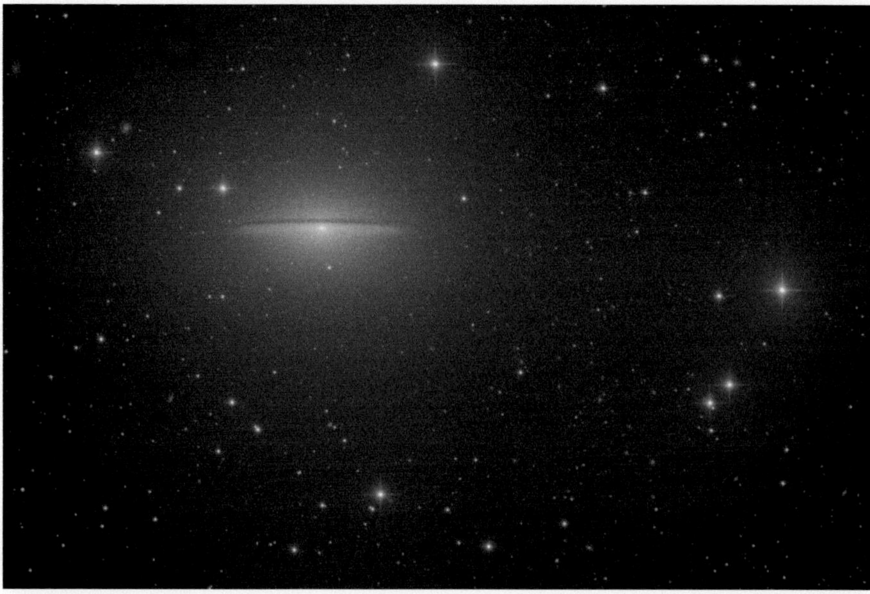

Fig. 8.22 The Luminance master stretched to show dust bunnies. (Image by the Author/ Telescope Live)

frames". This will initiate a new download request for the relevant Master Calibration Frames. You can then download the frames on your computer by navigating to Download Requests from the left sidebar.

The flat frames obtained are shown in Fig. 8.23.

It was simply a matter of re-running WBPP with the flat frames included. The repaired luminance frame is shown in Fig. 8.24.

Unfortunately, it was some time ago that I did this analysis, early in my learning curve. I preprocessed the L, R, G & B frames separately, and they were not correctly aligned. So I had to align them. The procedure is straightforward enough. I used the Star Alignment tool

$$Process \rightarrow Preprocessing \rightarrow StarAlignment$$

shown in Fig. 8.25.

Processing was otherwise straightforward. I used the channel combination tool

$$Process \rightarrow ColorSpaces \rightarrow LRGBCombination.$$

The idea is to do this in two stages, RGB first, then add the L channel in the second step.

My final image is shown in Fig. 8.26. The halo has been captured without overexposing the other detail.

Fig. 8.23 The luminance master flat frames have been stretched to show the offending dust bunnies. (Image by the Author/Telescope Live)

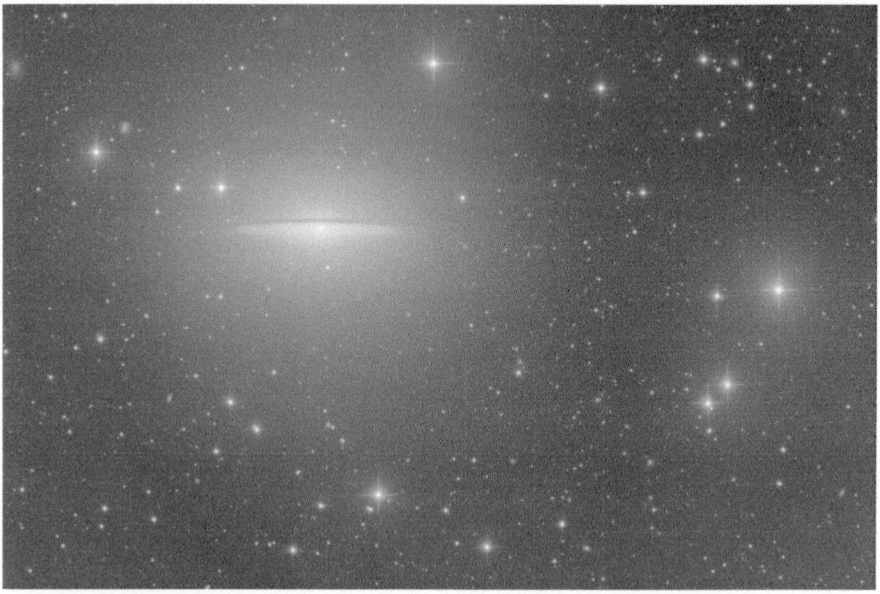

Fig. 8.24 The luminance master light frame has now been repaired. (Image by the Author/Telescope Live)

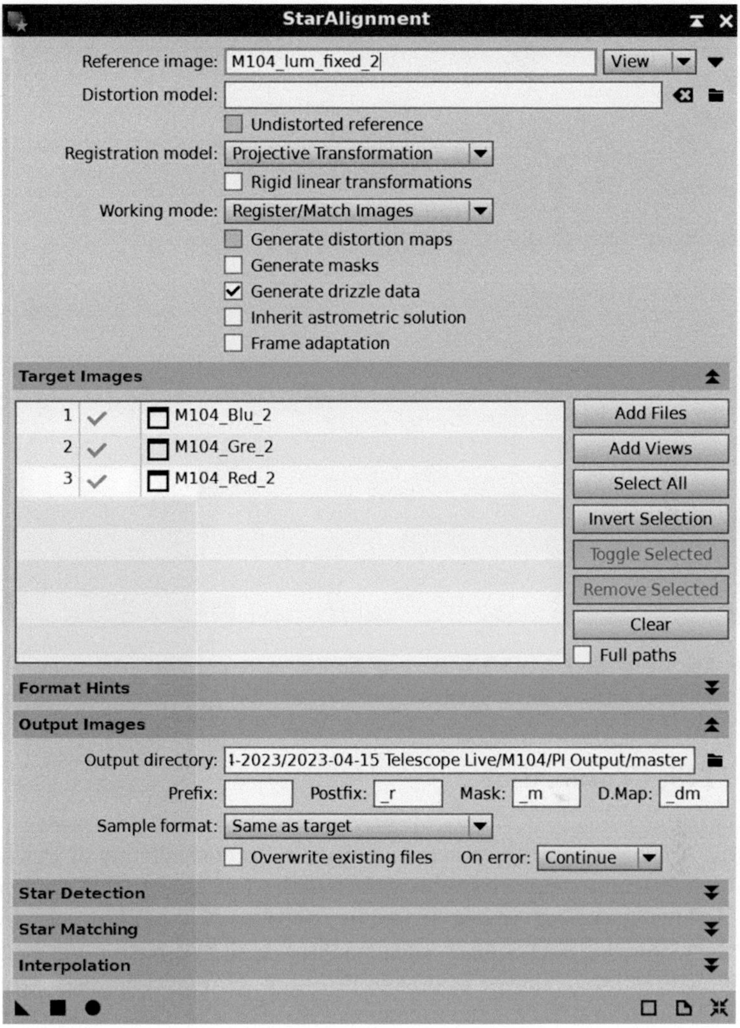

Fig. 8.25 Aligning the frames. (Image by the Author)

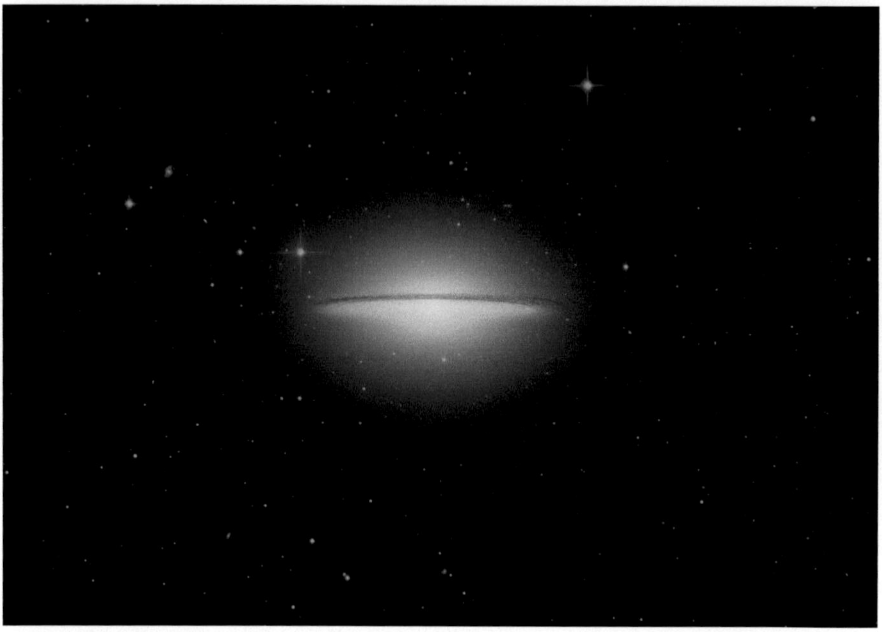

Fig. 8.26 The Sombrero Galaxy M104, showing the nucleus, disc and halo. (Image by the Author/Telescope Live)

NGC 5128, Centaurus A

Centaurus A is a galaxy with an active nucleus in the constellation of Centaurus. I took a very different approach to this galaxy. After processing the Telescope Live image, I downloaded an image from the website of the European Southern Observatory, showing data at three wavelengths, visual, 870-micron sub-millimeter data, i.e. near infrared, shown in orange in Fig. 8.29, Fig. 8.28, and X-ray data shown in Blue in Fig. 8.29. I then combined my Telescope Live image with that from ESO, which I also somewhat processed.

I have done this to feed you the idea that you can also do this with other galaxies, and increase your appreciation of the things that go on in galaxies. There is a great deal of publicly available data if you have the energy & stamina to look for it.

After preprocessing the Telescope Live data, I found that I could stretch for either the white halo around the galaxy or the dust band in the center. I therefore averaged these two images using Pixel Math. All you need to do is to use either of the two expressions below in the expression window

$$\text{mean}\left(\text{A,B}\right),\text{or}$$

$$0.5^{*}\left(\text{A}+\text{B}\right).$$

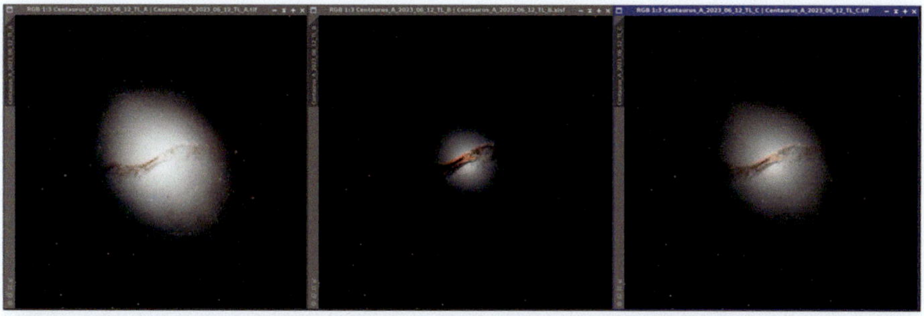

Fig. 8.27 The two images on the left were well-stretched for the halo and the dust lane respectively. The image on the right hand side is made by averaging the two images on the left. (Image by the Author/Telescope Live)

They are entirely equivalent. You check the "Create new image" box, and Pixel Math will create the averaged image. The two input images, and the output image, are shown in Fig. 8.27.

I then brought the European Space Agency's .tiff file into PixInsight, and tried to plate solve it. Unfortunately, I failed miserably, since I didn't have either the focal length or pixel size data, and did not manage to guess them. So, I could not use the Star align tool.

I did separate the galaxy from the stars, and used the Curves Transformation Tool to enhance the galaxy, and tone down the stars. The resultant image appears in Fig. 8.28. I had to make the colors stark in this one, because they would inevitably fade when superimposed on the Telescope Live white-light image.

I created a 4000-by-4000-pixel blank, black image in the free image processing tool Gimp 2.10.36. I dragged and dropped the Telescope live image (the right-hand image of Fig. 8.27) and the ESO Image (Fig. 8.28) onto the blank image. Gimp creates layers. In Gimp's jargon, you make the blank a "Normal" layer and the two dragged and dropped images, which automatically go into separate layers, as "Screen" layers. While working, you make the two Screen layers partially transparent. You then manually scale the images until they match, align them manually, and rotate the topmost Screen layer until it sits on the other Screen layer. It's a good idea to then save the image in Gimp's own .xcf format. Once you have done that, you "flatten" the image, which merges the layers. My final image is shown in Fig. 8.29.

You can now see how the active nucleus is driving matter out of the plane of the galaxy. For an account of active galaxies, I suggest "Galaxies in Turmoil" by Chris Kitchin [58].

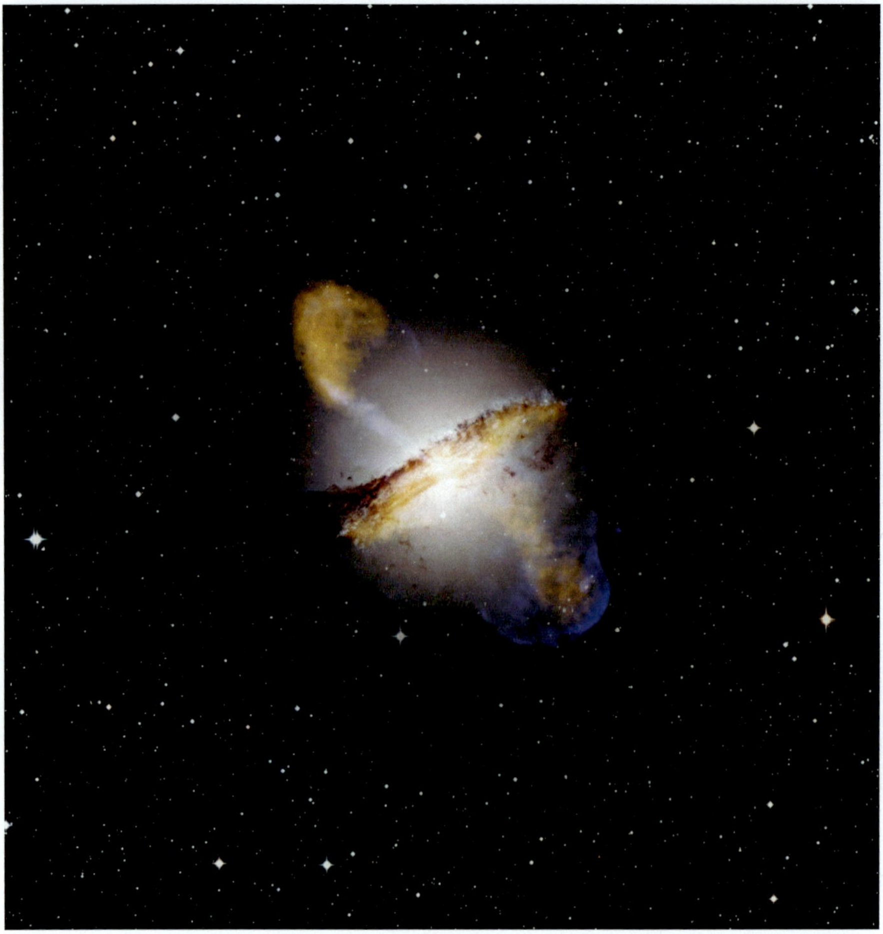

Fig. 8.28 The image from the ESO after I had stretched it. Credit: ESO/WFI (Optical); MPIfR/ESO/APEX/A.Weiss et al. (Submillimetre); NASA/CXC/CfA/R.Kraft et al. (X-ray)

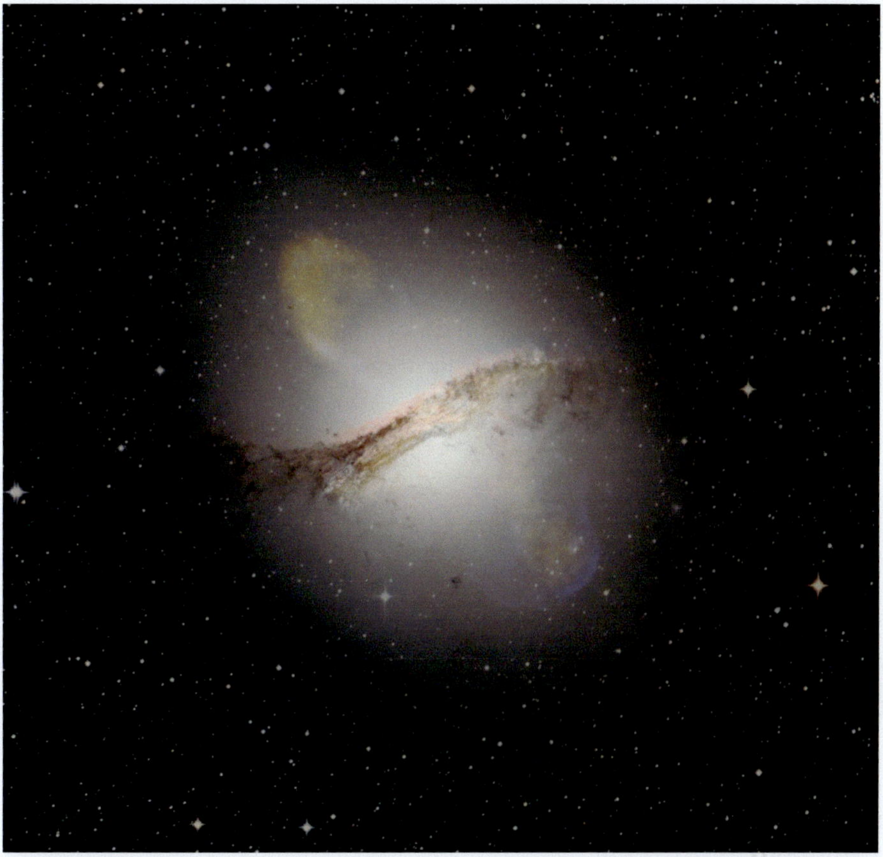

Fig. 8.29 Image processed by the Author, from data from Telescope Live and ESO/WFI (Optical); MPIfR/ESO/APEX/A.Weiss et al. (Submillimetre); NASA/CXC/CfA/R.Kraft et al. (X-ray)

NGC1365, the Great Barred Spiral

There's no question that this galaxy is one of the jewels of the southern sky, indeed of the entire sky. I can think of very open barred spirals in the northern sky, but none so impressive as NGC1363.

I set out with two aims. The first was that I knew from having processed this image about a year ago that there was more to it than just a giant S-shape. The second was to make a striking picture. Compositionally, as so often, my scope for playing around with nearby stars and other galaxies was almost non-existent. But I wanted to make a striking image anyway: that was my second aim.

First, I reprocessed the 80 minutes of good RGB and 25 minutes of good luminance frames I had from the CCD camera. I immediately found that this was a case where the luminance frame simply overpowered the others. With the L frame there it proved to be impossible to stretch the image. So I gave up on that and just processed the 80 minutes of RGB frames.

Nothing else worth noting as unusual happened, and my result is shown in Fig. 8.30.

Three things strike me about this image. The first is that I was not successful in my aim of showing more than the S shape; and the second is that the image is noisy because it has not been exposed for long enough. Third, it is a mirror image. I don't know why, but if I wanted to continue with this image, I could easily reverse it.

By the time I wrote this, there were about 9 hours of data on the Telescope Live site from Chile using a QHY 600 M Pro CMOS camera. That seemed to be likely to solve the problem of insufficient data. It was worth having a go with these data.

Once I had processed these and used the formula

Red: $R \times L$
Green: $G \times L$

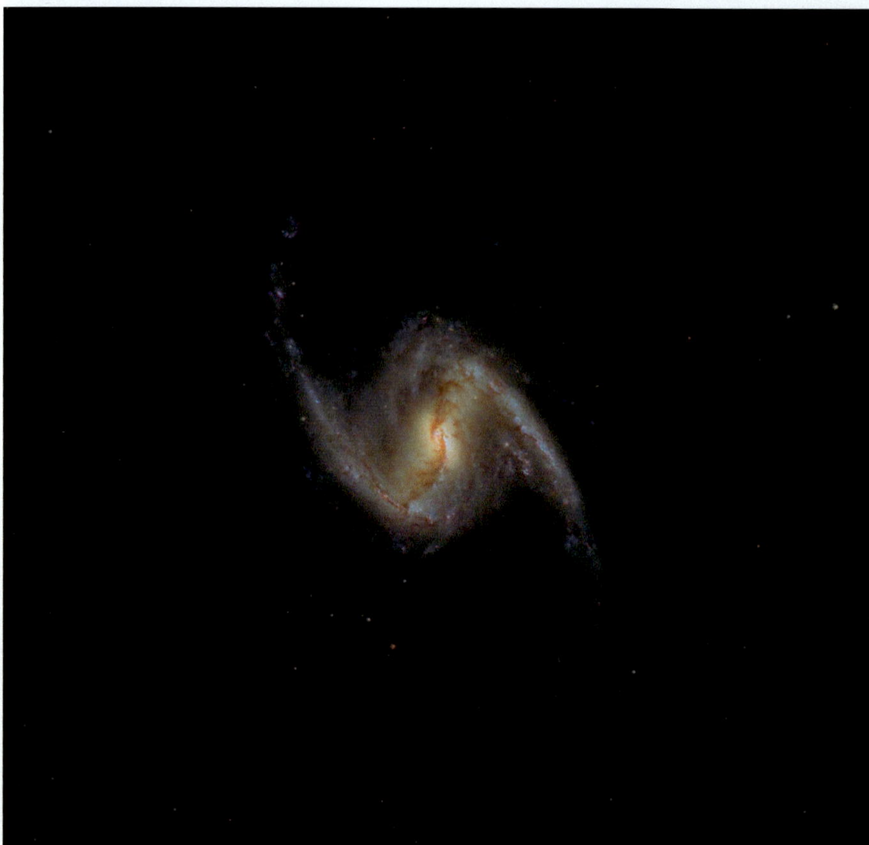

Fig. 8.30 Eighty minutes of NGC1365 taken with the CCD camera at Telescope live. (Image by the Author/Telescope Live)

Blue: B × L

I quickly discovered that, once again, the L frame overwhelmed the R, G & B layers. Having determined for the LMC that the luminance frame really does do something, I felt some reluctance simply to discard the luminance data.

It was time to experiment again, and try my luck. I figured that perhaps a combination of data with & data without the luminance master frame might work. I started out with

Red: NGC1365_R_2*(1 + (0.1*NGC1365_L_2))
Green: NGC1365_G_2*(1 + (0.1*NGC1365_L_2))
Blue: NGC1365_B_2*(1 + (0.1*NGC1365_L_2))

In other words, the image was about 90% without luminance, about 10% with. That image behaved itself, and seemed to be perfectly stretchable.

I upped the ante. By the time I reached the combination below, I got the feeling that, were I to increase the fraction with L frames, Then L frames would again become over-dominant.

Red: NGC1365_R_2*(1 + (0.4*NGC1365_L_2))
Green: NGC1365_G_2*(1 + (0.4*NGC1365_L_2))
Blue: NGC1365_B_2*(1 + (0.4*NGC1365_L_2))

This was the combination that made the LRGB frame with which I continued. It is shown in Fig. 8.31. Rather to my dismay, there were vertical stripes going right through the galaxies, which WBPP had not weeded out. Would I be able to make them go away with stretching?

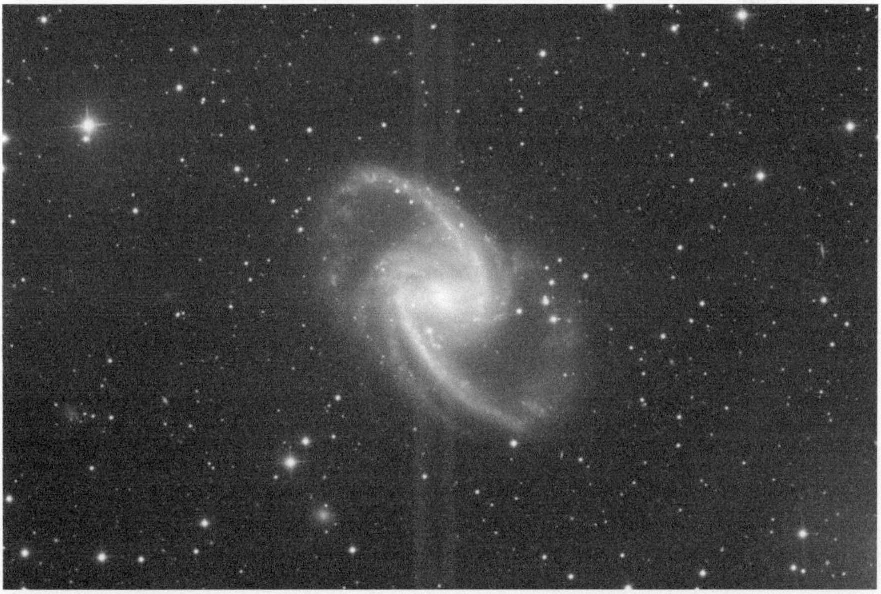

Fig. 8.31 Monochrome of the preprocessed LRGB image. (Image by the Author/Telescope Live)

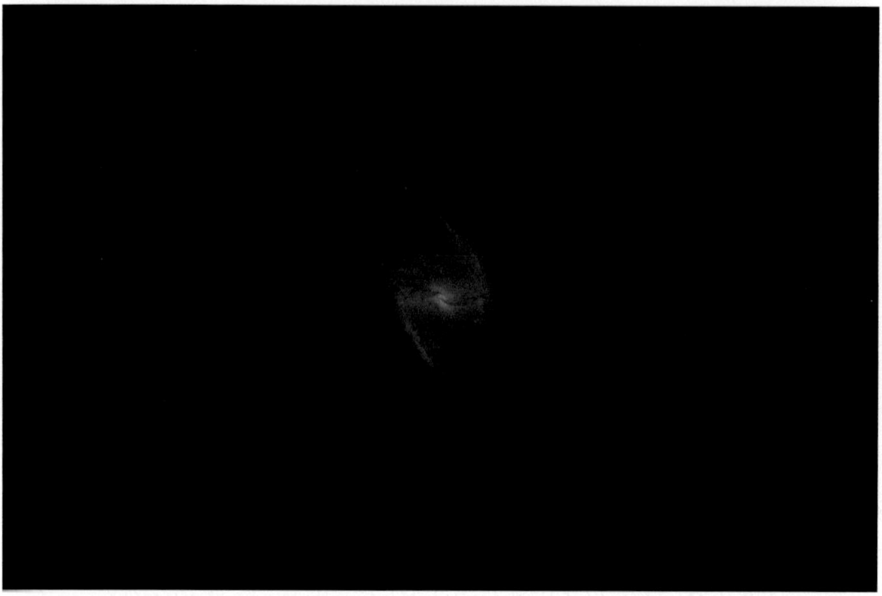

Fig. 8.32 Monochrome image of NGC1365 without stars after the initial Arcsinh stretch. (Image by the Author/Telescope Live)

Figure 8.32 suggests that the answer might be yes: stretching had suppressed the stripes. So far, so good. The stretching was done with

$$Process \rightarrow IntensityTransformations \rightarrow ArcsinhStretch.$$

The stars were removed with

$$Process \rightarrow Object\ Recognition \rightarrow StarXTerminator.$$

Further stretching got me to Fig. 8.33. I used

$$Process \rightarrow IntensityTransformations \rightarrow CurvseTransformation.$$

This is getting there, but it looks a bit monotonous to me. Time for

$$Process \rightarrow IntensityTransformations \rightarrow LocalHistogramEqualisation.$$

This gave me the image shown in Fig. 8.34.

This has more impact, but I've slightly overexposed the brightest parts of the image. I had a go at bringing these under control with Pixel Math.

The theory is that anything whited out will be at 100% brightness, i.e. a value of 1.000 in Pixel Math. In practice I defined "too nearly white" to be a pixel value greater than 0.975 rather than 1.000, and if the Pixel math tool found such a pixel as

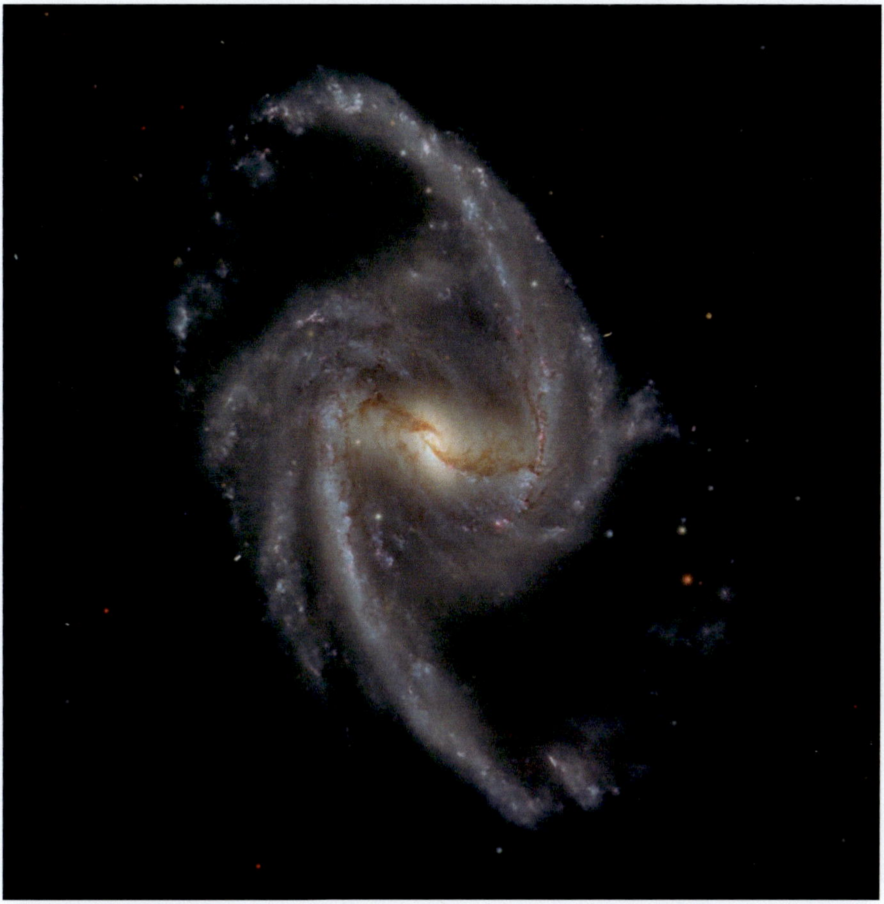

Fig. 8.33 The image after further stretching with the Curves Transformation tool. (Image by the Author/Telescope Live)

it worked its way through the image, it would reduce the value of any pixel greater than 0.975 to 0.975. The formula used is shown in Fig. 8.35.

That worked. I did not need to tune the 0.975 value, though I expected to have to. Sometimes you just get lucky with guesses.

Finally, I ended up with the image in Fig. 8.36. There are now no whited-out regions. There's no doubt that I found more than just the S-shape. Whether the image is punchy enough, which was my other objective, I leave to the reader to decide.

On the subject of composition, Fig. 2.20 is an example of how to make a striking composition when you don't have convenient trees and such like to frame your image. The angle is carefully chosen to make the subject of the photo appear to leap out at you from the background. The cropping is quite aggressive: no irrelevant, distracting background appears; and the subject fills the photo. Finally, it obeys a

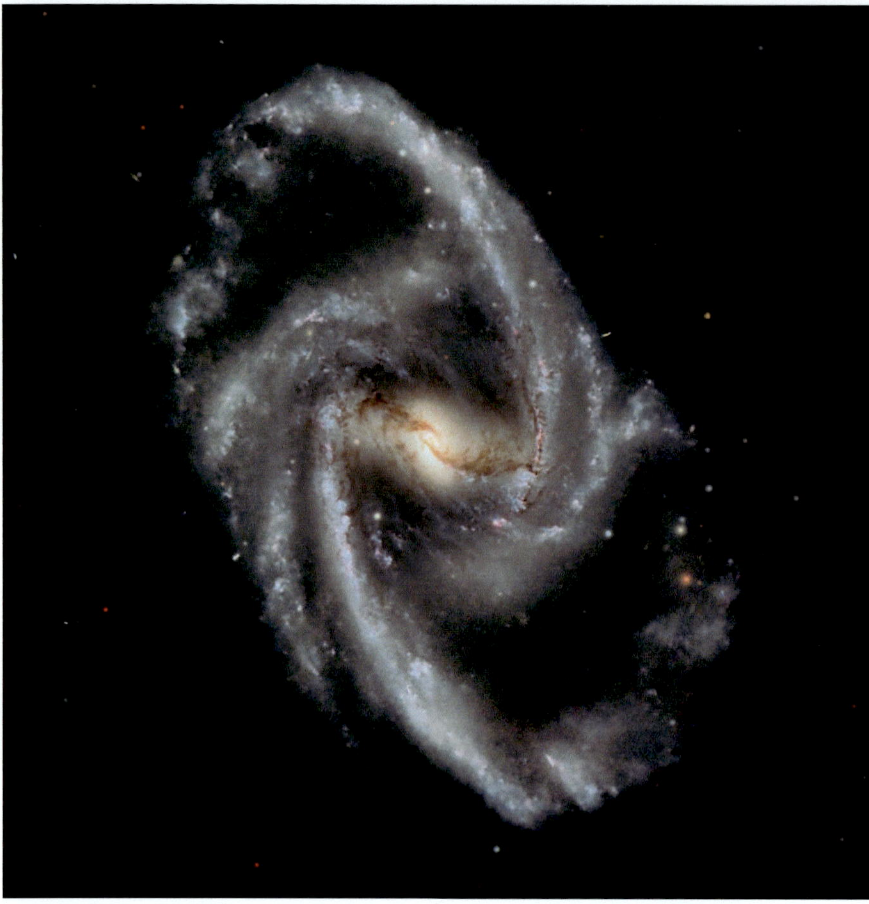

Fig. 8.34 The image after further stretching with Local Histogram Equalization. (Image by the Author/Telescope Live)

rule of thumb I learned as a teenager: the darkest part of the image should be barely black, and the lightest part should be barely white. That was in the days of black and white. A more modern version of that rule is that the brightest point can be any color as long is its numerical value in the pixel in one channel is close to 1.0 without over-saturating the image. As an aside, this especially applies to photos of the Moon. I often see those on social media with large areas of white. In other words, the image is over-exposed in the white regions. That's what you should try to avoid in images of the Moon or of galaxies, by careful stretching.

The art of stretching takes a little bit of practice. When you look back a year later, you will see how much your stretching has improved. In Fig. 2.20, I stretched in such a way as to bring out the detail in a camera which is mostly black. Fig. 2.25, in contrast, is not well-composed or exposed. The practical reason for this was that I didn't want to remove the camera until I had taken flats and bias frames, but that

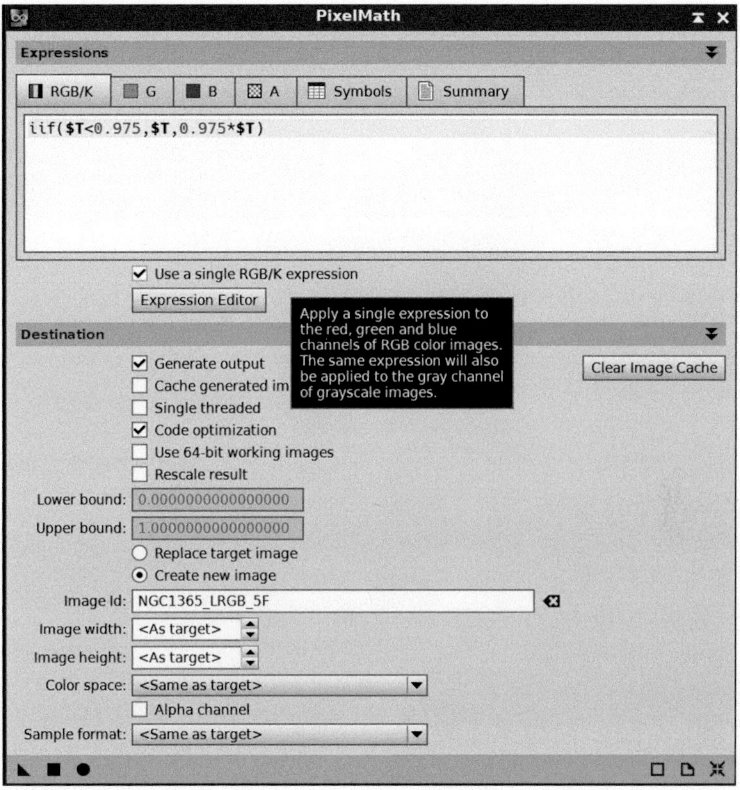

Fig. 8.35 The Pixel Math used to tone down any local areas of white-out. (Image by the Author)

doesn't make it a good photo. There are areas which are completely whited out, and a distracting white cable threading around. Those are the kind of errors to look out for.

Telescope Live has its limitations. You can rent time to take the images you want but it is expensive, and at your risk. Buying images that they have already captured greatly de-risks your investment, but they have chosen the images. For example, I would have liked to see how far the LMC actually extends. It clearly extends further than the area shown in Fig. 8.11.

The images you buy sometimes have to be rejected because of flaws. The most common such flaw that I have found is bright satellite trails that WBPP failed to weed out. Very occasionally you find a frame ruined by the telescope having been knocked. In practice I have never found this to be a serious limitation.

Balanced against this is the fact that I have had a chance to explore the galaxies of the Southern Hemisphere that I could never have afforded to explore by going there for an extended period of time. For that I am grateful.

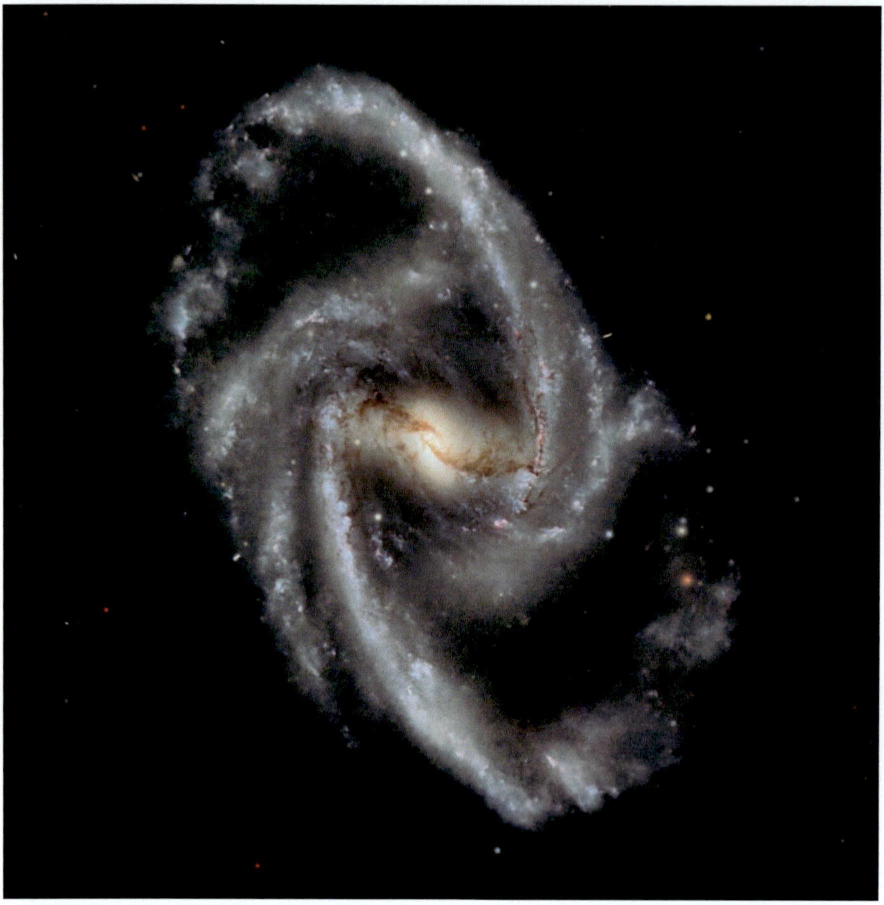

Fig. 8.36 My final rendition of NGC 1365. Notice the bar-within-a-bar in this galaxy. (Image by the Author/Telescope Live)

The earliest recorded example of wishing people clear skies was Edmund Halley of comet fame, who in 1724, finished the caption of a map showing a predicted solar eclipse with the words [59], "We wish our Astronomical Friends a Clear Sky".

Almost exactly three hundred years later, may I wish you all the same.

Appendix: Galaxy Naming

If only it were as simple as the list below. The lists contain errors and ambiguities, which often only became apparent as our observing capability increased.

The main galaxy catalogues are as follows.

- Messier catalogue, Mxxx
  - After Pierre Messier, eighteenth century
  - Of the 110 objects, 40 are galaxies.
- Caldwell catalogue Cxxx
  - After Patrick Caldwell-Moore, better known as Patrick Moore, twentieth century
  - Of 109 objects, 35 are galaxies
  - Intended as a complement to Messier's catalogue
  - Many GOTO mount hand controllers list these.
- New General Catalogue, NGCxxxx
  - Begun by J.L.E Dreyer, Armagh Observatory, 1888
  - 7840 objects
  - Appendix ICxxxx objects added later, 1895 & 1908
  - Further 5386 objects
  - Current edition contains 13,957 objects, 2019
  - Not all objects are galaxies.
- Principal Galaxies Catalogue PCGxxxxx
  - Lyon, France, 1989
  - 131,601 objects
- Uppsala General Catalogue of Galaxies (UGCxxxx)
  - Uppsala, Sweden, 1973

J. Clark, *Photographing Galaxies from Light Polluted Skies*, The Patrick Moore Practical Astronomy Series, https://doi.org/10.1007/978-1-0716-4506-2

  - North of Declination $-02°30'$
  - Min diameter 1 arcminute
  - Limiting apparent magnitude 14.5
- Arp's Atlas of Peculiar Galaxies, Arp xxx
  - CalTech, 1966
  - 338 objects
- The Abell catalog of rich clusters of galaxies
  - CalTech, 1958
  - 4076 objects

References[1]

1. Dreyer, J. L. E, *History of Astronomy from Thales to Kepler, Dover*, 1953 (a republication of a 1906 book), ISBN 0486600793
2. Williams, H. A., *Dutch Light: Christiaan Huygens and the Making of Science in Europe*, Picador, 2020, ISBN 9781509893317
3. Hoskin, M. *The English Background to the Cosmology of Wright and Herschel*. In: Yourgrau, W. & Breck, A.D. (eds) *Cosmology, History, and Theology*. Springer, Boston, MA, 1977
4. Wirght, T. *An original theory or new hypothesis of the Universe*, London, H. Chapelle, 1750
5. Kant, I., *Allgemeine Naturgeschichte und Theorie Des Himmels*, available online at https://web.viu.ca/johnstoi//kant/kant2g.htm, 1755
6. Herschel, W., Philosophical Transactions of the Royal Society of London, Vol. 75 (1785), pp. 40–126
7. Herschel, W., *Catalogue of 500 New Nebulae, Nebulous Stars, Planetary Nebulae, and Clusters of Stars; With Remarks on the Construction of the Heavens*, Philosophical Transactions of the Royal Society of London, Vol. 92 (1802), pp. 477–528, 1802
8. Herschel, W., *On the places of 145 new Double Stars*, Memoirs of the Royal Astronomical Society, Vol. 1, p. 166, (1821)
9. Herschel, W., *On the Construction of the Heavens*, Philosophical Transactions of the Royal Society of London, Vol. 75, pp. 213–266, 1785
10. Geheimen-Rath & Bessel, *Bestimmung der Entfernung des 61sten Sterns des Schwans (Determining the Distance of the 61st star of the Swan (61 Cygni))*, Astronomische Nachrichten, volume 16, Issue 5, p. 65, December 1838
11. Herschel, J. F. W., *Observations of Nebulae and Clusters of Stars, Made at Slough, with a Twenty-Feet Reflector, between the Years 1825 and 1833*, Philosophical Transactions of the Royal Society of London, Vol. 123, pp. 359–505, 1833
12. Dyson, F. W., *Measurements of the Distances of the Stars*, J. Royal Astron. Soc. Canada, **IX**, (9), 407-422, 1915; also published elsewhere

[1] I fully appreciate that many, if not most, of my readers will not be familiar with the little ways of researchers when referencing documents. I've avoided using well-known abbreviations for journals, because they are only well-known within the scientific "trade". I have, however, kept to the convention of writing volume numbers of journals in bold typeface.

J. Clark, *Photographing Galaxies from Light Polluted Skies*, The Patrick Moore Practical Astronomy Series, https://doi.org/10.1007/978-1-0716-4506-2

13. Pickering, E. C., *Periods of 25 Variable Stars in the Small Magellanic Cloud*, Harvard College Observatory, Circular 173, March 1912. Pickering states at the start that "The following statement…has been prepared by Miss Leavitt," but she was not credited as an author
14. Shapley, H., *Approximate Distance and dimensions of Large Magellanic Cloud*, Harvard College Observatory Bulletin No. 775, pp. 1–2, 1922
15. Nielsen, A. V., *Contributions to the History of the Hertzsprung-Russell Diagram*, Centaurus **9**(4), 219–253, 1964
16. Bravais, A., Journal de Mathématique, **8**, 435
17. Van der Kruit, P. C., *Pioneer of Galactic Astronomy: a Biography of Jacobus C. Kapteyn*, Springer, ISBN 978-3-030-55423-1 (eBook), 2021, pp. 126–9
18. Boss, L., *Convergent of a Moving Cluster in Taurus*, Astronomical J., **XXVI** (4), 31–36
19. Galli, P. A. B., Moraux, E., Bouy, H., Bouvier, J., Olivares, J. and Teixeira, R., *A revised moving cluster distance to the Pleiades open cluster*, A&A 598, A48, https://doi.org/10.1051/0004-6361/201629239, 2017
20. Kapteyn, J. C., *First Attempt at a Theory of the Arrangement and Motion of the Sidereal System*, Astrophysical Journal, vol. 55, p. 302, 1922
21. Van der Kruit, P. C., work cited
22. Paul, E. R., *The Milky Way Galaxy and Statistical Cosmology 1890-1924*, Cambridge, 1993, ISBN 0521027349
23. Eddington, A. S., *Stellar Movements*, London, Macmillan, 1914
24. Seeliger, H. von, Sitzungsberichte der Akad. Wiss. Zu Muenchen, 1911, pp. 413–61
25. Shapley, H., *Star Clusters and Their Contribution to Knowledge of the Universe*, Bulletin of the National Research Council, **2**, 171, 1921
26. Hubble, E., *A spiral nebula as a stellar system, Messier 31*, Astrophysical Journal, 69, 103, 1929
27. Easton, C. 1900, ApJ, 12, 136
28. Oort, J. H., Kerr, F. J., & Westerhout, G., The Galactic System as a Spiral Nebula, Mon. Not. Roy. Astron. Soc, **118**, 379–389, 1958
29. Blitz, L. & Spergel, D. N., Direct evidence for a bar at the Galactic Center, Astrophysical Journal, **379**, 631–638, 1991
30. Benjamin, R. A. and 20 others, *First GLIMPSE Results on the Stellar Structure of the Galaxy*, Astrophysical Journal, **630**, L149–L152, 2005
31. Hubble, E. P., *The Classification of Spiral Nebulae*, The Observatory, **50**, p. 276–281, 1927
32. Karen L. Masters, Chris J. Lintott, Ross E. Hart , Sandor J. Kruk , Rebecca J. Smethurst , Kevin V. Casteels, William C. Keel ,Brooke D. Simmons , Dennis O. Stanescu, Jean Tate and Satoshi Tomi, 2019, *Galaxy Zoo: unwinding the winding problem – observations of spiral bulge prominence and arm pitch angles suggest local spiral galaxies are winding*, MNRAS 487, 1808–1820, https://doi.org/10.1093/mnras/stz1153
33. Hubble, E., A relation between distance and radial velocity among extra-galactic nebulae, Proc Nat'l Acad. Sci. U S A., **15**(3): 168–173, 1929
34. Freedman, W. L., et al. , 2001, *Final results from the Hubble Space Telescope Key Project to measure the Hubble constant*, The Astrophysical Journal, **553**: 47–72
35. https://en.wikipedia.org/wiki/Kodak
36. Sandage, A. & Tammann, G. A., *A Revised Shapley-Ames Catalog of Bright Galaxies*, Carnegie Institute of Washington, 1981
37. https://ned.ipac.caltech.edu/level5/Shapley_Ames/frames.html
38. R. Berry & J. Burnell, 2nd English Ed., 2005, *The Handbook of Astronomical Image Processing*, Willmann-Bell, Inc., ISBN 0-943396-82-4
39. Sandage, A. & Tammann, G.A., 1987, A Revised Shapley-Ames Catalog of Bright Galaxies, Carnegie Institution of Washington Publication 635, 2nd Ed
40. Strateva, I. et al., *Color Separation Of Galaxy Types in the Sloan Digital Sky Survey Imaging Data*, Astronomical Journal, 122: 1861È1874, 2001

41. J. Brinchmann et al., *The physical properties of star-forming galaxies in the low-redshift Universe*, Mon. Not. R. Astron. Soc. 351, 1151–1179, 2004
42. Salim, S., *Green Valley Galaxies*, Serbian Astronomical Journal, vol. 189, pp. 1–14
43. Mutch, S. J., Croton, D. J. & Poole, G. B., *The Mid-Life Crisis of the Milky Way and M31*, Astrophysical Journal, **736**:84 (11pp), 2011
44. Gomez-Guijarro, C. et al., GOODS-ALMA 2.0: *Starbursts in the main sequence reveal compact star formation regulating galaxy evolution prequenching*, Astronomy & Astrophysics, **659**, A196, 2022
45. Heckman, T. M., in *Local Starbursts in a Cosmological Context*, eds. R. de Grijs, & R. M. González Delgado, **329**, 3, 2005
46. Gangopadyhay, G. & Kundu, A., 2024, *Rhythm in the Sky: Periodic Motions in the Universe,* Springer, ISBN 978-981-97-2588-5
47. Hubble, E., *A spiral nebula as a stellar system, Messier 31*, Astrophysical Journal, 69, 103, 1929
48. Toomre, A. & Toomre, J., Galactic Bridges and Tails, Astrophysical Journal, 178:623–666, 1972
49. Kreckel, K. et al., *A 50 pc Scale View of Star Formation Efficiency across NGC 628*, Astrophysical Journal Letters, **863**: L21 (7pp), 2018
50. Yun, M. S., Ho, P. T. P. & Lo, K. Y., *A high-resolution image of atomic hydrogen in the M81 group of galaxies*, Nature, 372, pp. 530–532, 1994
51. Telesco, C. M. et al., *Infrared Mapping of M82: A Starburst in an Edge-On Barred Galaxy*, Astrophysical Journal, **369**:135–146,1991
52. The Event Horizon telescope Collaboration, *First M87 Event Horizon Telescope Results. VI. The Shadow and Mass of the Central Black Hole*, Astrophysical Journal Letters, 875:L6 (44pp), 2019
53. Thornley, M. D., *Uncovering Spiral Structure in Flocculent Galaxies*, Astrophysical Journal, **469**: L45–L48, 1996
54. Rubin, V. C., *Kinematics of NGC 4826: A Sleeping Beauty Galaxy, not an Evil Eye*, Astronomical Journal, 107, 173–183, 1994
55. Toomre, A. & Toomre, J., Galactic Bridges and Tails, Astrophysical Journal, 178:623–666, 1972
56. Mirabel, I. F., Dottori, H. & Lutz, D., *Genesis of a dwarf galaxy from the debris of the Antennae*, Astronomy & Astrophysics, 256, L19–L22, 1992
57. Errani, R. et al., *Dark matter halo cores and the tidal survival of Milky Way satellites*, Monthly Notices of the Royal Astronomical Society, Volume 519, Issue 1, pp. 384–396, 2023
58. Kitchin, C., *Galaxies in Turmoil*, Springer, ISBN-13: 978-1-84628-670-4, 2007
59. Pasachoff, J., *Halley and his maps of the total eclipses of 1715 and 1724*, A&G, the magazine of the Royal Astronomical Society, 40, 2.18–2.21, April 1999

Index

J. Clark, *Photographing Galaxies from Light Polluted Skies*, The Patrick Moore Practical Astronomy Series, https://doi.org/10.1007/978-1-0716-4506-2